James Gulden

SPACE
OPERATIONAL
ANALYSIS

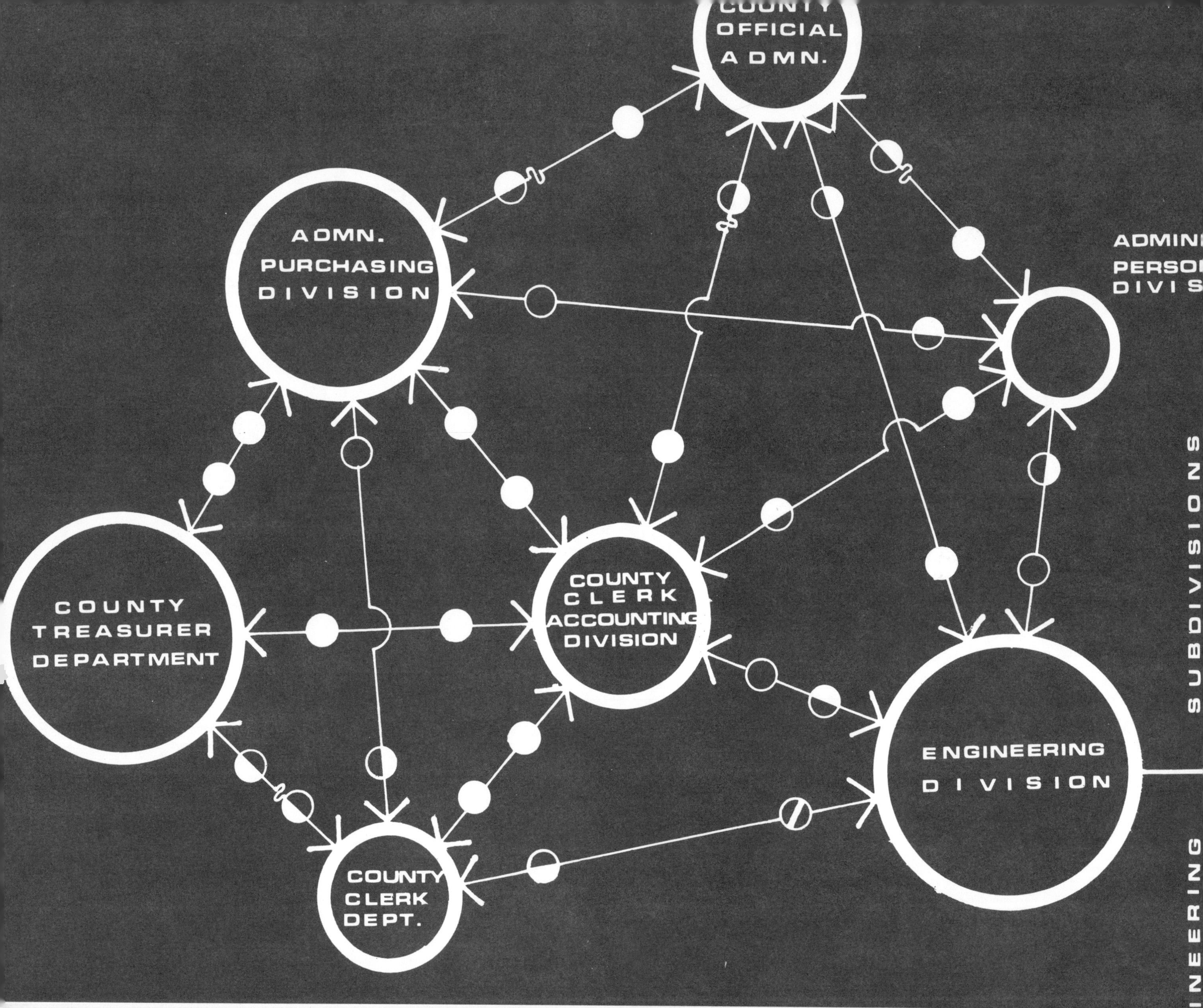

COUNTY OFFICIAL ADMN.
ADMN. PURCHASING DIVISION
ADMIN PERSO DIVIS
COUNTY TREASURER DEPARTMENT
COUNTY CLERK ACCOUNTING DIVISION
ENGINEERING DIVISION
COUNTY CLERK DEPT.
SUBDIVISIONS
NEERING

SPACE OPERATIONAL ANALYSIS

A SYSTEMATIC APPROACH TO SPATIAL ANALYSIS AND PROGRAMMING

MANUEL MARTI, Jr.

Member, American Institute of Architects
Member, Construction Specifications Institute

PDA PUBLISHERS CORPORATION

Additional copies may be ordered from: **P D A Publishers Corp.**
Box 3075
West Lafayette, Indiana
47906

Library of Congress Catalog Card Number: 80-81340

ISBN 0-914886-11-8

Printed in the United States of America

To Patricia:

friend, lover, wife, family
and guardian angel.
But above all... angel.

Table of Contents

Preface

The architectural programming techniques outlined in these pages were not created or developed by any particular individual or professional design organization, but rather constitute an updated selection of several widely accepted systems and methodologies presently used in the spatial analysis and programming of functions.

For this reason no specific system, data collection form, or table should be applied indiscriminately as a basic guideline without careful consideration of the advantages, disadvantages, and possible implications resulting from its use. Therefore, proper adjustments and revisions will be necessary before anyone can adequately apply the systems and techniques illustrated in this text to a determined project.

One of the principal objectives of this book is to offer a systematic approach to the spatial analysis and programming of functions in a generalized fashion and NOT to offer a series of loose systems and techniques to apply at random or time after time as the need arises.

The procedures illustrated are basically aimed at the design professions, those which deal primarily with the programming, designing and supervision of construction projects, and which therefore encounter, within the general scope of their professional services, the need for architectural programming technologies.

For this reason, and because of the innate diversity of the professional design activities toward which this text is directed, each procedure must be taken only in the context of its general meaning and adjusted properly to fit the needs of the specific design profession and definite project type it is intended to aid.

It has been assumed that the reader is familiar with general design and construction concepts and procedures; so, unless otherwise specified, the terminology used throughout is that generally employed by the professional design trades. A Glossary of Terms, at the end of the book, contains specific definitions of some of the concepts used in this text which may be new or uncommon or subject to misinterpretations.

None of the case studies and examples shown are exact reproductions of real projects but rather modified versions or adaptations of actual documents and formats which best apply to the specific situation they illustrate.

The main purpose behind the following pages is to awaken the consciousness of the design professionals to the need for developing their specific programming methodologies as clearly and distinctly as they do their own design trends and approaches, and NOT to establish inflexible patterns of programmatic techniques. It should never be construed that this book advocates ONE specific system above all others, because there is not, nor should there be, only one way of programming spaces by function, but many, as many as the innate individuality of the design professionals in charge of these procedures permits within the limits of proper operational performance and adequate professional achievement.

Therefore, it is essential that the reader submit each chapter of this book to a careful analytical scrutiny, questioning every paragraph and every sentence. Only in this manner will s/he reach his/her own conclusions, develop the proper value system and judgment, and arrive at the true meaning of his/her own programming approaches and techniques. Only in this manner can this text be of any use to its readers.

Only in this manner can it fulfill its purpose.

Madison, Wisconsin
March, 1980

Introduction

Problems Definition

The most essential element in the achievement of satisfactory operational performance is the proper recognition of problems situations as defined by a clear outline of their essential characteristics and the elimination of any peripheral or nonessential factors that might confuse or obscure the accurate definition of the problems in question.

The very act of problem solving implies much more than the actual development of an outline of adequate possibilities that might satisfy a set of goals or objectives. In fact, the act of problem solving — if traced back to its most essential roots — will invariably be found to depend on a core of "problem definition" for its foundation and support, no matter which case in point is considered.

Henri Poincaré once said, "The question is not, 'What is the answer?' The question is, "What is the question?"' But if such clarification of priorities appears simple, or even obvious on the surface, this veil of simplicity conceals many unanswered questions and variables which must themselves be clarified before any problem definition can even be attempted.

How should we ASK a question?

How can we outline a problem?

At first glance, it appears as if to ask a question all we must do is simply ASK it. But, is that ALL? And then, is that ALL we are actually doing or are we also establishing within its context guidelines and directrixes which will influence and alter the essential structure of its answer?

Furthermore, are questions and problems essentially synonyms whose differences consist only of formal characteristics, or are there more in-depth variables we should be aware of? Traditionally, a question has been defined as "a problem; matter open to discussion or inquiry."[1] A problem, on the other hand, has been defined as "a question proposed for solution or consideration."[2] Thus it appears that the main difference between the two terms lies in the fact that a question does not *necessarily* assume a great degree of difficulty in the pursuit or provision of a solution or answer, while a problem does. That is, when the answer to a question is not readily available, such question then outlines a problem. So while there are common denominators uniting the two concepts, the differences between them appear to be more formal than substantial, and the most important fact to be recognized in analyzing these terms is that either one represents a CLEAR STATEMENT OF GOALS. And this statement of goals, as the first statement of purpose for a specific action, will completely define, structure and direct the basic procedures to be followed in order to arrive at pertinent answers or solutions. Thus, the statement represents not only the description of goals or objectives but also the clear indication of the paths to be followed by subsequent programmatic developments. Such a statement of goals, therefore, is the factor which emerges from the problem identification action, regulates the problem definition process, determines all the following problem solving procedures, and serves as both foundation and yardstick for adequate and effective solutions.

Problem definition has been the subject of many studies and endless speculations ranging from the reaching of goals and objectives which involve more than mere actions, and must therefore recourse to THINKING procedures, as defined by Karl Duncker,[3] to the incorporation of "goal concepts" or satisfaction of "given descriptions" within a system by something existing or integrated within the system, as explained by W. R. Reitman.[4]

Actual statements of direction contained within problem definitions have also been defined as goals description and description of conditions which must be met by proposed solutions. These statements have been said to outline the operations, steps or procedures available for use in problem resolutions and to define the extent of resources and facts available for answers or targets development.

In addition, problem definitions have been grouped in elementary classifications based on their accuracy or extent, or on the strategies for development of their solutions. Problems have been called "well defined" if there is a TEST which can identify or verify their solutions or "ill defined" if the potential solutions, goals and constraints appear to be vague and confusing. By the same token, they have been classified as "underspecified," "uniquely specified" or "over-specified" depending on the amount of data they contain or on the number of potential solutions they suggest and "analytical," "search," "generative," etc. depending on the basic procedures employed in the development of their answers.

Problem definitions in professional design follow, in essence, the same patterns and classifications previously outlined, although their developments vary slightly in form from those of purely abstract problems' definition formats due to the factual characteristics of the concepts involved in their generation (as opposed for example to a purely mathematical or philosophical enunciation), as well as the basic nature of the elements involved in their outline. For example: Because the solutions of professional design problems can be represented by analogies, by merely establishing analogical similarities between corresponding elements of consideration, icons, or actual "look-alike" representations of concrete issues, and symbols, analogies in which words, numbers, mathematical operations or linear diagrams act as representatives of potential solutions; whether partial or complete, the problem outlines which call for such solutions should correspond with and refer to those same representational resources in the development of their own formats. Thus, the essential elements of design problem outlines will always be directly related to the basic language which structures and develops the design solutions. This reciprocity of values and characteristics should be maintained throughout the structuring of any system or methodology that might be developed for this purpose. The fact is that any design solution is simply the result of a primary definition of objectives which can only be properly illustrated when outlined by a system based on the productive interaction of common understandings with the subsequent design procedures by bearing within the systems' initial postulates the same concepts, values and definitions as well as any other viable means — whether abstract or graphic — of positive communication.

It is also because of this interacting correspondence among professional design procedures that the essential definitions of functional and spatial analyses presuppose and, at times, predetermine one another when observed in the light of design objectives. The spatial functionalism requirement developed for a specific design problem must include both a primary description of "function" as such, and an evaluative statement of the more specialized concept of "operational functionalism," such as that illustrated by operational relationships and interactions, in which the purely artistic license of aesthetic values and formal considerations may become only another determinant in the overall gamut of programmatic requirements, instead of being the essential force behind the definition of purpose and objectives for its practical development.

But if, on the one hand, it appears as if this approach undermines the role of aesthetic and formal values within the pattern established for executing functional programming, it should also be pointed out that this inevitable interaction of determinants weakens the argument in support of independent spatial programmers who would act as a specialized subdiscipline of the design professions. If to program a function properly one must, logically, be a specialist in that function, then to actually program the spatial requirements of any function one must not only become familiar with the function itself, but also be a specialist in the ways in which SPACE and SPATIAL RELATIONSHIPS, viably designed, affect the development and execution of such function.

Thus, any spatial programmer must not only be able to comprehend what spatial facts and interactions are all about, but also sense what their aesthetic implications and values are. In addition, he must possess the basic knowledge and understanding of the essential functional and environmental needs behind the development of specific operations. And these requirements — adjusted according to the specialties of the particular design activity involved — can only be met by qualified and active elements of the professional design disciplines; elements which not only realize the possibilities of spatial developments, but also understand their limitations, as well as the simple fact that one of the primary tasks of the programming activities is to supply information and establish guidelines for a designer, and that without full comprehension of what design is all about there cannot be spatial programming at all.

The question then arises: What special characteristics and which types and degrees of specialization(s) — if any at all — should the design professions allow or require of the elements and procedures in charge of the spatial programming of functions?

Actually, a question of that nature opens such a spectrum of possibilities and applicable answers that even a general outline could require considerable study in itself, and factors ranging from project or building type, size and specializations of professional design firms, and even regionality or territorial characteristics could play key roles in the development and conclusions of such a study. Ultimately, there is no simple and direct answer to the question we have raised.

In an article by John McLaughlin about architectural programming recently published in *Architectural Record* magazine, the actual programming and design boundaries are practically erased by basically considering the programmatic activities as intellectual procedures in which preconceptions and "predictions/experience filters" start limiting and defining specific design solutions for given projects at the outset. In other words, McLaughlin considers the beginnings of the design process to be what nowadays is normally called the "programming phase." The fundamental reasons for these conclusions are based on the premise that designers think and design in patterns, and just as the particular style of previously executed projects affects the client's selection of a specific designer, the subjective interpretation(s) of particular design trends, building prototypes or even formal stereotypes might shape or redirect programmatic scopes and stylistic determinants. McLaughlin goes on to specifically state that in many cases the generalized conception is:

[T]hat only the 'design' architect should design. The notion that the programmer is actively designing before the designer comes on the scene is threatening....[5]

Thus it becomes obvious that this argument places the activities that were traditionally considered to be "programmatic" within the scope of actual design activities and responsibilities. This concept is essentially correct if the traditionally accepted definitions which outline and determine the scope of design activities[6] could be modified to incorporate those pre-design activities which are currently regarded as falling within the scope of programming functions. On the other hand, as long as the problem definition stage of professional design procedures is considered an activity that precedes the design process and, while directly related to that process is clearly defined enough to constitute a completely independent development, the essential components and sequence of professional design activities can only be validly defined as follows: Programming, Schematic Design, Design Development, Contract Documents, Bidding and Construction Administration, a sequence which essentially underlines the same description that has been generally accepted to date.

Opposed to McLaughlin's point of view is that of W. M. Peña, who maintains in a recently published architectural programming manual that the essential differences between programmers and designers are the presumed discrepancies between analytical and synthetical minds and the intrinsic divergence of their innate approach to intellectual problem solving procedures. Thus he concluded that, generally speaking, it is very unlikely that one individual would be both a designer and a programmer.

There are valid points to this argument, but not enough on which to base generic descriptions, much less behavioral patterns or rules. The assumption that one individual cannot have the characteristics of both a programmer and a designer is inaccurate. If such a clear division between synthetical and analytical minds were accurate, no mathematician would ever have solved a single theorem, since the mathematical procedures of problem outline and resolution require both analytical and synthetical capabilities. Obviously, a programmer does not necessarily HAVE to be a designer. If that were the case, architectural programs would always have been executed by designers or even design professionals only, and, as we all know, such is not the case. But to maintain as a general rule that the same individual cannot be both a programmer and a designer because of the innate differences in their conceptual approach to the study or resolution of specific professional design tasks or problems is to carry a simple functional distinction of activities a bit too far.

We could speculate forever on this matter, but the essential fact is programming SHOULD and MUST be done by whichever individual within a professional design organization is capable of doing the best possible job of it. It is as simple as that.

Chapter 3 of this book contains more on the subject of programmers, their characteristics and professional qualifications. For now it is noteworthy that, whatever the speculations on the issues of programming and design interfaces, one fact remains clear: Whether due to firm size, specializations or professional developments, the need for programming has currently increased to the point where it is impossible to ignore the tremendous importance of its impact on the professional design achievements of today, as well as the essential role it seems destined to play in the professional design developments of tomorrow.

[1] Webster's New 20th Century Dictionary (1977) 2nd ed.

[2] Ibid.

[3] Duncker, K. (1945) "On Problem Solving." *Psychological Monographs*, vol. 58, no. 5.

[4] Reitman, W. R. (1965) *Cognition and Thought*. New York: Wiley.

[5] McLaughlin, Herbert, "Programming, predilection, and design." *Architectural Record*, Jan. 1979, p. 71.

[6] Such as those found in specialized dictionaries, reference manuals, etc. and even the descriptions established by article I of the A.I.A. document B141 "Standard Form of Agreement Between Owner and Architect."

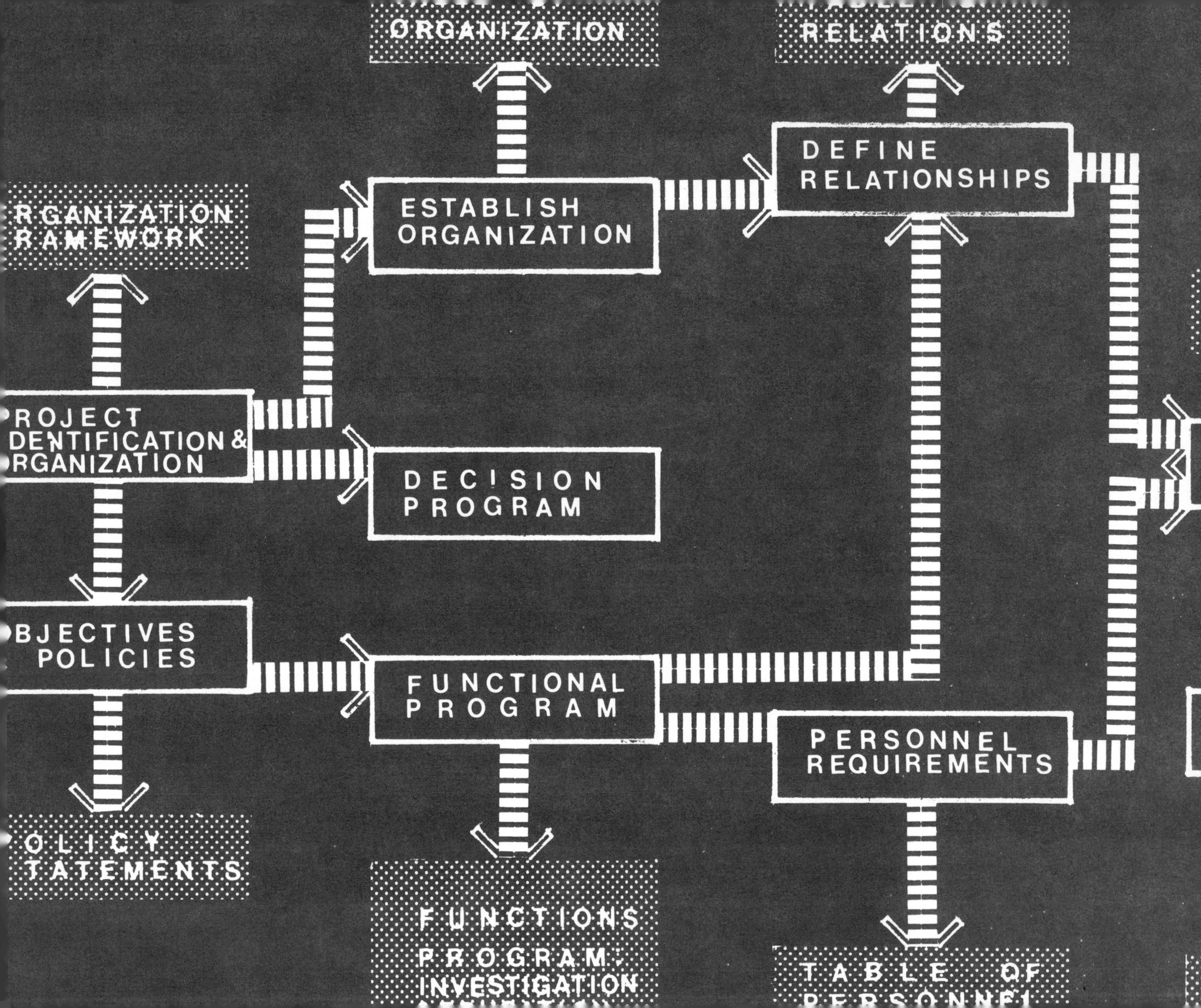
ORGANIZATION
RELATIONS
ORGANIZATION FRAMEWORK
DEFINE RELATIONSHIPS
ESTABLISH ORGANIZATION
PROJECT IDENTIFICATION & ORGANIZATION
DECISION PROGRAM
OBJECTIVES POLICIES
FUNCTIONAL PROGRAM
PERSONNEL REQUIREMENTS
POLICY STATEMENTS
FUNCTIONS PROGRAM INVESTIGATION
TABLE OF PERSONNEL

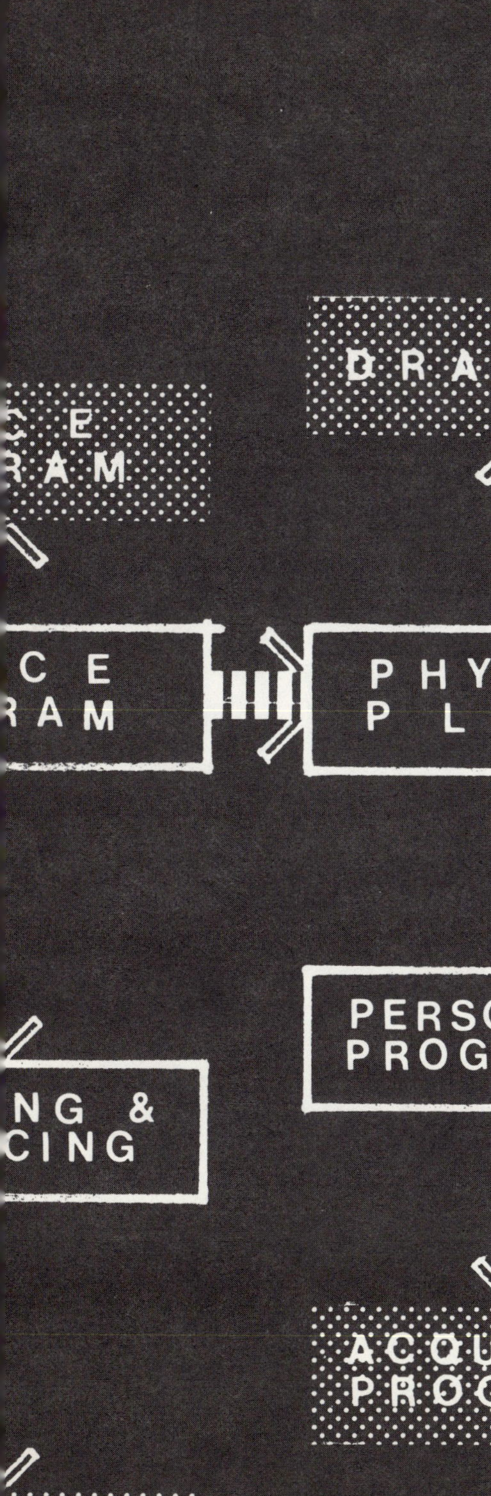

I

The Need For Spatial Analysis and Programming

1

Analysis and Programming

1.1 Building Analysis

The word "analysis" has, by definition, a dual meaning. On one hand it represents the separation of an intellectual or substantial whole into constituents for individual study, and on the other hand, the result of such study.

Because of this double meaning, the verb describing such action inherently represents a function of the logical thinking process in which the elements of cause and effect inevitably run parallel to each other and, in many cases, transpose their essential significance.

Analysis, therefore, cannot be construed as an isolated action, but as a process through which series of cause-effect relationships are established by means of a rational approach to the conditions of compliance with those universal factors which structure and regulate our complete framework of tangible concepts.

The analytical process begins in effect with the recognition of its own need — that is, the understanding of not understanding — continues with the translation of indeterminate elements into recogniz-able facts, and ends with a comprehensive evaluation of their characteristics.

Building analysis, like any other analytical process, follows the same outline, so in describing it the first step must be the establishment of its own need. Nothing is more damaging to the successful completion of an analytical procedure than those conclusions which derive from "assumed" facts.

Why is building analysis necessary? In order to answer that question, we must determine first what building analysis actually does. Based on our general definition, building analysis is the individual study of the constitutive elements of operations and their relationships, in order to determine their particular characteristics and spatial requirements as related to a building project. The primary function of this process is to provide the comprehensive layout of spatial needs and requirements necessary for a specific operation to take place within the limits of a project.

In actuality, the one process that effectively provides the physical layout in answer to spatial

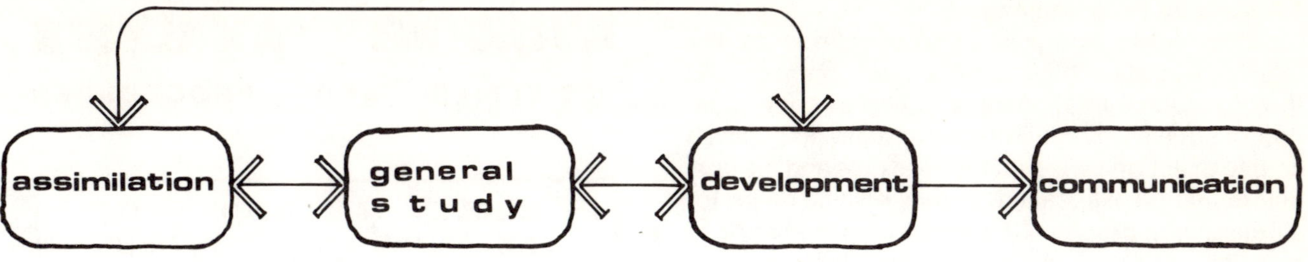

SUBJECTIVE   DESIGN   PROCESS

Figure 1.1

needs and requirements is the design process, through which all elements set forth as constituents of an operation are synthesized into a final solution. Still, in order to complete the design process, a basic summary of spatial conditions must be provided at the outset, and the development of this summary — which can only be formulated through an effective programming action — must inevitably be based on directrixes set forth by an analytical procedure.

The Royal Institute of British Architects, in its *Handbook of Architectural Practice and Management,* divides the design process into four parts, describing, with some variations, the subjective pattern shown in Figure 1.1

In this pattern, "Assimilation" and "General Study" represent the understanding of the elements and functions which structure an operation, their interrelationships, and the spatial requirements necessary for their functional performance.

Because of this, building analysis becomes an intrinsic part of that general process through which solutions to the requirements of spatial problems are found since this establishes the fact that, if we can arrive at such layouts only through the design process and the building analysis procedures provide those elements for assimilation and general study which form an essential part of the primary phase of that process, then building analysis plays a key role in the development and execution of spatial design solutions.

Figure 1.2 and the following list illustrate — respectively — the operations involved in the building anaiysis process and, in a generalized fashion, the activities involved in that process.

Building Analysis Process

    Description of Analyzed Operation

    Study of Existing Conditions
        What constitutes spatial occupancy?

    Research & References

    Evaluation
        What is right?
        What is wrong?
        What is provided?
        What is missing?

1.2 Building Programming

If building analysis can be defined as the individual study of the elements of an operation and their functional relationships as well as the evaluation of such study, building programming means a course of action based on those results and the final determination of spatial requirements and needs for a determined problem. If analysis means study and evaluation, programming means decision and determinations.

Cyril M. Harris defines "Program" in his *Dictionary of Architecture and Construction* as:

> A statement prepared by or for an owner, with or without an architect's assistance, setting forth the conditions and objectives for a building project including its general purpose and detailed requirements, such as a complete listing of the rooms required, their sizes, special facilities, etc.

Thus — if we are to place programming actions within the realm of professional design activities — building programming is the keystone of all professional design phases.[7] Furthermore, the basic etymology of the word *program* can be construed as "before or in front of the study of drawing."[8]

It is obvious, of course, that there cannot be building programming without some degree of building analysis preceding it. Still, there are some characteristics of the programming phase that cannot be found in the analytical process.

Analysis studies and evaluates, while programming ORDERS the evaluation, establishing patterns by which courses of action can be taken. Programming is thus the decision making process through which a conceptual layout of spatial requirements and their relationships will be accepted, modified, adjusted or even changed in order to produce a final composite of determinants making up the initial postulates from which any design process must derive.

If, for example, through a careful analysis of functional and spatial needs, we have determined that the clerical component of an operation will contain fifteen employees, and that a value of 75 sq. ft. minimum and 100 sq. ft. maximum per occupant must be assigned to that particular space, the

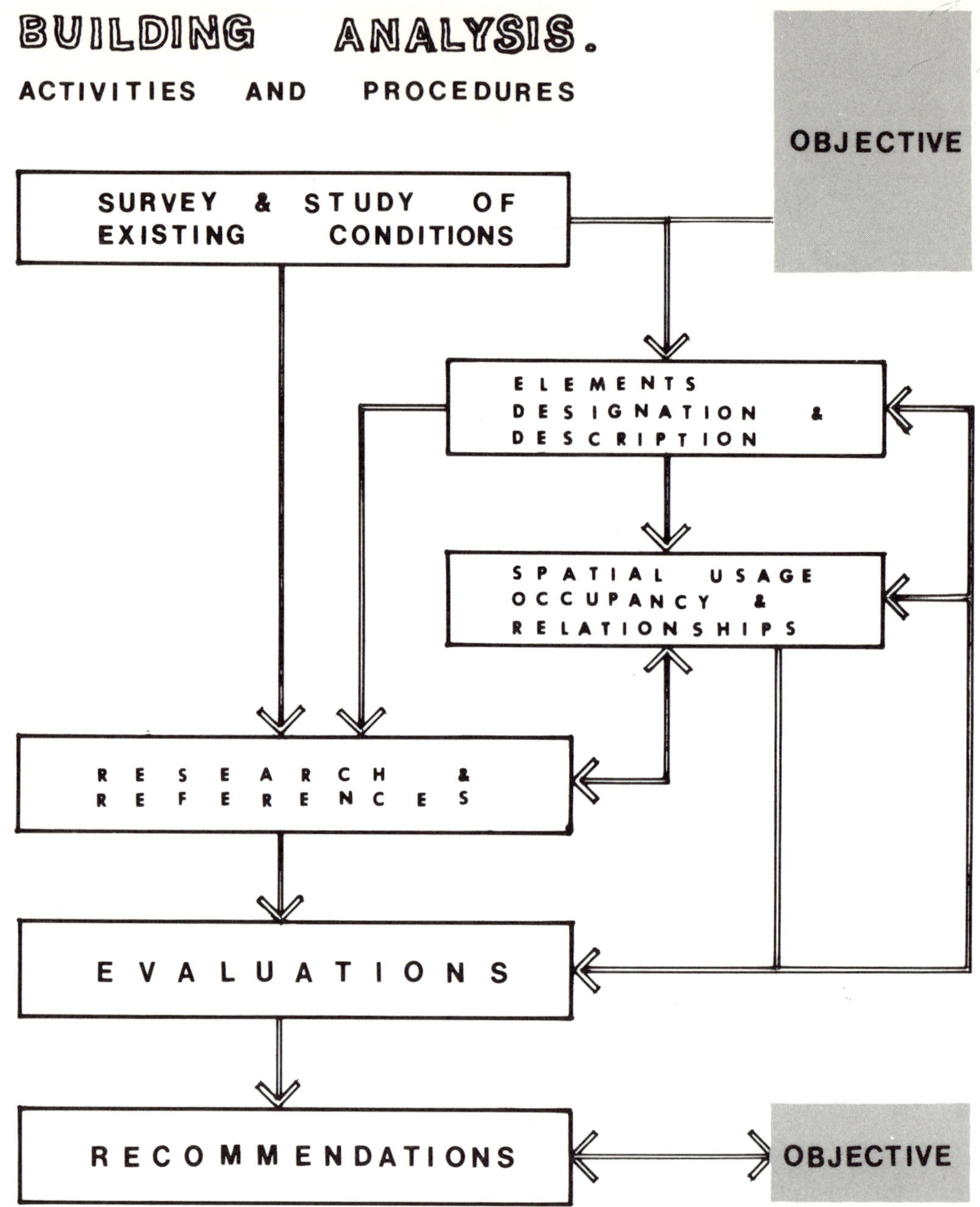

Figure 1.2

building programming action will determine one general office of 1300 sq. ft. or two offices (with or without the flexibility of being converted into one), one containing five employees with 500 sq. ft. of floor area and the other ten employees with 800 sq. ft., or any other possible combination; representing, by this means, the final decision based on the operational evaluation of the analysis results for the programmatic requirement of this particular clerical component. This requirement will include the directive of its statement as a program factor and, subsequently, a determining guideline throughout the entire design process.

One typical example of the difference between the analysis and programming processes can be easily seen in the fundamental structure of survey questions. While analytical questionnaires deal with the research and evaluation of existing conditions, the supporting data for programmatic concepts is obtained by formulating questions which present hypothetical situations and circumstances to those interviewed.

In 1968, Owens-Corning surveyed garden apartment residents' attitudes in complexes built within recent years in areas where such developments were booming. The results of questions related to noise and noise control appear in Table 1.1.

Notice that while question number 1 refers to existing conditions and is, therefore, strictly

1

Do you hear noise from any of these locations and what type of noise do you hear?

	Outside	Either Side	Below	Above	Halls	Importance
People talking	37%	31%	20%	21%	43%	1st
People walking	22%	13%	8%	45%	35%	2nd
Doors slamming	15%	28%	17%	19%	43%	3rd
Plumbing	2%	38%	25%	38%	3%	4th
Knocks on doors and doorbells	9%	24%	9%	10%	36%	5th
Children playing	40%	9%	7%	11%	17%	6th
Appliances	1%	23%	16%	24%	3%	7th
Traffic	51%	3%	3%	2%	3%	8th
TV sets	2%	19%	15%	17%	3%	9th
Telephones	3%	18%	10%	12%	3%	10th

2

In your next apartment, how interested would you be in renting an apartment where at least some of these noises have been reduced through the use of sound conditioning materials if you had to pay $5.00 more a month rent?

	Total	Singles	Married	Family
Extremely or very interested	42.1%	41.2%	39.5%	45.9%
Somewhat interested	22.9%	24.6%	21.2%	23.6%
Not very or not at all interested	32.7%	33.8%	36.2%	28.3%

analytical, question number 2 is designed to support "noise reduction improvements" as a logical reaction to the results of the previous analysis.

This analysis-programming balance, which can be noticed throughout most statistical studies, determines a very interesting characteristic. While in analytical research all questions are directed towards the study and evaluation of existing conditions, in programmatic research most questions are of a theoretical nature and are, therefore, less accurate. Because of this, it is extremely important to design such questions in the clearest and most concise manner, since the success of their result will depend greatly on the ability of those surveyed to project themselves into a nonexisting set of conditions and variables. This could result in more inaccuracies, speculative responses, and differences of opinion.

These characteristics will be found throughout both processes as a constant distinguishing element. When dealing with spatial programming, the first factor to judge is whether we are analyzing or programming a function.

1.3 Architectural Programming

There seems to be a confusion of terms in dealing with the use of the word programming when related to the construction industry, and this confusion should be clarified before we continue.

The terms project programming, architectural programming, and building programming seem to be used indiscriminately to signify either the same programming procedure, several interactive methodologies, or completely separate processes.

Without trying to set precedences or create controversies, and just as an explanatory note to this book, each of these terms will be used throughout this text as follows:

PROJECT PROGRAMMING. The programming process which involves the complete scope of a construction project, including — among others — Marketing and Sales, Legal and Contractual, Functional and Operational, and Architectural Programming, and Financial and Economic Conditions (see Figure 1.3).

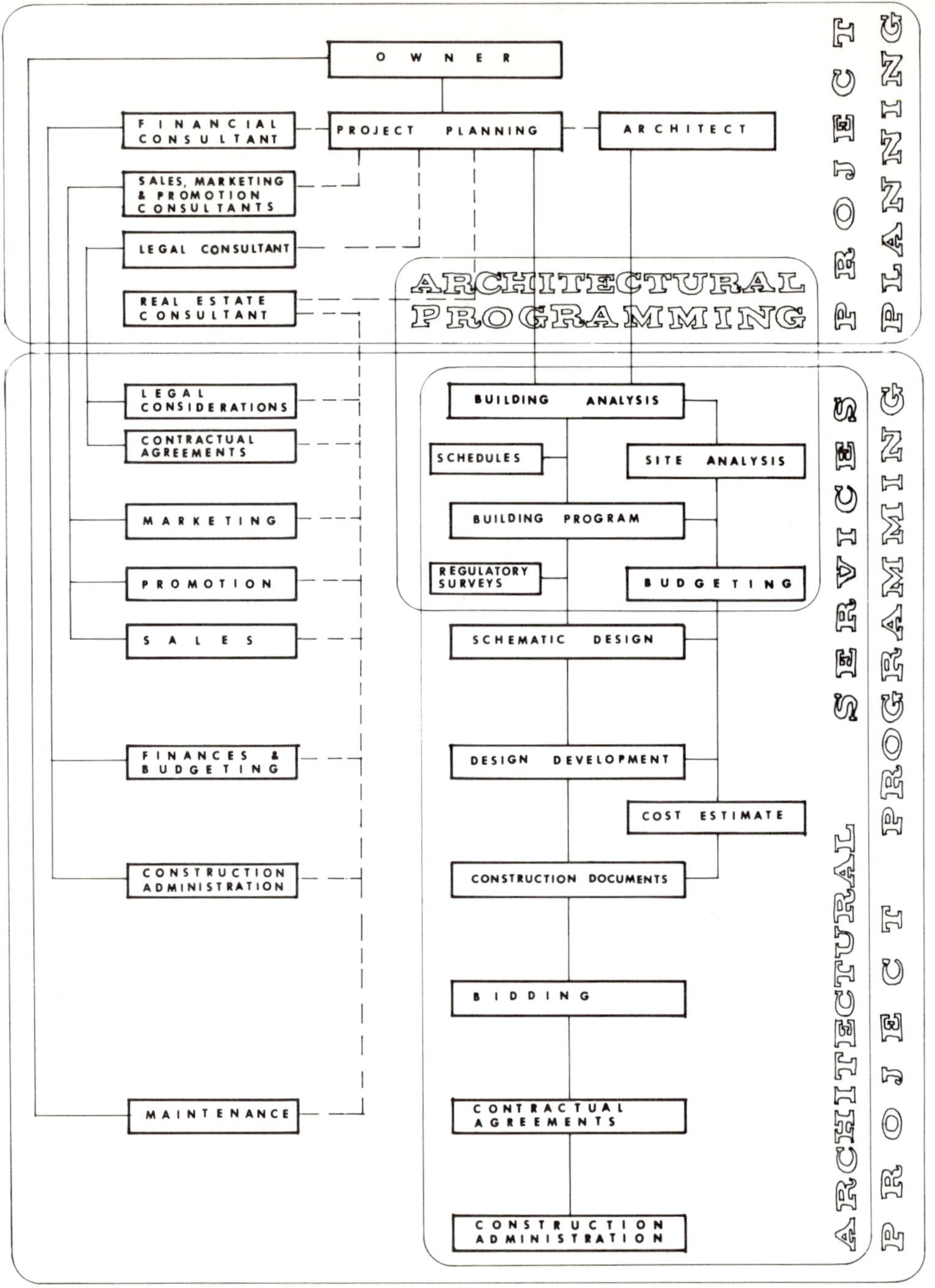

Figure 1.3

ARCHITECTURAL PROGRAMMING. That programming process which sets forth the requirements for the ARCHITECTURE of a specific project and which includes building analysis and building programming as its essential elements, as well as site analysis, budgeting, growth and change projections, etc. (See Figure 1.3)

BUILDING PROGRAMMING. The programming process through which the spatial and functional requirements for the structure or structures of a given project are set forth, based on a series of conceptual directrixes applied to the spatial analysis evaluation of all pertinent factors involved in the operational functions to be performed within the spatial parameters — whether interior or exterior — programmed for such structure.

Architectural programming will most likely be executed by an owner/architect team and includes the complete scope of those requirements and conditions affecting the architecture of a building project. (See Figure 1.3)

Project programming, on the other hand, can be done by an owner or a specialized consultant and might or might not involve the architect in a consulting or advisory capacity.

Finally, the building program, executed by an owner/architect team, represents the basic programming of the spatial functionalism for a determined structure by defining the conditions and objectives that structure is being designed to accomplish within a project, as well as the general outline and detailed requirements it must comply with in order to serve its purpose.

1.4 Elements of Building Programs

If the first two major constituents of an architectural programming procedure can be identified as building analysis and building programming, and if the essential objective of these constituents is to provide the design process with the necessary elements of assimilation and general study for its development, we must first consider the essential factors which structure the design process and their interrelationships with the subsequent architectural processes in order to determine the elements of an actual building program.

Spatial design, when related to construction projects, must deal with pre-established universal conditions (human, natural or temporal) which will determine every single phase of its development. Because of this, these factors which affect functional and operational performances become key issues of the intellectual procedure that is to create their scenario, namely: functional design. In order to achieve the goal of functionalism, the design process must not only develop schemes of resolutions for specific problems, but do so by means which will make such resolutions logically attainable. And this, by becoming the feasibility key to any solution, will underline the first statement of compliance with the aforementioned universal conditions. For this reason, any creative process should provide the following development and execution phases of professional design services with a more or less defined skeleton of operational data from which to derive their particular contribution to the final realization of a conceptualized construction project, as well as a message of INTENT within the forms and interrelationships illustrated in its designs. Logically, all factors contained within the universal conditions will intervene in this interaction. Chapter 7 contains more detailed descriptions of these conditions. For now, a simple listing of the major areas of interaction between the design procedure and the following professional design phases will suffice.

As a summary, therefore and in order to provide these subsequent stages in the complete framework of professional design procedures with their essential information and requirements outline, any comprehensive building program must consider the following compliance objectives:

1. Functional
2. Biological
3. Psychological
4. Sociological
5. Natural
6. Environmental
7. Temporal
8. Economic
9. Regulatory

This complete list of previously mentioned factors might not apply in every instance, but in each case every one of them must be considered and verified before the architectural programming process is completed.

Chapters 6 and 8 contain, in checklist form, information which must be provided by a comprehensive building program based primarily on the factors listed above.

It is important to note, however, that the architectural program for a specific building project, in becoming the layout of a determined problem, will always have special characteristics that will separate it from the rest. Nevertheless, because of the generalized nature of this chapter, we can only arrive now at a general outline of elemental factors, while stressing the fact that within the entire realm of professional design services there will always exist the need for a careful determination of the specifics and conditions which affect each one of its basic operating processes. Only through a careful analysis of the factors that intervene in each one of those processes will we be able to arrive at the proper courses of action to follow during their execution.

Any professional design activity, and in effect Architecture, begins inevitably with a program, basically structured by its building analysis and programming constituents, as the starting point of its primary design process, only to end with the completed project feedback data collection in the form of building analysis once more, as a measure of performance and verification of accomplishment and execution.

And this process, by successfully completing this specific circle of actions and fulfilling the total scope and objectives of a building project, underlines the primary function of its programming phase, defining within itself the goals of those very basic postulates which constitute the professional design services.

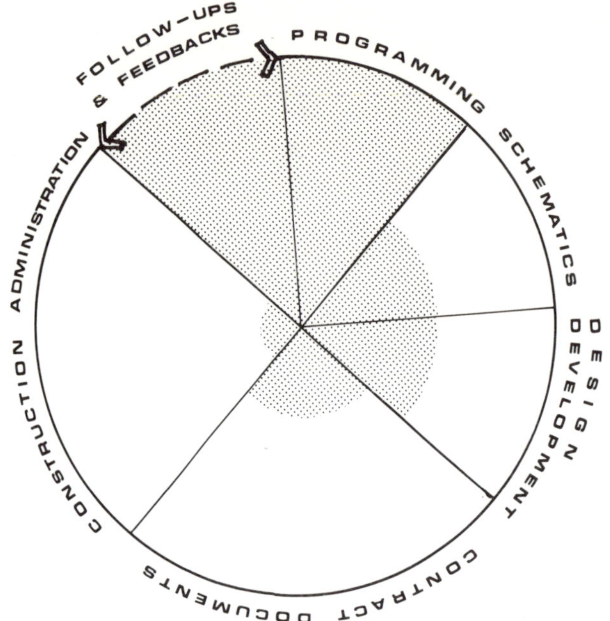

Figure 1.4

---

[7] In many cases the actual programming action has been considered the initial stage of the design procedure. McLaughlin, 1979.

[8] See, *The American Heritage Dictionary of the English Language*, 1969 and *The Oxford English Dictionary with Supplement on Historical Principles*, 1933, vol. VIII.

2
Professional Design Services

2.1 Layouts

Design, when taken within the context of its traditional definition,[9] cannot be presumed to mean an isolated action executed independently from any preceeding cause or resulting effect. Design requires the existence of an initial problem as primary cause of action and the intent of pursuit of an ideal target as solution to that problem.

Furthermore, design is actually a procedure structured by an elementary set of postulates which provides a framework for the principal causes of purpose and objectives. This procedure, when it is developed, produces an organized pattern of possibilities which illustrate, not only the response to a primary problem, but also a scheme of results as direct product of the initial precepts established by its elementary set of postulates.

In this light, a design problem statement contains not only the essential questions and tasks to be achieved by the design process, but also the rules which are to dictate and in many cases structure its courses of action.

All design professionals are well aware of this condition. In fact, such awareness may be one of the major characteristics which qualifies them as "professionals." Once the blind mask of idealism is stripped from the pure artistic conception and the inevitable limitations imposed by functionalism appear in the declaration of purpose for a determined design activity, the designer becomes a professional, no longer entirely subjective or conceptually pure in his artistic developments, but perhaps much more in touch with his fellow man. For this reason a design professional must always be recognized as one of society's most valuable assets, since s/he is by definition a person who has compromised a pure form of art in order to allow his/her artistic capabilities to maintain direct contact with and serve mankind.

A general listing of the professionals which offer artistic and aesthetically related design services as part of their essential functions must inevitably include the following:

Architects
Commercial Artists
Industrial Designers
Interior Designers
Landscape Architects
Urban Planners

For the purposes of this text, we will consider only those professions which deal directly with the functional and aesthetic ambivalence of construction projects. Therefore, the professional design services rendered by industrial designers or commercial artists in all their forms and specializations, as well as the strictly functional design service provided by engineers, will be referred to as "related disciplines or professions."

In describing the general layouts of the professional design services rendered by architects, interior designers, landscape architects and urban planners, and in trying to obtain a pattern of common denominators which will enable us to define the similarities and apparent differences of their programming activities, we will follow a procedure that separates the project analysis and programming functions from the remaining basic and complementary design services. This arbitrary outline seeks to filter and clarify the scope and definition of similar activities and functional equivalences. This procedure, which will use a common format throughout, will be based on the concept of comprehensive professional design services.

ARCHITECTURAL SERVICES. Between the years 1960-1962 the architectural profession underwent a dramatic change in the basic layout of its scope and responsibilities.

It was during those years that the concept of Comprehensive Architectural Services was established, defining all the corollaries and ramifications of what was once considered a simple four-step layout of total services: Schematic Design, Design Development, Construction Documents and Administration of Construction Contracts. Thus, the position adopted by the profession was a clear and direct response to the increasing demands and new requirements of a changing world.

The traditional architectural services were expected to change radically with the establishment of this concept; and today, less than two decades later, we can state that such change has indisputably taken place. The layout of Comprehensive Architectural Services has given the profession new directions and goals and has indicated basic courses of action in a world where the abilities to change and adapt are essential for survival.

In our day, any architectural firm that limits its services to those four phases which define the traditional basic services is inevitably destined to fail, not only because it lacks a more extensive and varied selection of services from which to draw income, but also because it is not fulfilling some of the essential requirements it is EXPECTED to meet.

Today, the practice of Architecture goes far beyond a simple layout of basic traditional services and includes an enormous variety of phases and procedures that extend from the very complex Project Planning and Programming processes to the specific Signage and Graphics design.

A general layout of these services can be summarized as follows:

Professional Services.

1. Project Analysis and Programming
    1.1   Project Planning and Programming
    1.2   Economic Analysis and Feasibility Studies
    1.3   Project Location and Site Analysis
    1.4   Architectural Programming

2. Basic Services
    2.1   "As-Built" Drawings, Studies and Reports
    2.2   Schematic Design
    2.3   Cost Estimating
    2.4   Design Development
    2.5   Contract Documents
    2.6   Bidding and Construction Contracts
    2.7   Project Supervision
    2.8   Construction Administration
    2.9   Promotional Design
    2.10 Postconstruction Services

3. Complementary Design Services

  3.1   Interior Design and Space Planning
  3.2   Engineering
  3.3   Urban and Regional Planning
  3.4   Landscape Architecture
  3.5   Site Planning
  3.6   Arts and Crafts
  3.7   Industrial Design
  3.8   Signage and Graphics
  3.9   Design-Built/Contracting Services

Related Services

4. Consulting Services

  4.1   Functional Programming
  4.2   Industry
  4.3   Business
  4.4   Specific Building Types
  4.5   Energy Conservation
  4.6   Prefabrication
  4.7   Marketing and Sales
  4.8   Legal Services
  4.9   Codes and Regulations
  4.10 Research and Testing
  4.11 Environmental Studies and Reports
  4.12 Real Estate and Land Use
  4.13 Construction Inspection and Certifications
  4.14 Surveying

5. Promotional Services

  5.1   Promotional Planning
  5.2   Public Relations and Communications

6. Educational Services

  6.1   Education
  6.2   Publications and Communications
  6.3   Historical and Preservation

In this outline the Professional Services sector encompasses all of those processes directly related to the architectural practice and essential to its structure, which services are to be provided by the firm itself or by specialized consultants acting as agents of the architect. The Related Services category covers those areas in which architectural organizations or individual architects act more as consultants than as executive agents. (See Figure 2.1)

INTERIOR DESIGN SERVICES. Interior Design, as a discipline, can be considered as old as Architecture itself, since until just recently both activities were executed by the same group of design professionals. Now, however, the increased proliferation of building types, functional requirements and specialized activities, as well as the enormous variety of furnishings, materials and building products, has led to the establishment of architectural interiors as an independent discipline, thus freeing the traditional scope of services of the architectural profession from the vast influx of architecturally related materials, products, constraints and specialized developments pertaining to interior spaces.

The American Society of Interior Designers defines a professional member as "one who is qualified by education and experience to identify, research and creatively solve problems relative to the function and quality of man's proximate environment." It also mentions fundamental design, design analysis, design of interior spaces, and space planning and programming, as being among the activities of its qualified professionals.

Based on the concept of comprehensive professional services, a general layout of Interior Design Services could be considered to encompass the following:

Professional Services.

1. Project Analysis and Programming

  1.1   Project Planning and Programming
  1.2   Economic Analysis and Feasibility Studies
  1.3   Architectural Interiors Programming

2. Basic Services
   2.1 "As-Built" Drawings
   2.2 Schematic Design
   2.3 Cost Estimating
   2.4 Design Development
   2.5 Contract Documents
   2.6 Bidding and Construction Contracts
   2.7 Purchasing, Contracting and Quality
        Control
   2.8 Construction Administration and
        Supervision
   2.9 Promotional Design
   2.10 Postconstruction Services

3. Complementary Design Services
   3.1 Architecture
   3.2 Engineering
   3.3 Landscape Architecture
   3.4 Arts and Crafts
   3.5 Industrial Design
   3.6 Signage and Graphics

Related Services.

4. Consulting Services
   4.1 Functional Programming
   4.2 Industry
   4.3 Business
   4.4 Specific Building Types
   4.5 Prefabrication
   4.6 Marketing and Sales
   4.7 Codes and Regulations
   4.8 Legal Services
   4.9 Research and Testing
   4.10 Real Estate

5. Promotional Services
   5.1 Public Relations and Communications
   5.2 Promotional Planning

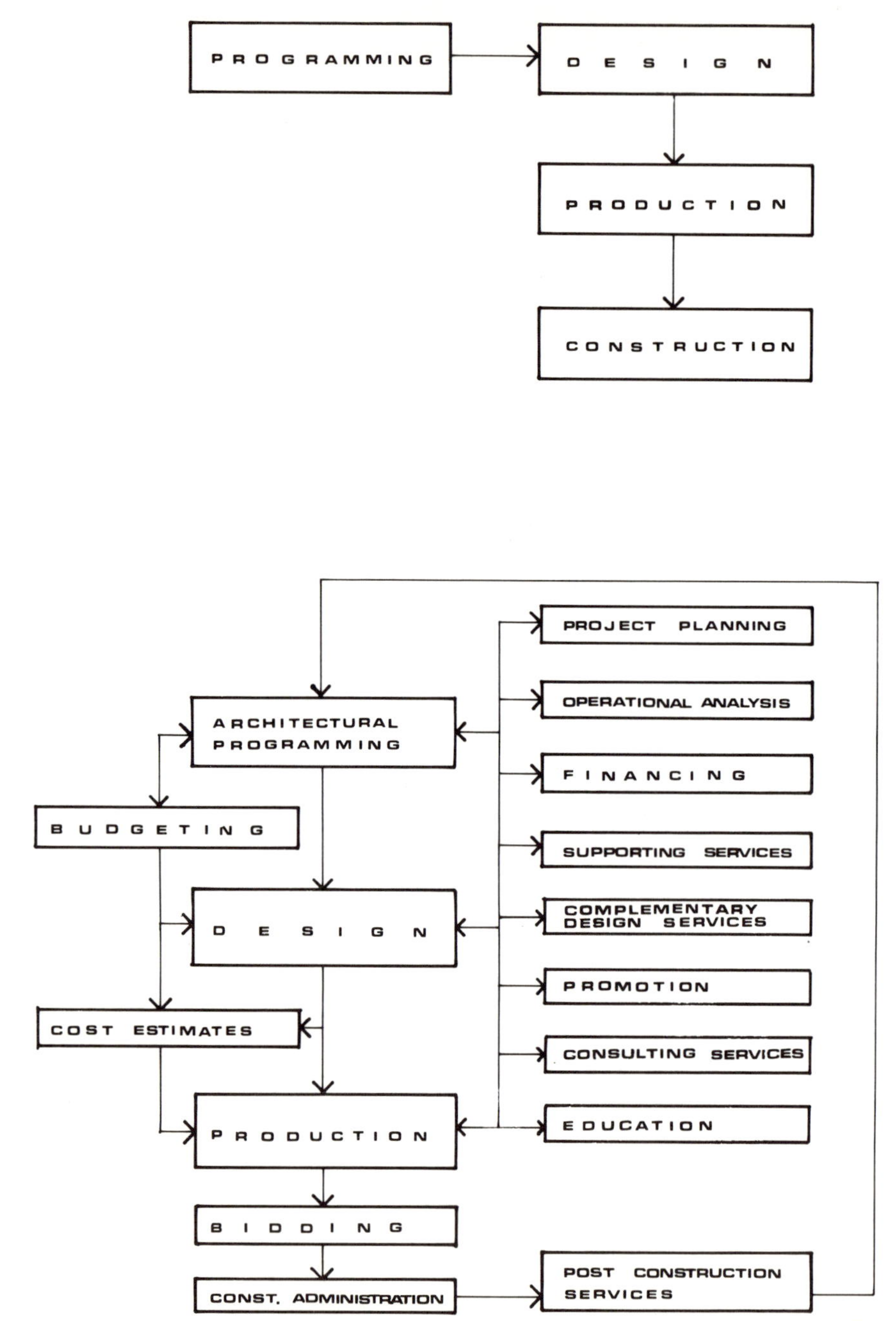

Figure 2.1

6. Educational Services
   6.1   Education
   6.2   Publications and Communications
   6.3   Historical and Preservation

As mentioned previously, the Professional Services encompass all of the processes directly related to the interior design practice and essential to its structure, services which are to be provided by the firm itself or by specialized consultants acting as agents of the interior designer. Related Services, on the other hand, cover those areas in which interior design organizations or individuals function more as consultants than as executive agents.

LANDSCAPE ARCHITECTURE SERVICES. In 1910, Charles W. Eliot wrote to the editors of *Landscape Architecture:*

> Landscape Architecture is primarily a fine art; but it is also concerned with promoting the comfort, convenience and health of urban populations with scanty access to rural scenery and urgent needs for the calm, beautiful and reposeful sights and sounds which nature can provide.

The American Society of Landscape Architects was founded in 1899 and the first degree of landscape architecture was granted in 1901. Since that time, the importance of this branch of the design professions has grown considerably and is now widely recognized nationally as a fundamental element of the professional design associations.

Today the role of landscape architecture can be observed in almost every urban development, where the most essential relationship, that of man to nature, is no longer ignored.

Thanks to the enormous developments and achievements of landscape architecture within the last few decades, we now accept its importance and seek its implementation in every setting of the urban profile; and just as the positive development of its basic goals has produced new approaches and solutions to the role of landscaping in the development of the human environment, so too the natural extent of its Professional Services can be expanded to encompass a comprehensive list of complementary design activities and related consulting, Promotional and Educational Services, as summarized below:

Professional Services.

1. Project Analysis and Programming
   1.1   Project Planning and Programming
   1.2   Economic Analysis
   1.3   Landscape Architecture Programming

2. Basic Services
   2.1   "As-Built" Drawings, Studies and Reports
   2.2   Site Planning and Schematic Design
   2.3   Cost Estimating
   2.4   Design Development
   2.5   Contract Documents
   2.6   Bidding and Construction Contracts
   2.7   Purchasing, Contracting and Quality Control
   2.8   Project Supervision and Inspections
   2.9   Construction Administration
   2.10 Promotional Design
   2.11 Postconstruction Services

3. Complementary Design Services
   3.1   Architecture
   3.2   Interior Design
   3.3   Engineering
   3.4   Urban and Regional Planning
   3.5   Arts and Crafts
   3.6   Industrial Design
   3.7   Signage and Graphics

Related Services.

4. Consulting Services
   4.1   Functional Programming
   4.2   Industry
   4.3   Business
   4.4   Specific Building Types

And, just as in the previous layouts, Professional Services are those involving landscape architects as primary agents; while Related Services encompass those in which landscape architects perform in a consulting or advisory capacity.

URBAN PLANNING AND DESIGN SERVICES. Urban planning, whether analyzed as a discipline in itself or as an architectural specialization, can be defined as that level or branch of Architecture which deals with the basic elements of urban development; or, as defined in the American Institute of Planning 1975 policy statement: "[T]hat level of design that deals with the relationships between the major elements of the city fabric."

The essential goals and objectives of its professional design services are concerned with better use of the land, more functional, economical and efficient environments, energy-conscious design, adequate and well-proportioned spaces, landscaping and architectural beautifications and compatibilities, as well as a sensitive and well-balanced planning of the urban scenario.

Among the many tasks undertaken by urban planning are the establishment of functional and efficient traffic patterns, the development of transportation facilities, the consideration of human safety and convenience, the provision of adequate utilities and services, and the balanced planning of well-situated buildings and open spaces, all requiring as little maintenance as possible. In short, the goal is well-designed urban environments according to projections of future needs and availability of resources.

Because of this discipline's basic concern with large scale developments, the planning and programming activities represent an essential element of its structure, as indicated by the fact that its descriptive title includes the word *planning.* And, as in the case of the other disciplines mentioned previously, the design activities of urban planning are entirely dependent on a program from which to derive and achieve positive results.

Based on a comprehensive view of professional design services, the following summary illustrates the Basic, Complementary and Related Services of urban planning:

Professional Services.

1. Project Analysis and Programming

    1.1 Urban Planning and Programming

    1.2 Economic Analysis

2. Basic Services

    2.1 Land Use, Development and Studies

    2.2 Environmental Planning

    2.3 Site Studies and Planning

    2.4 Urban and Regional Design and Evaluations

    2.5 Transportation Planning

    2.6 Traffic Engineering and Parking Studies

    2.7 Cost Analyses and Estimates

    2.8 Contract Documents and Project Supervision

    2.9 Codes, Regulations and Economics

    2.10 Signage and Graphic Design

3. Complementary Design Services

    3.1 Architecture

    3.2 Engineering

3.3   Arts and Crafts

3.4   Landscape Architecture

Related Services.

4. Consulting Services

4.1   Functional Programming

4.2   Industry

4.3   Business

4.4   Specific Building Types

4.5   Energy Conservation

4.6   Legal Services, Codes, Regulations
       and Ordinances

4.7   Research

4.8   Real Estate and Land Developments

4.9   Surveying

5. Promotional Services

5.1   Public Relations and Communications

6. Educational Services

6.1   Education

6.2   Publications

6.3   Historical and Preservation

The similarities observed in the layouts of the different Professional Design Services are not due to mere chance. The fact is that architects, interior designers, landscape architects, urban planners, engineers, etc. represent only particular components of one overall division of man's functional activities: Architecture and, therefore, tend to interact within common task forces when dealing with activities and procedures that pursue similar goals and objectives.

Therefore, a general summary of the Professional Design Services concerned with the project analysis and programming activities of the design professions can be described as follows:

1. Project Analysis and Programming

1.1   Project Planning and Programming

1.2   Economic Analysis and Feasibility Studies

1.3   Architectural Programming

Notice that in the individual listings established previously for each outline of professional design services the programming activities differ from one another according to the particular specialization involved. However, in all of them, programming tasks are listed as the basis for the subsequent design processes.

In view of this essential similarity, we will refer from this point on to the programming activities of the previously outlined design professions as ARCHITECTURAL PROGRAMMING, since, by reason of its own comprehensiveness, the scope of architectural programming will encompass all the programming activities considered in the preceding pages as specialized developments of its constitutive tasks.

2.2 Scope and Goals

The primary objective of professional design services as expressed in their general layouts is the successful execution of each phase of that layout so as to meet the requirements set forth by the operations the services are designed to serve.

Professional design services are not an end in themselves, but rather the means of finding solutions to the problems posed by human, natural, temporal, sociological or economic requirements.

Thus, these professions have responded to the ever increasing complexity of our times by making detailed analyses of the broad spectrum of architecturally related activities and then developing a comprehensive synthesis of directives to deal with every operational curve that might spring from those previously identified determining conditions and requirements.

During the analysis and programming phases of any professional design service, the function of the professional design specialist in charge is not that of operations analyst or programmer, but rather of space analyst and consultant. It is beyond the scope and responsibilities of a professional design consultant to take charge of changes or modifications of any kind in the actual operations he serves directly in his professional capacity. He is responsible only for those changes pertaining to the space

15

in which the operations take place. Thus, the professional designer becomes the link between the operations and their spatial environment.

In the role of spatial consultant and creator the professional designer fulfills its primary function. Architects, interior designers, landscape architects, urban planners and all others in disciplines directly involved in the spatial design of environments for human operations form the task-force behind the planning and execution of spatial developments.

In this area the importance of architectural programming has grown far beyond the limits established by any traditional conception, since to assume the responsibility of spatial design without a clear definition of the determining factors in a specific operation is as dangerous as having physicians proceed with the treatment of ailments without proper diagnoses, analysis and evaluation of all possible remedies, or having attorneys represent a client in a court of law without proper knowledge of all pertinent circumstances, applicable laws, precedences and facts.

The spatial design process is one of continuous exposure and isolation of alternatives so that the most viable solutions can be selected, studied and evaluated. Without a clear outline of tasks, it is impossible to formulate, much less execute, those alternatives.

In addition, once a specific commission that includes spatial analysis and programming as part of its services has been accepted, any professional design firm can become liable for errors and omissions in analysis and programming as well as in design. Needless to say, corrections of these particular stages after a building project has been executed will incur far greater expense than would result from an improperly designed planter or a faulty fascia; and unfortunately the discovery of such errors will most likely occur after the building is completed and occupied. Therefore, an accurate system of building analysis and programming is more than a desirable tool; it is an essential mechanism in the spatial design process and should provide the necessary elements for the development of proper design alternatives and thus serve as guideline and evaluative formula of schematic

proposals and final solutions.

2.3 The Future of the Design Professions

Perhaps the one isolated concept that most dramatically separates the professional design practice of our time from that of our predecessors is that of the "growth and change projections." The enormous number and complexity of human operations in our day has produced innumerable series of directions, theories and systems specifically designed to incorporate these new requirements as essential variables within the basic elements of analysis, programming and design. The concept of temporality, together with its flexibility and development derivatives, has probably had more influence on Architecture in the last few decades than any other single occurrence.

These socio-economic factors which began to appear during the last century have grown today to proportions far beyond those predicted by the forecasters of the human phenomenon and its spatial relationships. The technological advances of the Industrial Revolution added to the population explosion and all their logical consequences have probably contributed more to the transformation of Architecture than any other human development since the eclectic approach to the creation of forms and development of spaces which occurred during the Renaissance.

Only in recent times has the shortage of space become an issue of such magnitude that it has forced each one of the elements involved in spatial programming to establish the minimum-maximum values of areas and volumes in the determination of functionalism, usage, spatial density and dwelling characteristics to be attained.

This spatial crisis is, of course, a direct result of the population explosion and its increased demands for spatial development. It would not be an exaggeration to state that within the next 100 years it is very unlikely that single family dwellings will be contained within counties having major cities within their limits; or to predict that, based on the steady growth of the multifamily dwelling, within the next 100 years 99% of all residential structures will be of this particular kind.

Today more people than ever before were not born in a single family dwelling, have never owned one, and never will. Space has become more than a roaming environmental component; it is a tool and a very costly tool at that.

Alongside increased demand run the ever-present shadows of economic factors and inflationary trends. No longer do clients appear in our offices with scopes and objectives of lifetime permanence in their projects. The projects of our times invariably carry a strong suggestion of speculative intent that presents itself in the form of growth projections, change projections and economic considerations and budgeting, flexibility, functionalism, sales potential, marketing factors, image and appeal requirements, and movement and transitional needs.

The determination of how much the conceptual values of functionalism have changed within the past five decades, taking into account the determining factors apparent from our present standpoint, would be a subject too extensive to approach in this limited space. Still, it is imperative to note that the basis for the project requirements of our times are a direct result of those primary values interacting with the natural conditions which surround us.

Added to these human phenomena factors is the response of our natural habitat. This response, which comes to us in the form of natural crises, envelopes the functional requirements of human operations, curbs them and eventually affects the conceptual tasks behind them. Energy crisis, environmental considerations, natural preservation and availability of materials are just part of an extensive list of determining influences which modify and dictate the basic value judgment of function.

In analyzing the programming for this conglomerate of circumstances, one realizes three important facts:

1. Due to the complexity of operational requirements, a program becomes more than the simple layout of a specific problem, it will be the human response to the elements surrounding it and the germ of the solution outlined by its limiting charac-

teristics. Thus a program translates the first questions of "What?" and "How?" into clearly defined answers.

2. In evaluating the functional aspects of given requirements, a programming activity serves to measure a specific time; and that constant evaluative process not only provides a balanced outlook on present situations, but also serves as a means of forecasting with a logical degree of accuracy what can be expected to develop in the future.

3. At the same time, the programming process forces those agents involved in its execution to look at the future and project the development of their activities far beyond a simple day-to-day solution. Therefore, a program presents a clear and concise study of the situation as it exists today and offers the opportunity for effective actions, reactions and changes concerning tomorrow.

Faced with the unprecedented and continuing transformations of our days, any effort to make concrete predictions concerning the future of Architecture, professional design services, or professional design practice as a whole would be extremely dangerous, since it would be based on a world of extremely unpredictable events and circumstances. Yet some irrevocable trends that can be noted, if only to shed some light on a very limited portion of tomorrow:

1. Architecture is much more complex today than it has ever been before, and that is simply a direct result of the increased complexity of today's world.

2. The practice of professional design disciplines has diversified so that its responsibilities can no longer be covered by the traditional concepts embodied in any definition of basic services.

3. The functional values have departed from their traditional role of influences to become determining factors in the effective execution of architectural programming, design and development.

4. The natural crises are, to say the least, more a reality than a threat.

5. Therefore Architecture, because of the world in which it operates, has abandoned its traditional outline and finds itself enveloped in a constant

process of change, growth and renovation.

Thus, as we venture into analysis and programming in our time, we realize that the true value behind these processes lies more in the common denominator that unites them than in particular circumstances, problems or solutions; that is, in the development of a system or systems that would enable us to clearly formulate those very first questions of "What?" and "How?" and in defining them provide a framework for their answers.

---

[9]Design: 1. To compose a plan for a building. 2. The architectural concept of a building as represented by plans, elevations, renderings, and other drawings. Harris, Cyril M. *Dictionary of Architecture and Construction.*

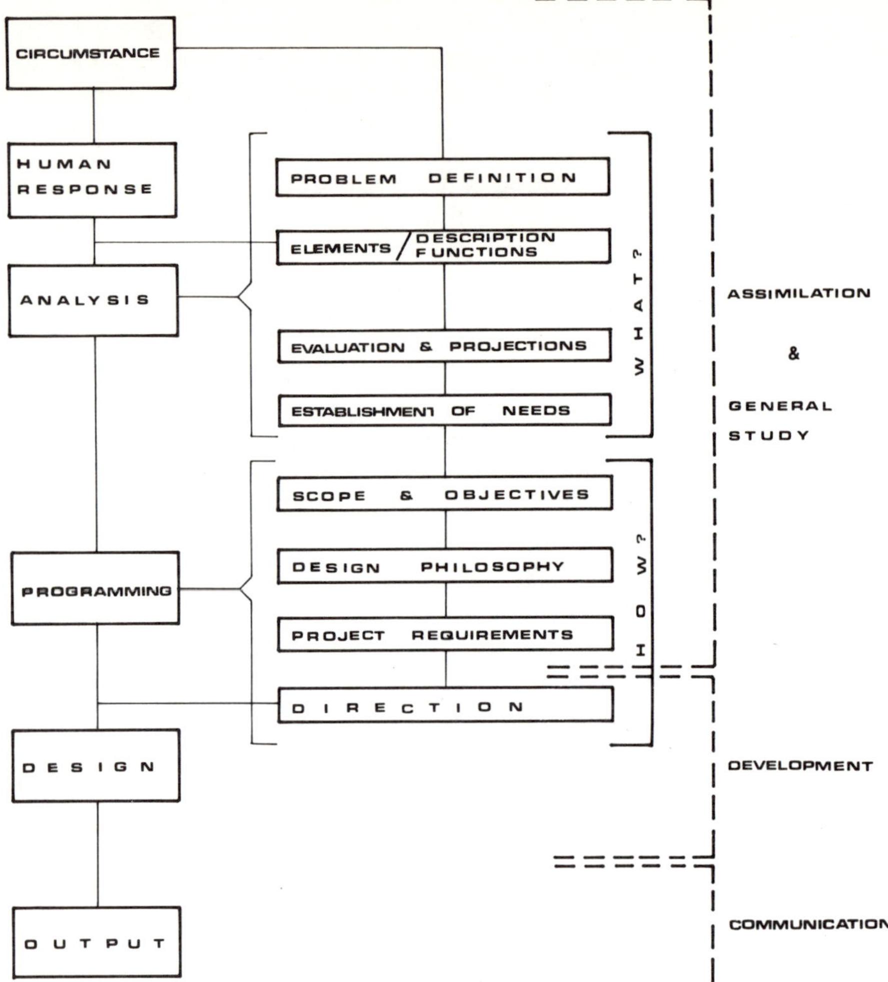

Figure 2.2

3

Designers and Design Professions

3.1 Education

Curriculums of professional design studies in the United States vary slightly from university to university, but essentially they contain the same block of basic subjects and the same general approach to the specialized training of students.

It is noteworthy, however, that in many other countries these curriculums differ from ours, and their graduates perform differently than we would normally expect once they are placed in their respective socio-economic circumstances.

In Latin America, for example, the term "general contractor" is unheard of. There, construction companies are founded, managed, and owned by architects and/or engineers, and our current "Design-Built" concept has been a standard component of their architectural practice for decades. By the same token, although widely accepted as specialties, we will not always find the definite separation of landscape architecture or urban planning as independently recognized professions.

Also, many foreign universities include in their first semester of design training a course called Architectural Drafting, which during that period takes the place of Design as we know it. Added to this, in their overall curriculum, is a three-year course in Building Analysis.

The idea behind these courses is, first, to provide the architectural student with the necessary tools to properly express his creative three-dimensional concepts on a two-dimensional and evaluative process of analysis and programming.

The direct conflict between the comprehensive professional design services layout and the basic concept of initiating a design project with a program furnished by the design professors is such that serious consequences may result. For example:

Several years ago a graduate architect started working for a firm somewhere in the southeast as a junior draftsman immediately after completing his architectural education. His duties consisted, basically, of running blueprints, lettering, etc. After about six months, he was finally given the more serious tasks of shadowing asphalt pavements on site plans, laying out surveys on larger scales, and even laying out buildings and parking lots on sites.

Within a year, he was dimensioning floor plans, tracing standard details, and copying typical wall section outlines from sketches provided by project architects.

After the second year, he was considered an intermediate draftsman by all standards. He was working directly under a project architect and instructing two new junior draftsmen on how to proceed with their occupations.

Throughout this time, this young architect had been studying for his State Board exam and, after the required two years of practice with a licensed architectural firm, he became eligible for examination. It took him two terms to pass the Board exam. At that point, he was able to run small projects by himself under the direct supervision of one of the managing partners of the firm, and he could even run the construction administration phases of a project while maintaining direct contact with the firm's construction administration and supervision manager. At that point, our young architect was, by all standards, a promising element in the architec=tural world.

Then it happened that his father, who was a prominent figure in his home town, convinced him to open up his own practice there. So with the qualifications described above, our young man embarked on his first independent architectural practice adventure.

The first project he became involved in was a small import/export operation which had overgrown its building area. The owner wanted a preliminary study from which to determine whether to proceed with alterations and additions to the existing facilities, or simply to purchase some land, erect a new building, and lease the existing complex. The spatial variations and modifications to the existing operations were a definite project requirement, since the company itself was undergoing a major restructuring process because of its growth.

Faced with a serious analysis and programming task for the first time in his professional career, our young architect applied a spatial analysis technique he had seen at the office where he had worked previously, which technique can be designated as the "Space Criteria Sheets," containing basically standard room requirements forms, space criteria sheets, and equipment checklist (Chapter 6 of this text contains a more detailed illustration of this type of technique); to these elements he added, on his own: flow diagrams, and room analysis charts in living color.

Following the completion of this study, two things became apparent: First, the restructuring process of the company made the study all but obsolete; and second, because of his inexperience in the field of programming he had used 75% of his professional fees at the time of its completion. Referring once more to Chapter 6, we will note there how some specific analysis and programming systems fail to incorporate past, present, and future conditions in their methodology.

In the particular case we are describing, although the obvious result of the first program was the recommendation for a new facility, a new study made AFTER the restructuring process of the company had taken place proved that the desirable approach was that of alterations and additions to the existing facilities, while a newly created sales branch would be relocated in a rental office complex near the downtown area.

Furthermore, since our young professional was advocating a new building at this point, if only to break even in his unfortunate venture, the client opted for another architectural firm from a nearby city and left our friend with an enormous amount of useless paperwork, a bad reputation, and a stack of unpaid bills.

The end result of this case cannot be considered an unfortunate story altogether because while he failed in a programming activity in this particular instance, the architect DID apply an analysis technique with a certain degree of success in spite of its inaccuracy. The extreme case would probably be most common in situations similar to this one; that is, NO ARCHITECTURAL PROGRAMMING KNOW-HOW AT ALL, very few references on the subject, and no previous education on the basic elements and procedures of this phase.

Based on the knowledge accumulated during his college years, if one of the first tasks our young friend had, when he began working for the architectural firm he was employed by for three years, been one related to building analysis and programming, chances are he would have been fired within two weeks. The problem here is not one of lack of methods or systematic approaches, but one of misconception or ignorance of principles.

Architectural programming does not consist of simple formulas which can be applied time after time without fear or risk of failure but of very involved processes which must be understood at their most basic level, since each project will have essential variations, developments and adjustments that must be carefully identified before a proper programmatic approach can be determined.

To provide a student with a program for a specific design project in the form of a typewritten sheet setting forth the room count and a sketchy description of operational function is just as grave as to lay out projects with no economic considerations, budgeting or construction requirements. This approach assumes the participation of a non-existent client who knows exactly what his needs are, what his wants are, and how to distinguish between the two and who, at the same time, can assist you in the design evaluation of his project. Needless to say, nothing could be farther from the actual conditions under which the real professional design practice takes place. If physicians were trained in such a manner, a visit to a young doctor's office would be as safe as playing Russian roulette.

Undoubtedly, just as there is a need for proper programming in the successful execution of professional design services, there is a need for proper training of the people responsible for the execution of those services during the education and learning stages.

3.2 Professional Design Firms. Their Structures and Components.

Behind every successful professional design operation there is a person or group of persons with very definite characteristics.

Professional designers are multi-faceted people. In order to effectively determine the environment in which specific operations will take place, designers find themselves empathizing continuously with all the constitutive elements they are designing for. And this characteristic, which stretches from the path a secretary must follow from his/her desk to the storage room in order to get a fresh supply of stationery, to the arrival of 80,000 people at a football stadium within a matter of minutes, places the design professional in a position where performance means more than just service. Much more. For a design professional performance means a way of life.

Furthermore, the design phase of the architectural world is one where today is inevitably structured by tomorrow. While this is true of any profession in which planning procedures constitute more than incidental activities, any architecturally related design practice must LIVE this duality throughout every minute of its existence in order to develop. In the design process every idea is evaluated, judged, and considered as real as if it were not just an idea but a finished project. Only in this way can we establish a clear conception of its value.

This holds true not only in the design process itself, but throughout the design development and contract documents stages, in which dimensions, details and material selections are all brought in from TOMORROW and forced to interact TODAY.

One of the fascinating aspects of this group of professionals is, although most of them possess the same basic qualifications and characteristics, they still perform under an organized framework of corporate structure. The concepts of employer-employee and qualified personnel are not only maintained but desired. Specialization, although it has increasingly become a part of professional design practices, has yet to find a comfortable niche in which to rest within their particular structures.

For example, today we may find Promotion Departments, Design Departments, Production Divisions, and others within the structure of an architectural firm; but we do not yet find a Heart Architect, a Mercantile Architect, or a Mechanical Architect. In the field of Architecture they are still just Architects. The reason is simple: the operations contained in the architectural execution framework are so interrelated that major fragmentations would become a hazard rather than an advantage. This also holds true in the fields of Interior Design and Landscape Architecture, where we will rarely find firms entirely dedicated to a specific building type or operational function, much less limitations like Restaurant Interior Designer or Playground Landscape Architect.

Such fragmentation tends to diminish the inventiveness and creative reactions which follow the primary assimilation process by overshadowing its developments with preconceived postulates and directrixes, which might or might not apply in each instance, and by predetermining basic layouts of scopes and objectives which might not correspond entirely to the specific conditions surrounding a problem. Such direct effects and their immediate ramifications must inevitably raise questions about the true value of excessive specialization when applied to the design professions. We must not forget that the scope and objectives statement of a client in the programming stages can determine the approach to take towards the solution of a specific parapet detail, just as the design layout of a series of rooms can essentially affect the construction sequence of their execution.

Therefore, the programming activities of a diversified professional design firm will most likely be done by one of its members rather than by an operational analyst, functional programmer or specialized consultant.

That being the case, the most important considerations in assigning these responsibilities would be those pertaining to the characteristics and objectives of the position itself as well as the necessary qualifications to fill it. That is, the description of the programming job itself and the person behind it.

To illustrate this, let is describe three theoretical architectural firms[10] and their respective approaches to this position. These imaginary firms will be referred to as:

Beige and Associates, Architects, Planners, A.I.A., P.A.

Gray, White and Green Inc., A.I.A.

The Black and Brown Corporation

In describing each firm, we will note three overall approaches to the actual programming job. This does not mean that only three major categories are possible, but that the averages will most likely be found somewhere between them.

In a recent A.I.A. Newsletter, a study of firm sizes by total personnel number demonstrated that while 55.8% of the architectural offices were of a size that only encompassed one through four employees, and 39.5% fell in the five through twenty-four employee range, only 4.7% of the firms had twenty-five and over members.

By the same token, while 26.7% of the registered architects were members of the one through four employee firms, and 49.6% functioned as members of firms with five through twenty-four personnel, only 23.7% of the registered architects operated from offices with twenty-five and over employees. Most of these firms, just as other professional design organizations, follow the general layout of operational functions illustrated in Figure 3.1.

In the illustration of our imaginary spaces and their architectural programming activities, we have structured each one in a different manner. Here is a brief description of each:

BEIGE AND ASSOCIATES, ARCHITECTS, PLANNERS, A.I.A., P.A. There is little doubt in this four employee firm about who is in charge of programming, since Mr. Beige is the head of the office and his duties cover promotion, design, office management and general project supervision. Most of the working drawings portion is handled by Mr. Beige's associate who, in conjunction with an intermediate draftsman, is in charge of construction drawings and specifications.

ARCHITECTURAL OFFICE OPERATIONAL FUNCTIONS

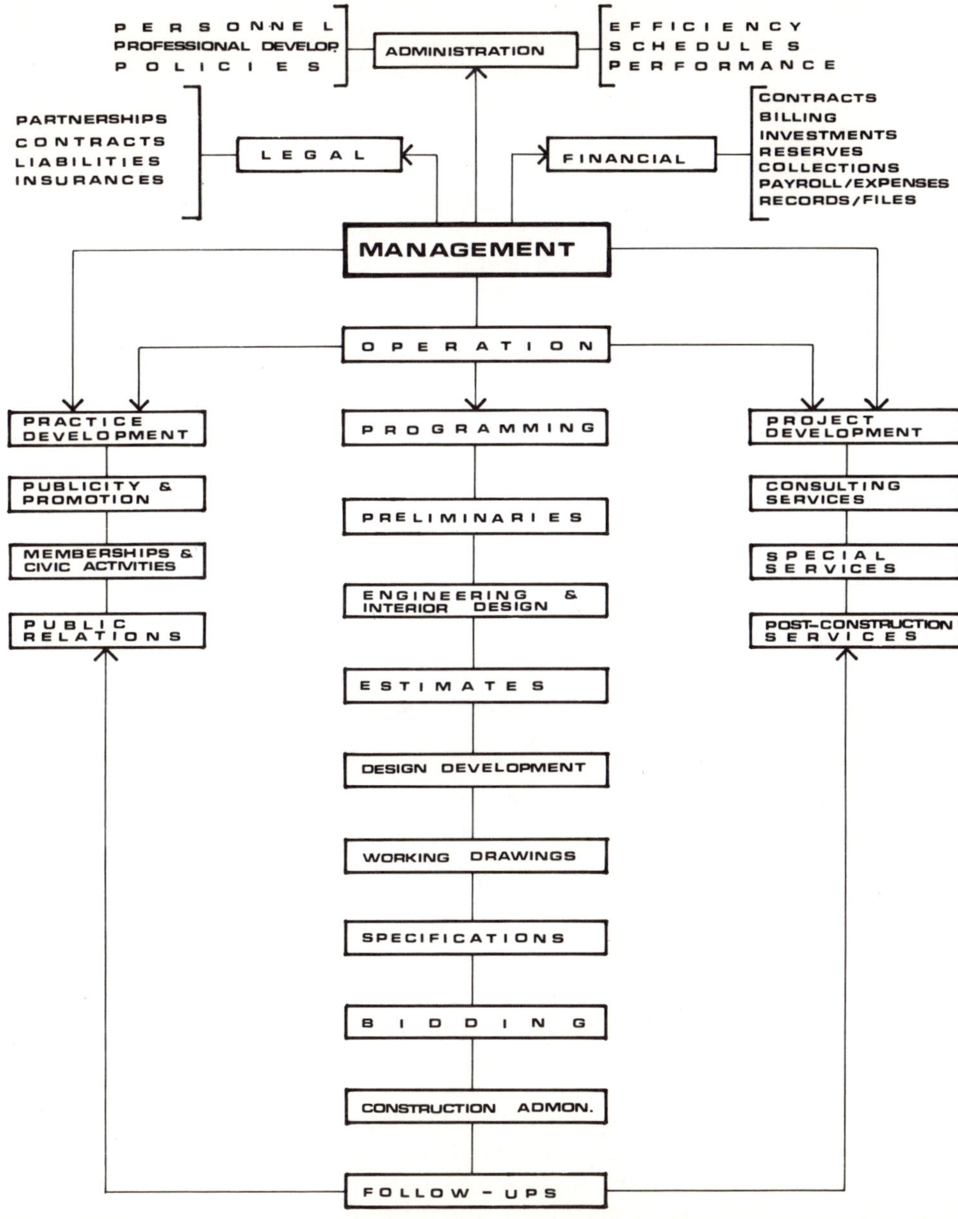

Figure 3.1

One advantage to this type of arrangement is that, since the programming and design responsibilities fall on the same person, chances are that whatever system is used, their programming activities will answer most of the questions the designer might have. Since one individual is both programmer and designer, there will be a much clearer understanding of the problem at the beginning and throughout the design process. And if any questions or difficulties arise because of incomplete investigation, the programmer/designer will become aware of them immediately through the promotion architect, who by now is on a first name basis with the client. Still, there are some drawbacks:

Mr. Beige's programming system is specifically designed to provide the designer with all necessary answers regarding a project's requirements outline, but it does not render any information to his associates concerning construction requirements, spatial relationships description, material selection, etc. In addition, most of the firm's consultants, who never get involved in the programming activities, will inevitably have numerous questions to be directed to Mr. Beige's associate, since he is the one in charge of production and specifications. This results in several errors; the wrong material selection is made once in a while, and Mr. Beige complains continuously that no matter what the situation, he always ends up being a "plan checker" and his twenty-four hour days never seem to be long enough to effectively accomplish his tasks. What happens here is that Mr. Beige has the only first-hand knowledge of the project's scope and objectives as well as its general design requirements, and he has not bothered to include the full breakdown of information pertinent to those activities he is not completely concerned with himself; so upon his return to the office after a full day of meetings, supervision, and promotion, he inevitably finds himself spending valuable hours in conferences and telephone conversations with his associates and consultants. At the end of which time he must return to the office that evening in order to either review what has taken place during his absence or simply

"

try to design another project in a peaceful and quiet atmosphere.

In situations like this one we will find typical conditions like reflected ceiling plans improperly designed, thermostats and telephone outlets placed at the wrong locations, window and hardware schedules improperly selected, the wrong door swings and types chosen, and many more.

The fact is if during the building analysis and programming phase all pertinent disciplines are taken into account, the subsequent margin for errors and questions is greatly reduced, and answering, at the beginning, the questions that will inevitably arise later on is faster and more sensible. This way, it is not necessary to spend time on the correction of errors that occur due to lack of information.

If the problems of Beige and Associates seem relatively simple, let us take a look at the firm of Gray, White and Green Inc., A.I.A.

GRAY, WHITE AND GREEN INC., A.I.A. The corporate structure of this firm is as follows: Mr. Gray, the senior partner, handles promotion, office management and overall design concepts; while Mr. White is the partner in charge of design and some promotional tasks. It is his responsibility to follow up on every project until the completion of the design development phase. Mr. Green, on the other hand, is the partner in charge of working drawings, specifications and construction administration. Their office has, on the average, fifteen to twenty employees, including two secretaries, five registered architects (two of them associates) and approximately three intermediate and two junior draftsmen.

Building analysis and programming are basically done by either Mr. Gray or Mr. White, although the project architect assigned to a specific building sometimes gets involved in these stages.

Because of the different qualifications of the employees and associates, the design process is handled by two or three people working under the direct supervision of Mr. White.

This is a design-oriented firm whose members take great pride in their performance. For this reason they have developed a comprehensive building analysis and programming system that is implemented whenever possible.

This organized approach to analysis and programming has several advantages: the element in charge of design relates directly to the establishment of the program, as in the previous example; one of the managing partners is actively involved during the first working session with the clientele, and most important of all, they have a SYSTEM.

Still, some disadvantages can be noted. In many situations, Mr. Gray develops the program, but he is primarily a promotion and managing architect, not a programmer. Programming, like any other architectural discipline, requires constant practice and development; it should not be done by just "anyone," no matter how simple the approach might seem. Mr. Gray is very well qualified to resolve matters of design and general construction in conferences with clients; but because of his performance, he has lost contact with the everyday production requirements, codes and regulations, specifications, etc.

Mr. White, on the other hand, is much more knowledgeable in these areas; but because he is essentially a designer, his approach to analysis and programming is not totally impartial. He emphasizes those areas which deal directly with design in his analysis and programming activities. Furthermore, when exercising influence on a client's decisions, he finds himself advocating those which will benefit the program in favor of his design predilections. This brings us to a second disadvantage: If, during the building analysis and programming stages, the programmer must function as an actively directive element while maintaining total impartiality towards all subsequent architectural phases, it may be risky to place the partner in charge of design in this position.

Another problem is that the designs themselves are executed by the other project architects who do not always sit in on the programming sessions. If they do not attend these sessions, the firm is definitely making a terrible mistake. The second hand, word-of-mouth information passed on to these people will inevitably result in variations or interpretations of the actual response. Having the other project architects sit in on the programming sessions will help, but they may still find themselves confused because programming is not their specialty, and they are unfamiliar with the system being used, since this practice has relegated them to the role of mere spectators throughout these procedures. While there should always be only one programming department in professional design organizations, every element of a firm should be familiar with the method or methods employed by that department.

This brings us to Mr. Green, who very rarely gets involved in the programming stage, and who inevitably ends up arranging conferences with the client or his representatives to resolve unanswered questions regarding hardware schedules, construction procedures and costs, materials selection, engineering requirements, bidding and contractual procedures, etc.

These conferences are not only necessary but advisable in most cases, but the frequency with which they take place could very well raise the questions: "Are we doing something wrong?" or "Is there any way we can reduce the number of problems?" These questions can be resolved, either completely or to a great extent, by utilizing the project analysis and programming stages to better advantage.

To criticize or rectify a poorly executed program is, in a case like this one, extremely difficult, since the program is carried out by one of the bosses and very few people will confront their employers with the fact that they are not doing their job properly. But even if it goes unmentioned, the fact remains that they are not doing their job, because in establishing the program, they function as programmers and their work should be evaluated only on that basis.

It can be argued that generally the processes involved in building analysis and programming are so extensive that the managing partners of the firm have no time to handle them. If that is the case SOMEONE in the firm ought to be given that responsibility, someone with the necessary qualifications, knowledge and experience and, most important, someone who will answer to management for the performance of these duties.

Because of their size and complexity, as well as their departmental organizations, we will not find this disadvantage in The Black and Brown Corporation.

THE BLACK AND BROWN CORPORATION. This 80 to 100 person office is basically headed by Mr. Brown, Jr. with Mr. Brown, Sr. acting as chairman of the board. The name of the firm's founding partner, Mr. Black was retained in the corporation title following his death.

One of the senior partners of the firm is the man in charge of architectural programming, who has held this position for several years. Not only are these processes his responsibility, they constitute his basic duties in the corporation.

The board of directors of The Black and Brown Corporation has already realized that the first thing an architectural office sells a client is not ideas, but information. They are no longer stressing the importance of design over any other portion of their services, since design is not their only strong point. Stressing the importance of any firm's strong point is not, by any means, an error; what is an error is to ignore the importance of other phases, or to consider them as mere functions of one's own strengths.

The success or failure of the programming activities in The Black and Brown Corporation hinges on the system or systems used by its programmer and on how well-implemented these methods are within the personnel structure. The advantages or disadvantages here are those of the system itself rather than of the allocation of responsibilities.

As long as their method works, there will be no problem. If their method fails, there is no need to say what will happen, and on what scale. The main advantage of The Black and Brown Corporation is the fact that building analysis and programming have been given their definite identity as an architectural function; and as such, this phase has been studied, analyzed, planned, programmed, and executed.

The only disadvantage of this situation is a somewhat theoretical one, namely that, because of the many elements involved in the subsequent phases, and the need for all of them to become familiar with the program results, the method's basic approach is fairly rigid and allows little room for adjustments or variations should the need arise. That is, before varying his analysis and programming method, the programmer at The Black and Brown Corporation must be very careful not to disrupt the natural flow of events to follow.

Still, any problem of this sort could be easily corrected.

Summarizing, we can draw a general sketch of what the programming position encompasses.

First, it serves as the client's primary source of guidance and information.

Second, as a direct result of this characteristic and because of its early contact with the clientele, the programming position becomes the first professional service rendered to a client, and thus sets forth the first positive or negative influence in the client-firm relationship. It should give assistance and build confidence, as well as establilsh the rules which are to govern the relationship.

Third, it defines scopes and objectives for specific projects.

Fourth, it collects data and facts as well as pertinent references regarding existing or theoretical conditions, and studies and evaluates their entire layout. In other words, it analyzes them.

Fifth, it provides a framework for working with the proper agents to establish concepts and determining factors, as well as specific needs and controllable variables or "wants."

Sixth, it determines courses of action regarding these elements, reaches conclusions, and programs them.

Seventh, it then projects itself towards the remaining professional design procedures and provides them with the necessary information to effect their tasks.

To fulfill these duties, a person must have not only the appropriate personality, with capabilities for listening and understanding, and an overall disposition for dealing with clients, but also the necessary educational background, construction knowledge, and design sensitivity. It is obvious that the description fits many design professionals. Therefore, the fulfillment of the requirements CAN and SHOULD be achieved.

3.3 Professional Design Practice

THE OUTER OFFICE. Since the architectural programming process involves the two major procedures of analysis and programming as essential constituents, it is very surprising that professional design firms are not promoting building analysis services more often then they do. If we take building analysis and building programming as two separate entities, we will find that, while the second process presumes the existence of a problem, the first one does not.

If space is currently considered as much a tool as machinery itself, it is very strange to find no servicing activities for that specific portion of human operations,

A physical check-up does not necessarily involve an ailment, but rather an analysis and evaluation of present conditions in order to offer suggestions that will either minimize or prevent major problems. With this approach, professional designers could avoid eliminating possible projects; and, they could also help their clients diagnose their spatial conditions and predict their development. As a matter of fact, a careful analysis might result in an altogether new project if previously unnoticed problems became apparent before their effects became critical.

By the same token, once these activities are explained, the client will probably realize that there are as many benefits to be garnered in this process as in servicing and maintaining operational equipment or accounting records. Space is as operational as anything else, and the relative cost of these services, as compared to those of full professional fees, is minimal.

The problem in many cases is one cannot sell what one cannot deliver; and since building analysis and programming, as elements of an architectural practice, have generally been considered (and in many cases are designed to be) supporting services for the basic design and production disciplines, they have very rarely stood by themselves as independent processes. But even that circumstance does not mean that spatial diagnoses cannot be accomplished. They most definitely can be. All it takes is the interest, the concern, and the effort to take greater care in the layout and establishment of an effective programming method.

THE INNER OFFICE. Try to imagine for a moment the evaluative process of a preliminary sketch. Isn't it done as a function of the program it serves? How can we logically evaluate design without programming?

We can not; it is as simple as that. If we do not know to what end something is done, we cannot pass judgment upon it.

Now, if the essential tool for design evaluation is programming, there is little doubt that the more clearly we define this last process the better we will achieve the first. Throughout the design process, just as in any other creative discipline, a dynamic transformation operates which can only be described as a conceptual understanding of sorts and which, by reason of its own structure, determines many architectural programming approaches.

It is no secret that if a program is given to a designer in the form of typewritten sheets which state room names, sizes, and occupancy designations, his/her preliminary solutions will attempt to establish the primary relationships between those spaces. But without a clear understanding of how these elements interact, chances are those first attempts will only be preliminary efforts to understand how individual spaces are to perform within the whole.

The process would be similar to trying to create a dynamic entity out of a series of inanimate cells with complete disregard of the proper arrangements and flow patterns necessary to develop, sustain and distribute its energy forces. Furthermore, in professional design spaces are not "inanimate cells" but living operational elements.

After working on design solutions for awhile, the possible spatial arrangements and alternatives always become less complex. The reason for this is very clear: by then, all the inert elements have become active through subsequent research and discarded interactions. In other words, through a long process of trial and error in the determination of something that could have been clearly simplified beforehand.

This study, which constitutes the essence of the spatial relationships investigation, does not correspond to the design process at all, but rather to the precluding architectural programming phase.

That is why, from the traditional approach to the objective design process (see Figure 3.2), we can derive the revised procedure shown in Figure 3.3.

Thus, in contrast with the simple program statement in Figure 3.2, where we force the design process to interpret, evaluate, and classify its operational factors, in Figure 3.3 we provide the design process with both a clear definition of the problem and also a strategy to follow as indicated by that definition.

3.4 Architecture and Perspective

Before closing this chapter, let us make one final remark concerning the fact that Architecture is generally not viewed from the same perspective by its users as it is by the team of professionals who execute it.

For all of those occupations involved in the design and construction of a building project, its completion means an end, while for its occupants this represents only a beginning. In general, the functional characteristics of a specific structure will be more important in determining its value for its inhabitants than for its creators.

This does not mean that aesthetic values and form/function compromises are undesirable; but generally, when a portion of a building does not operate correctly, chances are we will not only have to face the complaints and their consequences, but also find the necessary remedies for the problems that occur.

It is of the utmost importance in programming to realize this fact at the outset and maintain this attitude throughout the entire procedure. If a program does not clarify all the functional requirements for a building, as well as its spatial classifications, needs, purpose and objectives, it is not fulfilling its function. And if to omit is grave in these situations, to be indeterminate is even worse; because while omission may lead to questions and delays, indetermination may lead to misinterpretations and errors.

Every element involved in the architectural processes, from owner to electrical subcontractor, must learn to compromise at a certain stage. The ability to compromise is essential to any individual who wishes to succeed in any field related to the construction industry and, probably, in most other areas of life, but improper knowledge of situations and circumstances can only aggravate the compromising action, never facilitate it. That is why Architecture, as such, should never be viewed as either means or end, but rather as both. Because while its programming activities DO outline an end, the completion of its physical execution only marks the beginning of another of its phases: its performance.

---

[10]Although all three firms will be assumed to offer essentially comprehensive architectural services, similar structures and functional characteristics can be found within other professional design organizations.

[11]The A.I.A. Document D200, "Project Checklist" Pre-Contractual Part, provides a comprehensive checklist of basic data which is highly recommendable for these purposes.

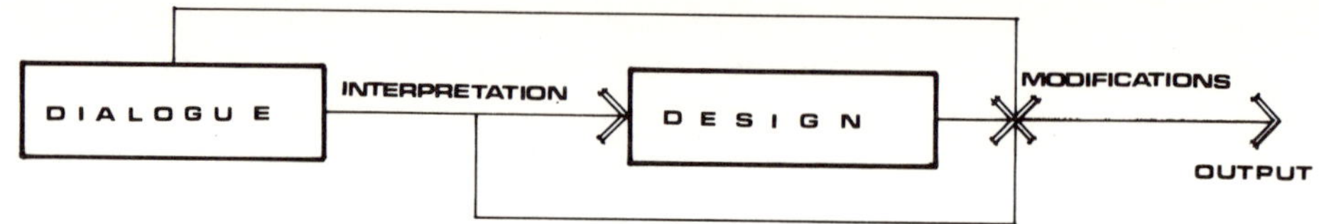

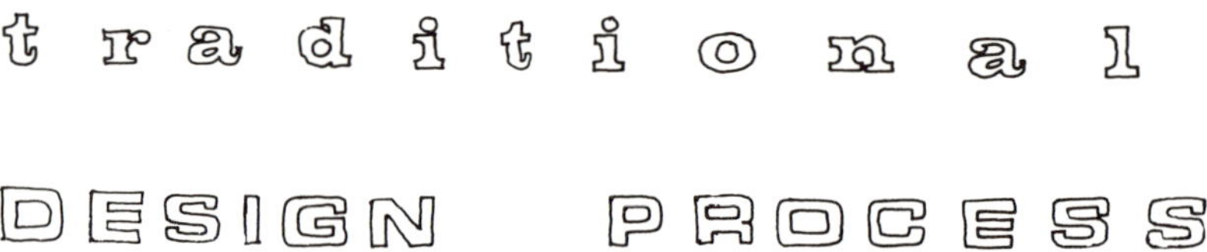

Figure 3.2

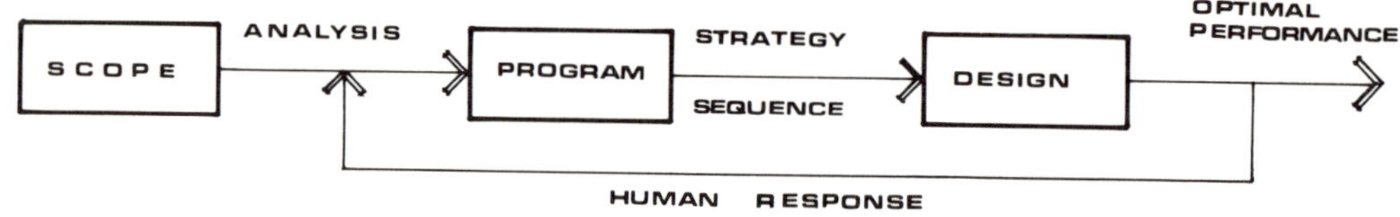

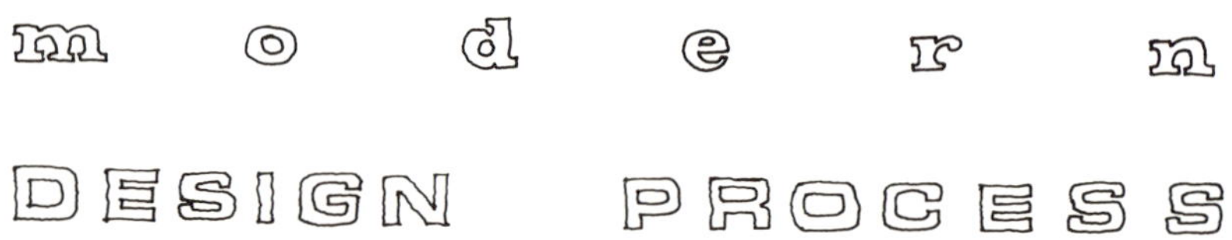

Figure 3.3

4

Clientele and Related Professions

4.1 Clientele

The elements included in the term "clientele" as used by our professions are so numerous that any attempt to tabulate them would probably result in an inaccurate outline. The word *client* sometimes represents not only the owner, the owner's representative, or the user, but also a wide variety of elements ranging from single individuals to building commissions.

Among those who might function as clients of a professional design firm at a given point, we can include the following:

Individuals
Families
Professional associations
Business officials
Construction or business consultants
Corporate officials and consultants
Government officials
Building commissions
Developers
Institutional groups or associations
Owner's agents or representatives

We can add to this list variables based on their function, regionality, local, state or federal levels, as well as socio-economic class, nationality, structure, race, number, creed, etc. Obviously, a comprehensive chart would end up classifying mankind.

Still, when any of these elements approaches a professional design firm, they become "clients" and, as such, there are certain common characteristics which apply to most of them.

First, any client approaching a professional design organization probably needs the design services such an organization provides. While this might appear to be an obvious statement at first glance, it does not hold true for other professions in which the range of activities has been fragmented into much more specialized tasks, such as physicians, dentists, attorneys, engineers, etc.

Second, generally, a client's first inquiry is usually a request for information in a field s/he knows little or nothing about. Actually, except for those who have been previously involved in construction projects, the majority of clients will approach a professional designer with nothing more than a general problem outline and a series of basic elementary questions.

Third, the client almost always has to be briefed on what professional designers do and what their roles and responsibilities are, as well as on what part the client himself will play in the overall professional design framework.

As we have mentioned before, it is during this educative stage that certain concepts should be clarified, but more often than not the careful explanation of what architectural programming represents is omitted.

Since a contractual agreement will be made, it is of the utmost importance to explain to the client the role this professional will play in the programming, design, and execution of the project. Many clients believe that building analysis and programming are part of the basic professional design services they are contracting for. Moreover, many of them have no concept of what these two procedures are all about, much less any understanding of the role they themselves are about to play during those phases.

In facing this educative process, the professional designer should concentrate more on principles than on details. Once a client has been informed of the importance of these phases, it is very unlikely that he will choose to circumvent or eliminate them.

Furthermore in stating principles, scope and responsibilities, it would be wise to establish a series of clearly defined Programming General Conditions, by virtue of which each party will be responsible for the acquisition of pertinent information and determining factors, as well as decisions, directions and outlines regarding alternatives, tasks and developments.

One particular problem which we might refer to as the "free sketch syndrome," can be corrected if not totally eliminated by emphasizing the vital importance of the architectural programming phase. Most clients who appear in a professional design office with the hope of obtaining, by means of "some sort of sketch," an idea of the designer's talent and capabilities and approach to a problem, or just outright competency, will probably learn for the first time that design is not an isolated action dependent only on talent, experience or knowledge, but an involved process which can only be realized effectively once these factors have been applied in the proper direction. Direction which can only be outlined in the comprehensive study of requirements and courses of action contained within a building program.

At the same time, a free sketch executed without proper knowledge of a problem can be extremely misleading. This also holds true for ideas or suggestions ventured in the course of preliminary conferences which prove inadequate after the analysis and programming processes have been completed, causing embarrassing corrections of statements or worse yet, becoming established guidelines that, instead of helping the project, result in erroneous approaches and create obstacles in the search for the final solution.

Morris Lapidus, in his book *Architecture: A Profession and Business*, tells the story of a personal friend and client of his who in spite of his desire to give a particular project to his friend, ended up hiring an architectural firm other than Lapidus' after a misunderstanding concerning the architectural style proposed for the project. If we "assume" that any project carries, within its scope and statement of objectives, a clear definition of style or design approach, we leave ourselves open to this type of pitfall. Moreover, we make the terrible mistake of assuming that the selection of architectural styles is our domain and no one else's when the first postulate of architectural programming has always been: Do not assume anything, or take anything for granted.

One noteworthy system employed very successfully by a southeastern architectural firm not only reduces the possibility of a free sketch syndrome, but, at the same time, clarifies at the outset several items which generally go unnoticed at first. This system can be defined as the *Introductory Questionnaire* or *Telephone Interview Data Sheet*. The idea behind this system is not only to gather as much preliminary data as possible during the first

OWNER / REP.	NAME	
	ADDRESS	
	PHONE	
PROJECT	NEW	
	EXISTING	EVALUATION
		REMODELING
		ADDITION
PROFESSIONAL SERVICES	BASIC	
	COMPREHENSIVE	
	OTHER	
SITE	SELECTION	
	EXISTING	EVALUATION
		ANALYSIS
		IMPROVEMENT
		EXPANSION
		EXISTING STRUCTURES
		SIZE
	SURVEY	
	UTILITIES	
	PARKING	
	LEGAL CONDITIONS	
BUILDING TYPE	TYPE	
	FUNCTION	
	SIZE	
ECONOMIC DATA	BUDGET	CONSTRUCTION COST
		SQ. FT. COST / QUALITY
		PROJECT COST
	FEASIBILITY STUDY	
TEMPORAL DATA	SCHEDULE	
	PHASING	
	GROWTH / CHANGE	
REGULATORY CONDITIONS		
PROJECT CONSTRUCTION	BID	
	NEGOTIATED	
	DESIGN / BUILT	
	OTHER	
REFERRALS	SOURCE	
	MEDIA	
REMARKS		

Figure 4.1

phone call about a prospective job, but also to pinpoint, with all the precision the situation permits, the trouble areas and preliminary definition of programming elements as listed in a comprehensive checklist: Questions and items which might otherwise have been omitted due to lack of order and continuity or simply due to forgetfulness. (See Figure 4.1)

Notice that although the questions are very generalized, they still cover all the basic areas to be determined during the early stages of a project.

Further developments of formats of this type include more detailed questionnaires or checklists[11] based on building types, size, etc. Such developments, although advantageous in certain cases can sometimes become so involved that their implementation is awkward in contrast with a simple and direct generalized checklist. One thing to remember is that these introductory questionnaires are not designed to be presented to a prospective client over the telephone as a question and answer quiz, but rather as a checklist through which the interviewer moves gradually from one area to another, *leading* the CONVERSATION rather than *conducting* an INTERROGATION.

Another phase generally downplayed is that of the follow-up of professional design projects. This portion which should definitely be included in the building analysis phase, if only to double check its conclusions and establish an Analysis Reference Library, can also serve as a useful tool in clientele surveying.

For example, a single follow-up questionnaire that would gather the data needed for such feedback could be summarized in the following list:

Office Performance
    Professional Services
    Technical Personnel
    Clerical Personnel
    Correspondence
    Contracts
    Appointments, Meetings, Presentations
    Deadlines
    Executive Personnel

Design
  Aesthetic Value
  Functionalism
  Performance

Materials
  Adequacy
  Performance
  Care and Maintenance

Equipment and Furnishings
  Selection and Performance
  Maintenance
  Operating Costs

Economic Considerations
  Budgeting
  Maintenance
  Taxes

Overall Comments: _______________________

____

____

Chapter 10 contains a much more comprehensive follow-up/feedback format; for now, this general outline will suffice to illustrate the basic analysis areas in question.

Surprisingly, professional design organizations have not yet surveyed in depth the performance of the design professions on a national basis. That is, no survey has been conducted that would enable design professionals to improve certain facets of their services, or even their overall performance. It would actually be very interesting to learn some public opinions regarding functionalism and aesthetics as related to building types from owners, users or even passers-by. And just to prove the extent to which research in this area has been neglected, try for a moment to determine which building type the general public would select as the most aesthetically unpleasing? Or which as the most attractive? This research section of our profession, which to our general knowledge is virgin land, could not only produce startling results, but also improve, redirect, or even radically change the course of some present trends.

Peter F. Drucker, in Chapter 7 of his book *Management*, makes an excellent presentation of Business Purpose and Business Mission by clearly outlining the following questions:
1. "What is our Business?"
2. "Who is the Customer?"
3. "What is value to the Customer?"
4. "What should our Business be?"
Just by substituting the word "Profession" for "Business" and substituting "Client" for "Customer" we will find the entire context of that chapter directly applicable to the professional design practice of our days.

Through an impartial evaluation of the answers to these questions, as well as a comprehensive study of existing conditions and their ramifications, we can produce and develop an unlimited number of benefits. In doing so, we will also benefit our clientele and everyone affected by our professions and their accomplishments; in short, society as a whole.

Summarizing: The involvement of clients throughout the architectural programming stage can be defined as follows:
1. Determination of scope and objectives
2. Procurement of all pertinent data
3. Establishment of spatial requirements
4. Establishment of spatial relationships
5. Determination of special characteristics
6. Establishment of design philosophy
7. Determination of budgeting and economic considerations
8. Procurement of marketing and sales studies results
9. Establishment of temporal aspects, project deadlines, time tables, etc.
10. Forecasting of growth and change projections, implications and flexibilities
11. Determination of operational and maintenance requirements
12. Verification of construction requirements

Thus, the involvement of the client can range from simple assistance in the establishment of a program to participation in major decision-making processes and data collection activities, with or

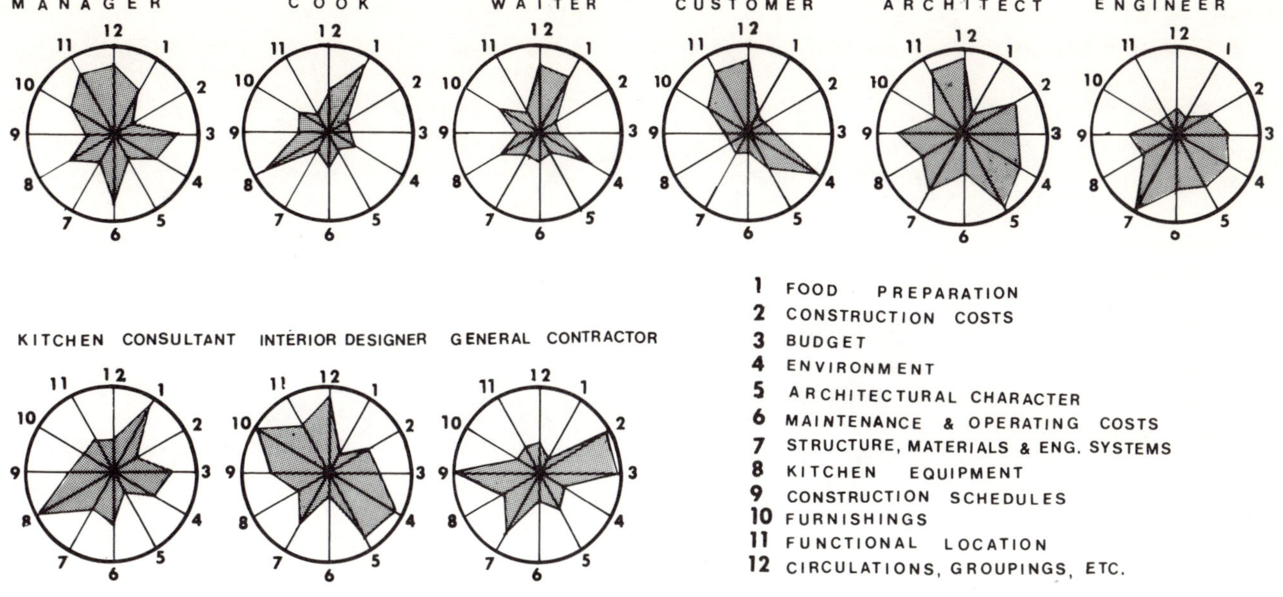

Figure 4.3

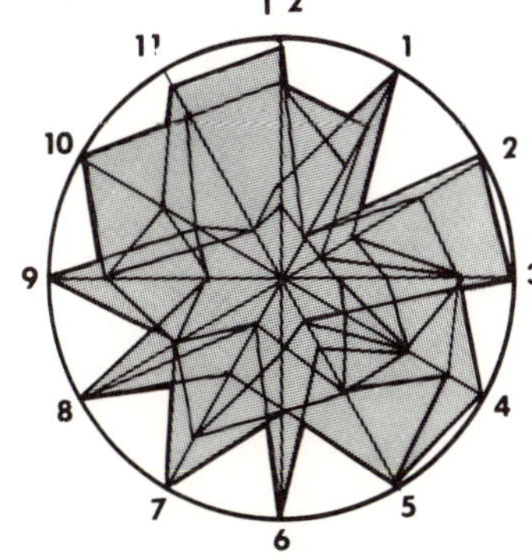

Figure 4.2

without the direct involvement of a specialized programmer as the case may dictate. Needless to say, the role played by a client during this stage is by no means an easy one; but then neither are the building analysis or programming phases of a building project.

4.2 Related Professions

Because of the complexity of operations included in the construction industry today, its planning processes involve a series of specialized disciplines which sometimes act as supporting services and sometimes as basic elements of execution, and which cannot be ignored if we are to establish a complete framework for professional design accomplishments.

Some of these disciplines, even when functioning as mere aids in the initial stages of the professional design process, can become determining factors in the final programming of a project; and in many cases assume the importance of essential factors from the very beginning.

If we chart the specialized knowledge of nine basic professions with twelve tasks directly related to the architecture of a restaurant, we get the results shown in Figure 4.2.

These specialized areas of knowledge, once incorporated into a team, produce the diagram which appears in Figure 4.3.

As we can see, only through such superimposition of specialties can we satisfy all the necessary demands for the successful execution of the project. Not all of the services rendered by all design-related professions will be consistently employed in analyzing and programming architectural projects; but in some cases the considerations of each one or several of them could be extremely useful or even essential to establishing a proper list of project requirements.

A concise breakdown of architecture and construction-related professions and their possible involvement in the programming stage of building projects could be summarized as follows:

ARCHITECTS.
1. Research and analysis of architectural project requirements
2. Project programming
3. Site planning recommendations
4. Operational and maintenance requirements and recommendations
5. Improvements
6. General construction cost analyses and life cycle cost studies
7. Growth and change projections, phasing and schedules
8. Construction systems, materials, procedures and recommendations
9. Project design philosophy requirements
10. Regulatory surveys.

ENGINEERS.
1. Research and analysis of specialized systems
2. Evaluation and recommendations of engineering systems
3. Engineering cost analysis and life cycle cost studies
4. Operational and maintenance requirements and recommendations
5. Improvements
6. Growth and change projections, evaluation and comments
7. Spatial requirements data
8. Related needs and requirements
9. Regulatory surveys

INTERIOR DESIGNERS.
1. Analysis and research of interior design philosophy and requirements
2. Spatial needs, recommendations and programming
3. Cost analysis of interiors and furnishings
4. Recommended improvements
5. Growth and change projections, evaluation and comments
6. Materials and systems recommendations
7. Project design philosophy requirements

URBAN PLANNERS.
1. Research and analysis in site and urban planning
2. Community characteristics, evaluation and comments
3. Recommended improvements and programming
4. Growth and change projections, evaluation and recommendations
5. Environmental considerations
6. Regulatory surveys
7. Cost analyses and life cycle cost studies

LANDSCAPE ARCHITECTS.
1. Research and analysis of project requirements
2. Site planning recommendations
3. Materials and systems recommendations, upkeep and maintenance studies
4. Landscape architecture cost analyses
5. Improvements
6. Landscaping spatial and functional requirements
7. Growth and change projections
8. Landscaping availability and regionality
9. Project design philosophy requirements
10. Regulatory surveys

BUILDERS AND SUBCONTRACTORS.
1. Materials and construction systems recommendations
2. Project location and site analysis recommendations
3. Growth and change projections
4. Construction time, costs and work schedules
5. Related trades effects, designation and comments

ESTIMATORS.
1. Costs analysis and projections
2. Materials and construction systems availability, evaluation and comments
3. Performance/Cost evaluation of products, materials and systems, evaluation and recommendations
4. Consideration of growth and change projections

REAL ESTATE CONSULTANTS.
1. Analysis of market demands and existing facilities evaluation and recommendations
2. Site recommendations
3. Growth and change projections
4. Leasing and operational requirements, evaluation and recommendations
5. Costs, interest rates and contractual factors

LEGAL ADVISORS.
1. Project legal requirements
2. Contractual conditions
3. Growth and change projections input
4. Legal recommendations

ECONOMIC CONSULTANTS.
1. Financial analyses and feasibility studies
2. Research and recommendations
3. Consideration of growth and change projections, evaluation and comments
4. Taxation, depreciation, maintenance, etc.

SPECIAL CONSULTANTS.
1. Analysis and recommendations of the items covered by their particular discipline, as well as those directly or indirectly related to them.
2. Spatial, functional and operational requirements
3. Site planning and building programming recommendations
4. Growth and change projection analysis and consideration of effects

BUILDING OFFICIALS.
1. Project analysis, definition and classification
2. Applicable codes, regulations and requirements
3. Recommendations
4. Growth and change projections, evaluation and recommendations

PRODUCTS REPRESENTATIVES.
1. Complete product data
2. Availability, costs and recommendations
3. Maintenance, guarantees and performance information
4. Product spatial, environmental, operational and servicing requirements
5. Installation procedures and instructions

As we can see, the coverage goes beyond the limits of a simple breakdown of spaces and their overall relationship to one another. Obviously, each of these professions could, when applicable, determine directions and requirements for a given building project. Thus, the involvement of these trades in the primary stages of architectural programming is not only desirable but, in many cases, absolutely essential.

---

[11]The A.I.A. Document D200, "Project Checklist" Pre-Contractual Part, provides a comprehensive checklist of basic data which is highly recommendable for these purposes.

5

Effects of the Lack of Proper Building Analysis and Programming

5.1 Professional Design Services

Perhaps the most damaging result of the infrequent use of proper building analysis and programming methodologies is the overall lack of awareness of the intrinsic advantages of those procedures. In other words, their inability to prove their own value due to their scarce application. We can cite numerous reasons why these procedures are often relegated to a mere preliminary activity without definite structure or identity, reasons which will be based on the inadequacy of the system of systems employed for a specific project, lack of proper execution by the personnel in charge of these activities, or simply the common rushing through undefined programmatic objectives towards one definite design activity with limits and tasks roughly defined according to the generalized concept of basic professional design services only.

Whatever the reasons, the fact is unless a programming system is adopted by a professional design firm, its benefits will never be known. Still, while said advantages may not become apparent at the outset, the disadvantages resulting from the absence of a programming system can definitely be felt and recognized. If we were to consider the lack of building analysis and programming as a "cause," it could very easily be blamed for a series of adverse effects which affect the design professions performance from their very basic services to their practical applications, as well as some of the related developments of the design disciplines, such as marketing and promotion activities, related professions and building trades, and the construction industry in general.

Professional design services, whether better or best, depend entirely on the successful execution of a series of consecutive tasks rather than on the particular achievement of isolated actions or procedures. To assume that a poor design solution can be overcome by brilliantly executed contract documents is as erroneous as to presume that qualilty drafting can compensate for dimensional errors.

Considering only three basic constituents of the design professions practice, Schematic Design, Design Development and Production, as elements of analysis, we can easily determine how flaws, errors and omissions can occur due to improper analysis and programming.

SCHEMATIC DESIGN.

1. The absence of improper execution of architectural programming will result in the lack of adequate information to successfully start the design process with a clear and comprehensive understanding of the problem it must resolve. The improper execution of the programming stage will tend to confuse the designer, especially during the very critical conceptual stages of creative tasks. Also, because of this, the completion time of the design process will inevitably increase.

2. Although improper programming results may not eliminate the best design solution, they will definitely become an obstacle in its development because they will complicate the inevitable "trial-error" approach of the design procedure and its performance. Moreover, the absence of building analysis and programming results in the failure to provide pertinent information regarding the identification and structure of programmatic flexibilities, issue which is a key factor in attaining viable solutions for professional design problems. Furthermore, a poorly executed program creates confusion and disagreements between designers and programmers since the former are receiving only incomplete information from which to proceed. Obviously, the lack of building analysis and programming creates frustrations and unnecessary expenses through the repeated submittals of possible solutions as new requirements appear with each new scheme presented for evaluation.

3. The absence of a clearly defined program prevents the design phase from transmitting sufficient information to the following professional design phases. Moreover, the client is not provided with the necessary information and value system to understand and assimilate why certain things are done in a certain way; nor is s/he helped towards the full comprehension of the project's requirements, constraints and objectives, and their reflection in the proposed solution. Without a program there is no adequate basis for judging and evaluating a specific design solution objectively.[12] The absence of programmatic statements also imperils the positive effect of possible feedback information from which to develop spatial and operational standards, new design philosophies and functional solutions.

DESIGN DEVELOPMENT AND PRODUCTION.

1. The communication between the design activities and the production phase will inevitably become much more complex and, as it does so, the margin for errors and omissions will increase if the analysis and programming procedures are ignored. Without a program no information whatsoever is provided concerning the intent of the design, and this can lead to misinterpretations regarding material, furnishings and equipment selections, sound and visual barriers and construction procedures or dimensional and overall adjustments.

2. Detailing, hardware specifications, the location and sizes of doors and windows, and many other miscellaneous specialities cannot be pinpointed logically and accurately if, during the first analysis and statement of intent, the corresponding data has not been properly defined. Such absence of architectural programming will provide no feedback data from which to develop new systems, techniques and procedures which can help improve the contents and effectiveness of contract documents.

5.2 Professional Design Practices

As a direct result of the effects on the design and production services previously noted, the generalized concept of professional design practice is affected by the lack of building analysis and programming at its roots, whether through the basic structure or essential activities or through their ultimate performance factors.

Thus the practice of professional design is dependent on these two primary processes either in the general conception of its goals or in a particular set of circumstances that may develop in a specific firm.

The effect of inadequate building analysis and programming on professional design practices in general can be outlined as follows:

1. The absence of a program prevents clarification of the design intent, especially when a client's original statement of scope and objectives happens to be inaccurate, obscure or simply wrong and unapplicable. Since without the aid of analysis and programming procedures, factors essential to the successful execution of construction projects, such as socio-economic problems, psychological considerations, energy conservation, cost control and budgeting, etc., are not studied properly at the outset and consequently their effect cannot be weighed or determined accurately during the planning stages. Furthermore, improper execution or lack of building analysis and programming methodologies will inevitably create the need for specialized references, which, in most situations, will act as strict guidelines. What initially can be construed as a positive development, becomes in the majority of cases a stumbling block.

2. The lack of properly executed programmatic activities diminishes the role of the design professional as a spatial specialist, making him appear to be only a designer and construction technician. The question is whether form and function are living organisms or simply building blocks with which to play and reorganize the human environment. Obviously, without proper programming, professional design cannot accomplish its primary task: performance. In professional design almost every factor and solution is a simple matter of choices which must be backed up by sound reasoning and careful scrutiny. The lack of building analysis and proper programming imperils both the reasoning and scrutiny of such alternatives and, therefore, their proper selection.

As mentioned previously, the professional design practice is also affected on a concrete, practical level by improperly conducted analysis and programming. Among those facts which may affect the practice of professional design in a more concrete manner, we find that a properly executed program will help maintain satisfactory cost control over the entire professional design process; but even if cost control proves unsatisfactory, we will find that projects whose cost exceeds the original budget by 10-15% go back to programming reviews and the design table for cutbacks. In most cases of this nature, production alone cannot (by means of material substitutions, construction alternatives, etc.) decrease costs sufficiently to keep the project within its budgetary limits. The lack of architectural programming increases the amount of design and production time and, therefore, their cost and directly affects the construction administration and project supervision phase by jeopardizing a clear interpretation of the design intent. Regarding clarification as a factor in itself, we also find that an improperly prepared program can cause serious errors in design and production by reinforcing the assumption that, since there are no clearly defined requirements, previously used references and standards "work well all the time" when the fact is NOTHING works well "all the time."

Improperly executed programs will result in the need of numerous meetings with the client for review of possible schemes and alternatives, with the direct effects of time consumption and increased execution costs. Furthermore a program, when properly executed, produces a much clearer picture of schedules, scope and extent of services and research procedures, since the clear definition of any problem contains the basic outline of those tasks necessary for its resolution. To this we must add that inadequate analysis and programming activities result in the failure to pinpoint responsibilities, objectives and statements of intent, not only of the design professionals and clients, but also of specialized consultants and the various elements of the professional design firm itself. Thus there is no definition of a clear schedule of activities.

Like the basic disciplines directly related to the professional design practice, other phases of a project are also affected by improperly prepared programs. Marketing and promotion, the construction industry in general and other related professions will suffer from the lack of adequate analysis and programming.

A brief summary of some of the problems that might arise can be outlined in the following manner:

MARKETING AND PROMOTION. The lack of a building analysis and programming system will obviously preclude the marketing of these particular services to clients. Brochures, promotion, etc. cannot refer to analysis and programming services unless a proper architectural programming method has been developed, executed and proven effective. Also, when analysis and programming are needed as back-up material or specialized reference in the bid for a construction project, there will be no examples or actual case studies on the basis of which to prove or justify expertise and experience.

CONSTRUCTION INDUSTRY. As a direct result of the effect on the production of contract documents, improper planning and programming will prevent the trades involved in the execution of construction projects from obtaining pertinent information regarding the basic intent of a set of drawings. This can also discourage product manufacturers from creating, developing or improving new products or existing construction systems, since they will not be provided with requirements for specific conditions of functional programming and levels of performance for construction products. Standardization and stagnation are dangerous corollaries of this state of affairs.

Furthermore, suggestions and variations regarding stipulations found in a set of working drawings from trades not directly related to the programming procedure will most likely be based on cost or simplicity of execution, without regard for the intent of specific requirements as applied to functional programming. This generally results in undesirable changes or quality control variations in the contract documents, if the suggestions are not brought to the attention of the programming professional but are simply suggested to the owner without proper knowledge of the "whys and hows." Just as standards of performance are based on specifications, the intent of those specifications is directly determined by the standards of performance outlined by programmatic requirements.

In addition, energy conservation and attempts to deal with today's environmental crisis will inevitably fall back on the designer's table once everything that can be achieved through existing construction systems has been done. Without a clearly defined method of dealing with these problems, the solutions will consist only of "effects" treatments rather than "causes" remedies.

RELATED PROFESSIONS. The lack of building analysis and programming can adversely affect the overall design criteria and conceptual development approaches of each related discipline involved in a building project, since analysis and programming procedures are the basic elements which provide specialized consultants with the spectrum of informative data for a project so they can determine and clarify the scope and extent of their collaboration. Undoubtedly, the lack of understanding of a problem inevitably leads to the lack of adequate suggestions for improvements of solutions.

Hours of use, general timing of activities, type of occupants and many other facts, which any comprehensive building program must contain, are essential factors not only to the fundamental design approach of a project, but also to the basic operational approach of each discipline involved in the actual execution of construction phases. Elements which rank from the simple detailing of a window to the essential phasing and scheduling of an entire building complex can facilitate execution of the viable solution in each instance. More than that, they can determine such solutions. To this add the fact that the basic interest in a project will clearly increase as the understanding of its requirements and intricasies grows, and this familiarization is facilitated by a clear definition of programmatic requirements while it would be inhibited if there were no clear outline of facts from which to draw a subjective comprehension of scope and intent.

5.3 Comments

The effects noted in the preceeding pages do not necessarily occur as a direct consequence of improper architectural programming, but many of them CAN result from that deficiency, if only indirectly.

It would be useless to point out the effect of a total absence of analysis and programming procedures since these activities are *always* carried out in some degree no matter what the circumstances. Because of the inevitability of these realizations, a clearly defined measure of judgment becomes essential in any objective evaluation of the system or systems applied in each set of circumstances. In this light, it would be convenient to analyze the effects of the lack of programming systems more as "symptoms" than as actual crisis; that is, more as the symptoms of a tendency towards standardization and mechanization instead of accurate programming by function, tailored solutions and humanizations.

The common phrase, "The client does not know what he wants," often serves to justify the application of systems or solutions proven effective in previous situations involving similar conditions. A variation of the same phrase is the common observation: "The client is not too sure of what he wants." If the latter statement is generally inaccurate, the former is always incorrect. The fact is the client KNOWS what he wants; what the client might not always know is what he NEEDS.

In Chapter 7 we will study the relationship between "wants" and "needs" in more detail. At this point we will only establish that the concrete determination of needs is primarily the responsibility of programmer/client teamwork when applied to the program requirements. In this role, the programming activity becomes the translating process between ideal operational requirements and the factual realism of Architecture, technical realizations and construction requirements. Thus the programmer acts as the interpreter of "wants" and "needs" into those actual requirements which constitute the basis for any building project no matter what the circumstances, and the starting point for the determination of performance guidelines, value judgments and professional fees and compensations.

Regarding the establishment of professional fees and compensations, it is noteworthy that, since the results of the analysis and programming activities can be as basic as the determination of how many stories a specific building should have, which construction systems are desirable and why, etc., the information these processes provide is essential to the correct establishment of fees, contractual agreements and schedules.

In many instances we find preliminary agreements between the design professional and his client for the execution of analysis and programming phases with a set of figures structured against the overall professional design services contract. When properly executed, this arrangement can be extremely advantageous for both parties. However, we often find architects, interior designers or landscape architects entering into contractual agreements for commissions which turn into much more complex projects than anticipated after the architectural programming phase has been executed. This common mistake is unforgivable. The design professions have at their disposal not only the basic capabilities and qualifications to avoid such situations by employing an adequate analysis and programming system, but also a great many sources from which to derive effective architectural programming methodologies.

A fee structure based on size, cost or building type without regard for the actual degree of complexity of the proposed facility can be one of the worst mistakes any professional design firm can make and, so far, the only way known of determining these complexities is through a carefully executed architectural program. However, usually the structure of an Owner-Professional Designer Agreement is based solely on the objectives and requirements set forth in a project program, and the evaluation and performance measures of its result derive entirely from that program. There lies the importance of its comprehensiveness.

Finally, during the analysis and programming activities, the design professional has the best opportunity to awaken an aesthetic consciousness in his client. Since it is the first working process in which both design professional and client are involved, it is during this phase that the importance of image, appeal, and character can be brought into play. By this it must not be understood that these activities should be used to "sell" the client ideas or "award-winning designs" which could prove misleading, confusing or even detrimental to the functional performance of a project or its budget, in exchange for aesthetics, since that would be nothing but glorified deceit. Our purpose is rather to awaken, orientate, and stimulate our clientele in this field. The proper distribution and allocation of funds can not only produce a much better project in the end, but can also become the determining factor in the project's optimum performance.

Thus we will benefit our client and the project as well as our own practices while performing, at the same time, a service for our entire community and, through it, for society as a whole.

---

[12]One of the greatest causes of disagreements and disputes between designers and managing design professionals is the degree of objectivity employed by the later in evaluating and judging specific design schemes. In the majority of cases, the objectivity displayed becomes a function of pre-establilshed subjective approaches to form/function relationships or simply personal interpretations regarding character, image or style concepts. While it can be argued that such subjective filtering can help the style unification or the design cohesiveness of a specific firm; there is no real value in the argument that professional designs MUST bear the seal of its creator as prerequisite of excellence. Further comments on this can be found in Notes on Interfaces.

6

Existing Methodologies and Techniques of Spatial Analysis and Programming

6.1 Introduction

No programming system can physically provide all the necessary data to develop, complete and answer every task or question of the design procedure. In most cases, an effective design process will automatically include the exploration of alternatives which will lead to new questions and decision-making processes. In actuality, it is not surprising to find a sequence of activities as shown in Figure 6.1 following the initial development of a program.

Significantly, most design procedures will maintain a constant interaction with the data base and postulates established by the program, but this relationship functions more as the reciprocal interaction and adjustment of outlined variables than the statement result layout of inflexible determinants.

With this in mind, and based on their performance, we can easily establish two basic categories of architectural programming methodologies: the ones which supply the design process with the information required for its effective development, and the ones that do not. It is as simple as that.

From there we can derive several categories depending on the degree of completeness of the information provided which simply make up variations in the second kind of programming.

Using a computer, a pocket calculator or a deck of cards would all be equally useless unless all the primary applicable angles were covered in the basic system and all pertinent data were introduced to the design phase and processed in the proper manner. Otherwise the process is incomplete, which is not to say that it is wrong; it is simply incomplete.

Actually, the programming process can be complemented by some of those variations; it can be restructured and in many cases even revised; but no matter which order or programmatic development is followed in the course of these interactions, one fact will always remain unchanged before the design process can be completed, the programming phase must be terminated.

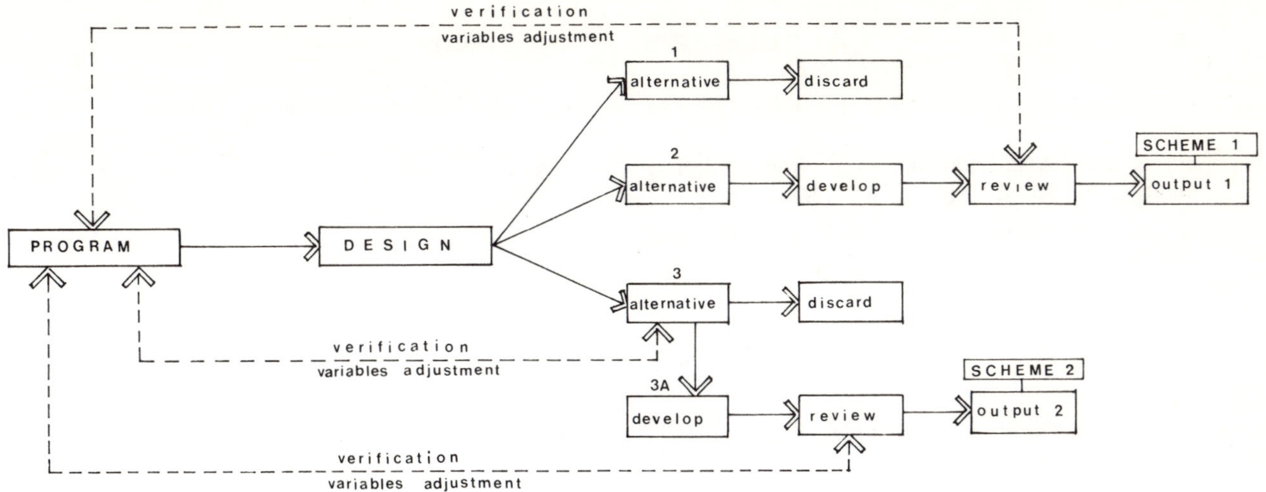

Figure 6.1

As originally stated, it is undesirable to view the programmatic activities and results as part of an inflexible system that does not allow room for adjustments, variations and revisions. In architectural programming, the ideal methodology is one that offers flexibility of use in the generality of cases and still contains within its structure characteristics which enable it to become a specialized format of programmatic data and determinations.

A method which creates sets of requirements without room for adjustments or revisions as the foundation for design activities is definitely wrong. On the other hand, any method which does not allow the establishment of a comprehensive outline of needs and requirements as determinants of a specific design problem layout is definitely worse. To employ a programming method which serves as guideline and problem definition *up to a point* can be as damaging, in certain cases, as establishing one which sets inflexible rules and constraints. As a matter of fact, the essential characteristics of any viable programming system are its flexibility of use and contents and its inherent capability to indicate when and where there is room for adjustment or variation and when not.

Therefore, the first positive corollary of this primary characteristic can be defined as the system's ability to provide a comprehensive layout of degrees of importance by developing a process for evaluating the programmatic requirements it contains. The system must have a self-evaluative scale based on a clear and concise definition of priorities.

As far as the actual procedures are concerned, we can clearly distinguish several possible situations, such as those illustrated in Figure 6.2.

For the purpose of this Chapter, we will not consider any computer-aided programming procedure, simply because this subject is covered more extensively in Chapter 10. Proceeding to the other categories, here is a brief description of each:

PRE-ESTABLISHED PROGRAMS. Sometimes a professional design firm is approached by a client with a set program. Then the programming activity is actually an analysis of the existing outline of tasks.

Possible procedures are (A) The assumption of the pre-established program as an absolute to be followed only by schematic designs; (B) The computer-aided review of pre-established programs

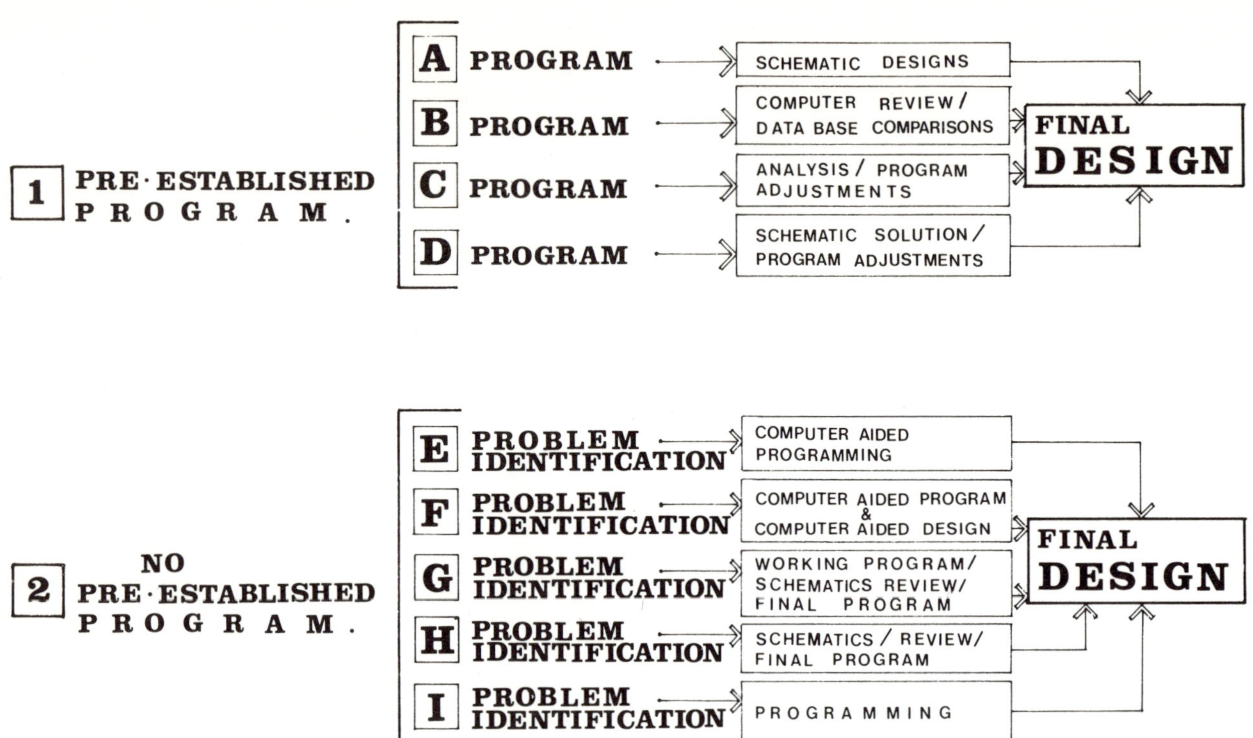

Figure 6.2

as compared to existing data base standards (C) The given program analysis and adjustments without the aid of a computer to a final determinate; or (D) The preliminary execution of schematics in lieu of an analytical procedure, only to clarify and evaluate through an actual preliminary design process, the programmatic requirements (whether strict determinants or controllable variables) and review, adjust, and define the initial program's content, if necessary, in order to determine the realistic or viable scope for a final solution.

NO PRE-ESTABLISHED PROGRAM. When the client has no pre-established program, we might (E) have the programming phase done with the aid of a computer, (F) have both programming and design executed in this manner,[13] (G) establish a preliminary or working program which will be completed with the aid of schematic designs, (H) have the programming activity executed through a sequence of schematic designs on a trial and error basis, or (I) determine the program independently, without any previous schematic design's interference and without any computer-aided process.

Any system which does not encompass all of these possible methods of execution is incomplete, because it is impossible to maintain that only one system works all the time and certainly not for all cases. In addition to the self-evaluative scale mentioned previously, a second characteristic in our definition of programming would be a comprehensive outline of flexibilities in its layout.

As we examine each one of the possible cases mentioned previously, many partial methods of programming could satisfy the definite circumstances of a given case. For example, there would be no need for any spatial relationships data collection

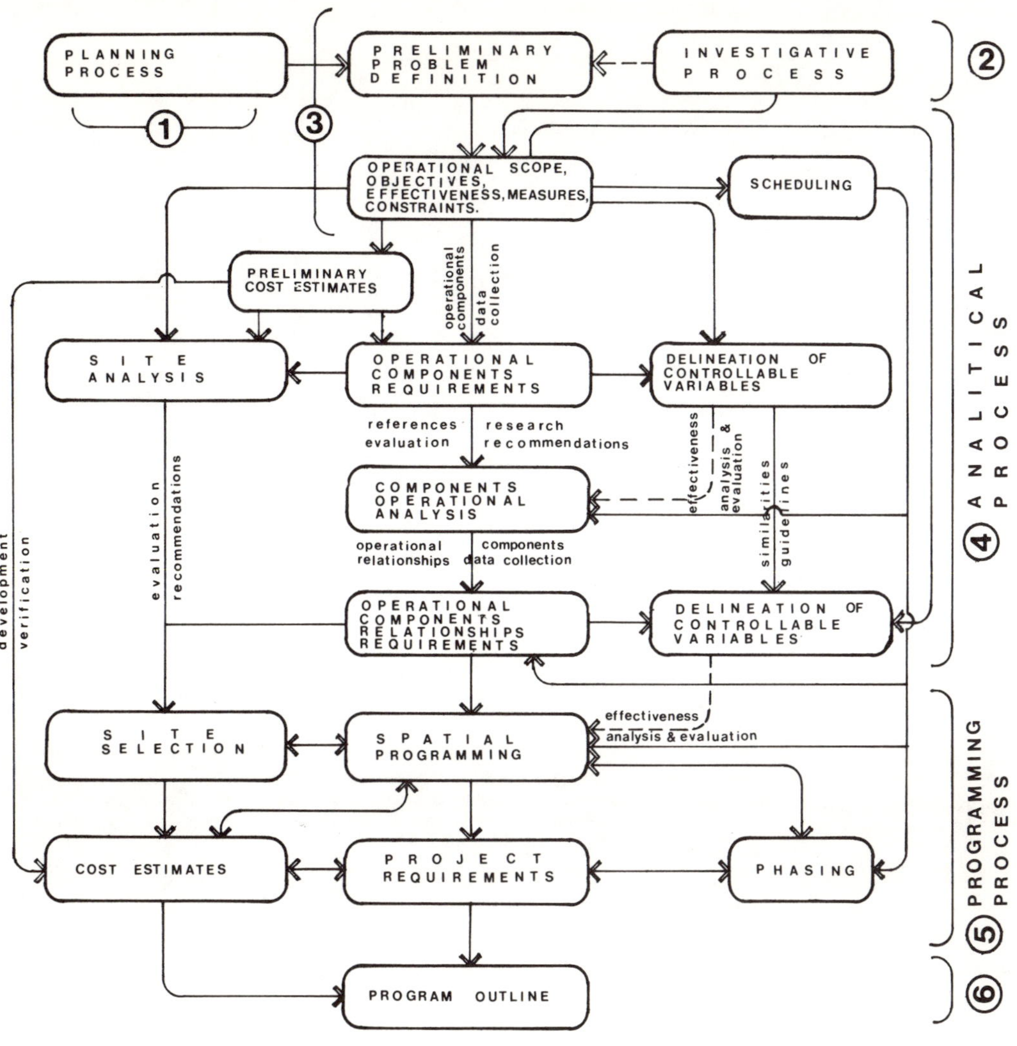

Figure 6.3

forms in the cases of (D) or (H) above where the links among spaces are established through a sequence of schematic designs, and no need for spatial data collection or determination of activities by the professional design firm in the case of (C) where the program supplied by the owner is taken literally as final determination of his project's needs and requirements.

However, it is extremely important to note in cases like (B), (C), (E), (F), (G), and (I) the application of a comprehensive architectural programming system by the professional design firm is essential.

6.2 Existing Systems and Techniques

Recently, we have seen the development of several systems which define programming steps in direct interaction with determining factors, and others which establish programming steps in a linear sequence covering everything from investigation and analysis procedures to volumetric design. Although the general outline varies in each case, the sequence Objectives-Analysis-Programming-Requirements is still the same, and the direct dependency of these steps on universal factors involving man, time and space determines the basic structure of all existing architectural programming methodologies.

The range of activities involved in architectural programming systems can be represented as illustrated in Figure 6.3. It includes several individual processes as interdependent or isolated constituents of the system's overall development and results, such as Planning Process, Investigative Process, Objectives Definition, Analytical Process, Programming Process and Program Outline.

The following examples illustrate some of the techniques employed in effecting and developing each one of the previously outlined steps. These examples should never be confused with actual programming systems; they are simply techniques employed to effect or illustrate separate segments of those systems.

1. PLANNING PROCESS. A basic outline of the

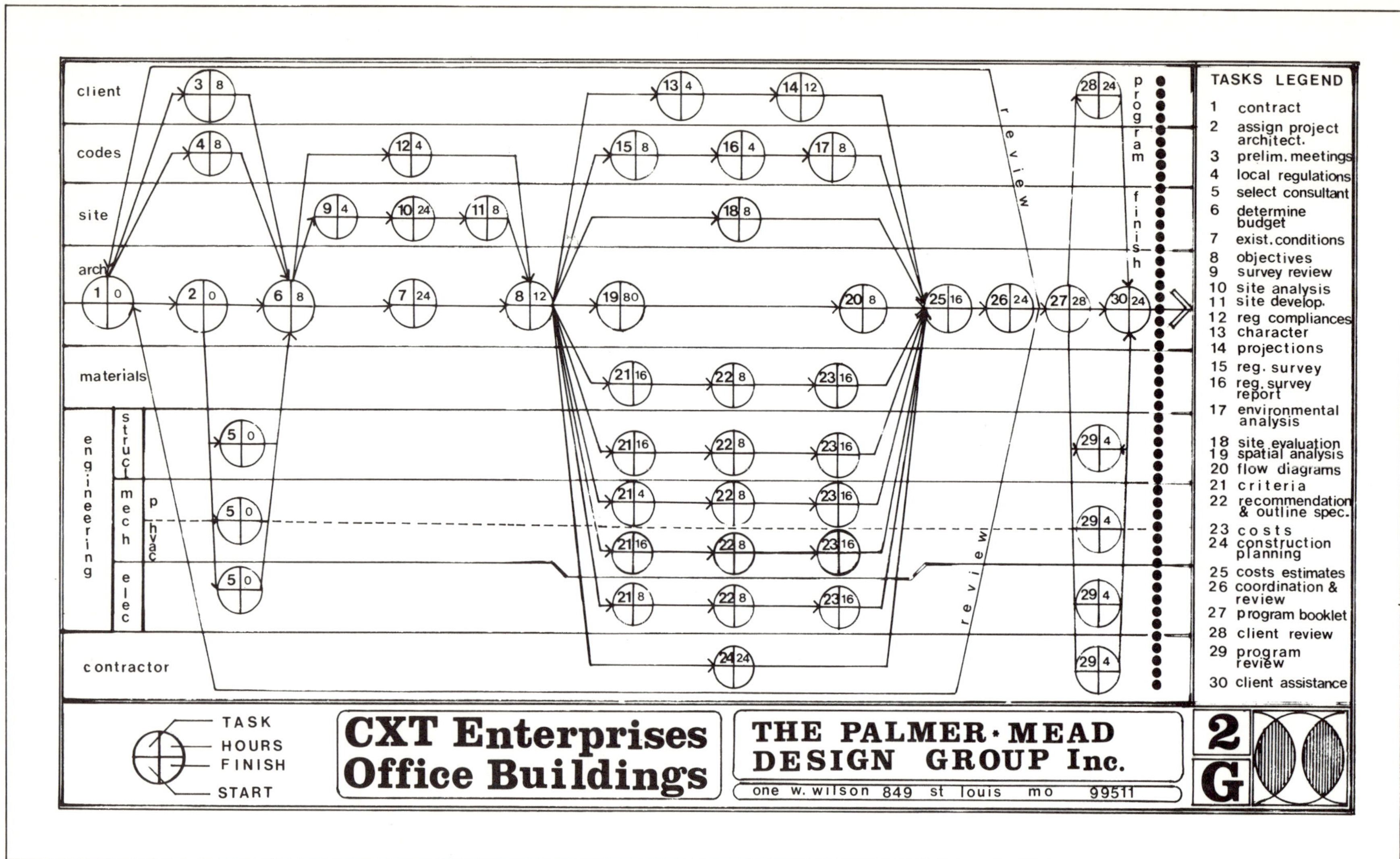

Figure 6.4

breakdown and definition of the programming process itself. (See Figures 6.4, 6.5, 6.6)

2. INVESTIGATIVE PROCESS. Preferences gathered either through feedback information from previously executed projects or existing facilities of similar characteristics or through a simple data collection process to gather existing references. (See Figures 6.7, 6.8, 6.9)

3. OBJECTIVES DEFINITION. The determination of scope and objectives and the establishment of a sequence of operational activities. (See Figures 6.10, 6.11, 6.12)

4. ANALYTICAL PROCESS. The establishment of data collection forms, checklists, criteria sheets and spatial relationships formats. (See Figures 6.13, 6.14, 6.15)

PHASE 1
DATA COLLECTION
EXISTING TRAVEL PATTERNS
LAND USE & ECONOMIC DEVELOPMENT
EXISTING TRANSPORTATION FACILITIES
PHASE 2
ANALYSIS
TRAVEL PATTERNS CHARACTERISTICS & DEVELOPMENTS
TRIPS DISTRIBUTION CHARACTERISTICS & DEVELOPMENTS
IMMEDIATE ACTION PROGRAM
FUTURE LAND USES ACTIVITY & DEVELOPMENT
FUTURE TRAVEL PROJECTIONS
NEW & COMMITTED FACILITIES
FORECASTING
FORESEEABLE DEFICIENCIES
PHASE 3
ALTERNATIVES
OBJECTIVES & STANDARDS
DEVELOPMENTS
EVALUATION & SELECTION
IMPLEMENTATION PROGRAM
PHASE 4
CONTINUING PROGRAM
IMPLEMENTATION

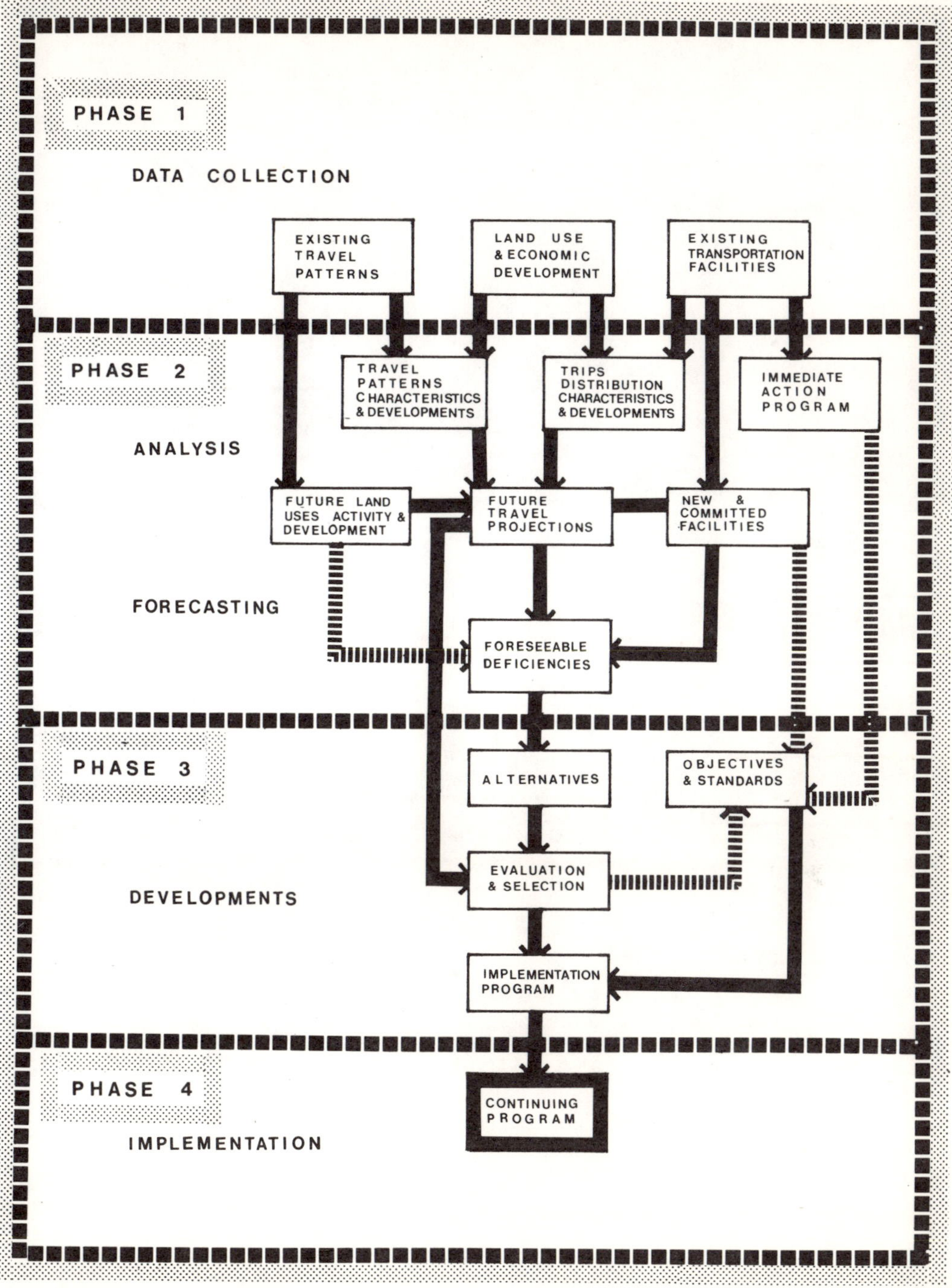

Figure 6.5

TABLE OF ORGANIZATION
TABLE OF RELATIONS
ORGANIZATION FRAMEWORK
ESTABLISH ORGANIZATION
DEFINE RELATIONSHIPS
SPACE PROGRAM
DRAWINGS
SCHEDULE
PROJECT IDENTIFICATION & ORGANIZATION
DECISION PROGRAM
SPACE PROGRAM
PHYSICAL PLANS
STAGE CONSTRUCTION
OBJECTIVES & POLICIES
FUNCTIONAL PROGRAM
PERSONNEL REQUIREMENTS
FUNDING & FINANCING
PERSONNEL PROGRAM
CONSTRUCT & OCCUPY
POLICY STATEMENTS
FUNCTIONS PROGRAM: INVESTIGATION & EDUCATION
TABLE OF PERSONNEL
APPLICATIONS
ACQUISITION PROGRAM
EQUIPMENT PROGRAM
ACQUISITION PROGRAM

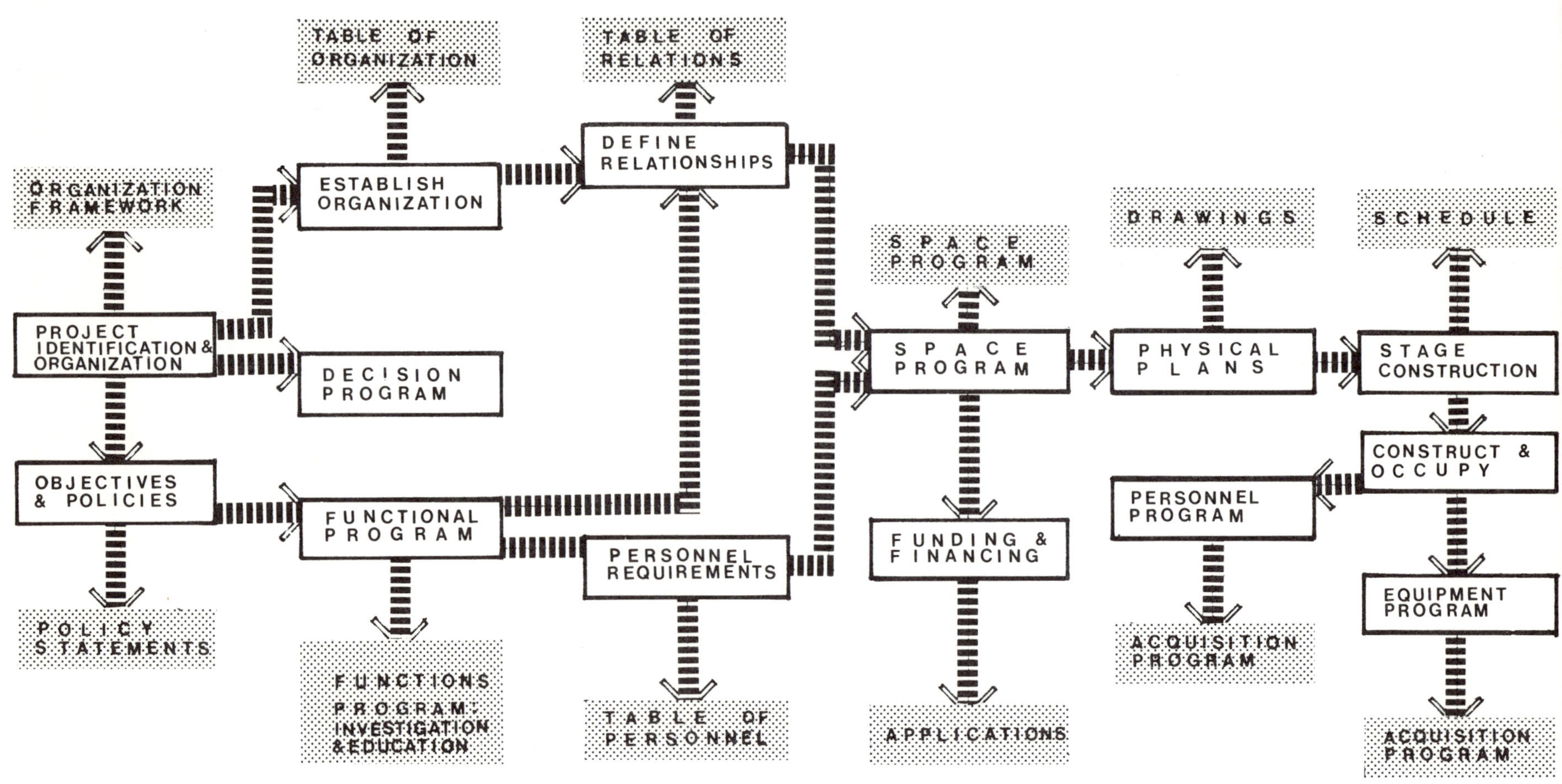

 Figure 6.6

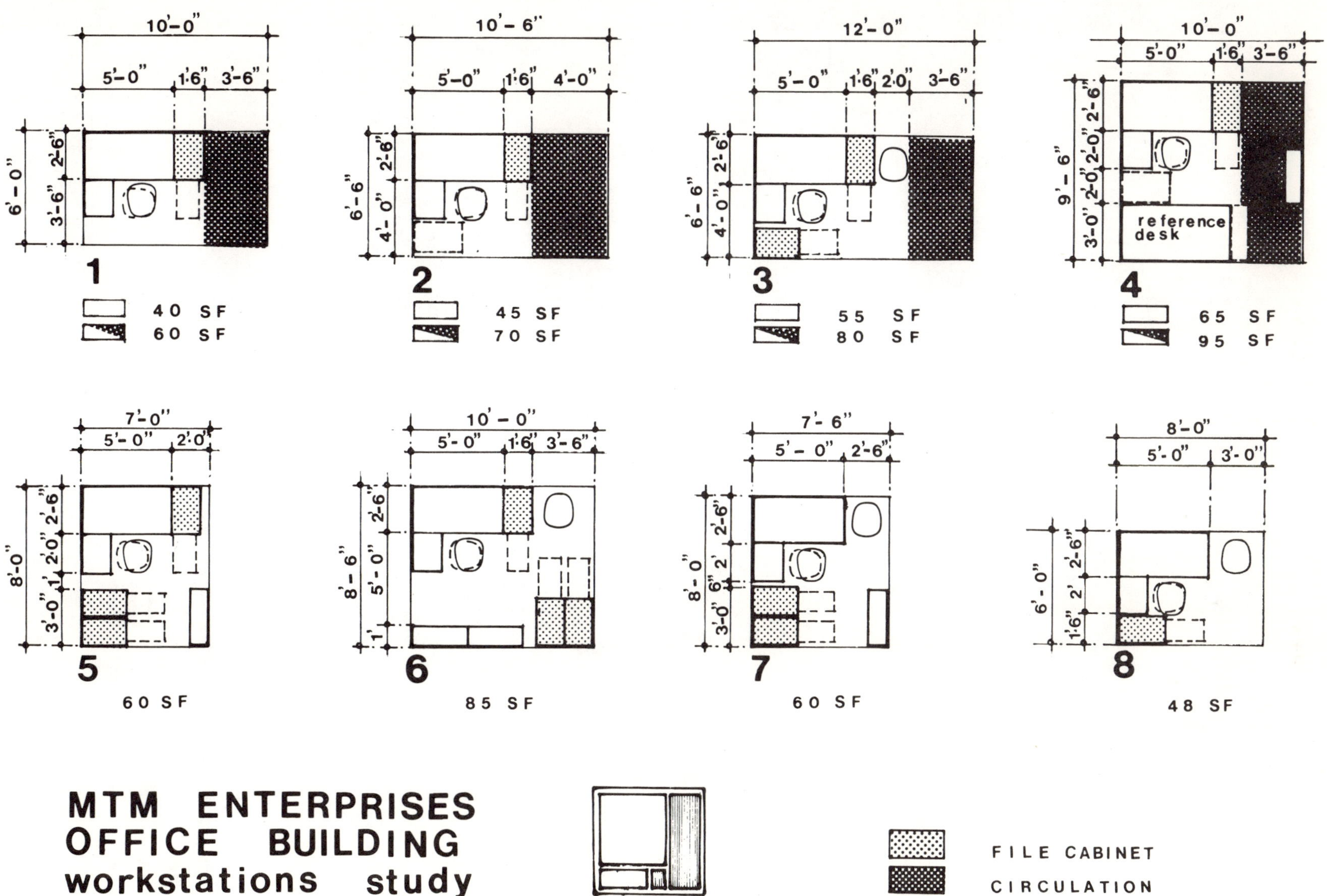

MTM ENTERPRISES OFFICE BUILDING
workstations study

Figure 6.7

9 — INTERCOMMUNICATION & PERSONNEL MOVEMENT PATTERNS — STUDY 2

Legend: f = frequent · I = irregularly · s = seldom

#	DEPARTMENT AREA	2	3	4	5	6	7	8	9	10	11	12	13	14	15
1	Reception, offices, conferences	f	f	f	f	f	f	f	f	f	I	f	I	s	s
2	Students		s	I	s	s	s	I	s	I	I	s	I	s	s
3	Chambers			f	I	I	s	I	f	f	I	f	s	f	I
4	Test labs				I	f	s	I	f	f	I	f	s	s	s
5	Chemistry					f	I	I	I	f	f	I	s	I	s
6	Specialties unit					s	I	I	I	I	I	I	s	s	s
7	Research unit							I	s	I	I	s	s	s	s
8	Shop								s	I	f	s	I	f	I
9	Medical unit									I	s	f	I	s	s
10	Data processing										I	s	f	s	s
11	Chemical unit											s	I	I	s
12	Preparation units												I	s	s
13	Biomechanics complex													s	s
14	Mechanical equipment														I
15	Penthouse														n.a.

NOTES

Figure 6.8

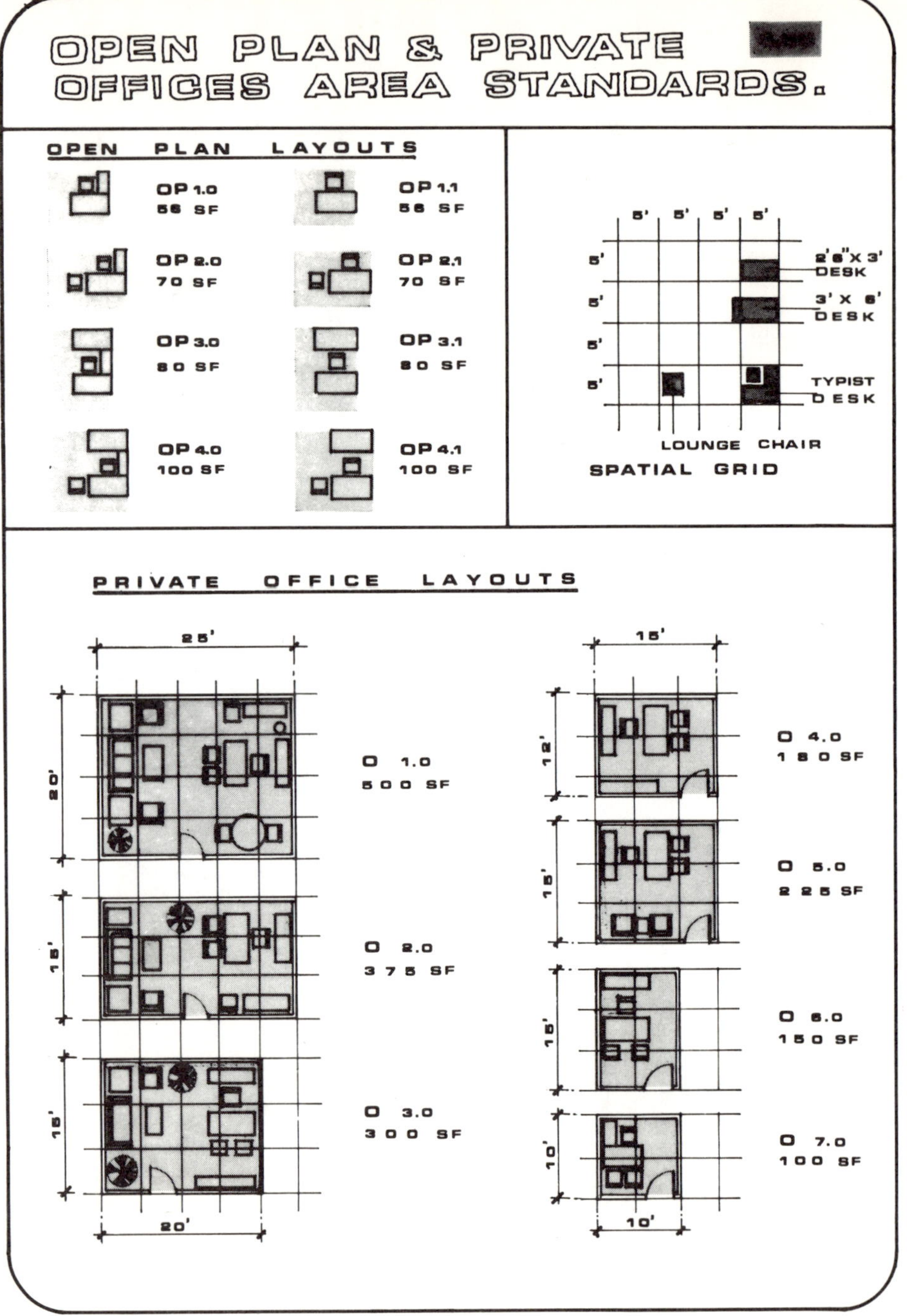

Figure 6.9

OBJECTIVE:

INCREASE OPEN AREAS & MAINTAIN DENSITY

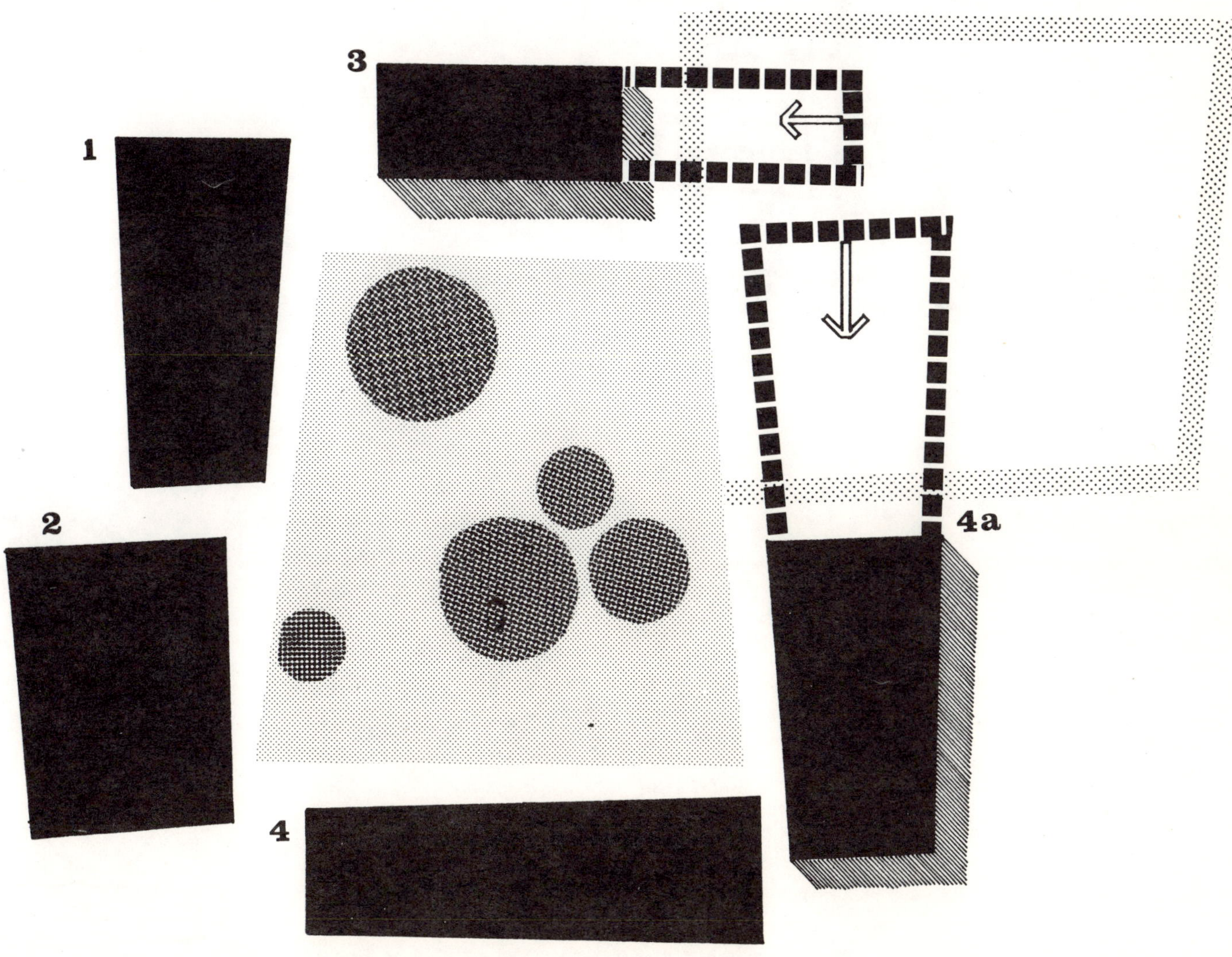

Figure 6.10

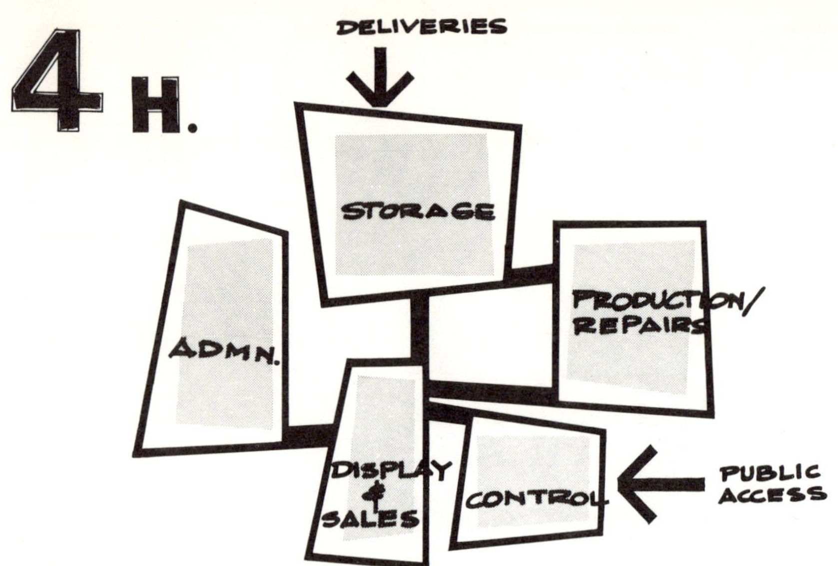

Figure 6.11

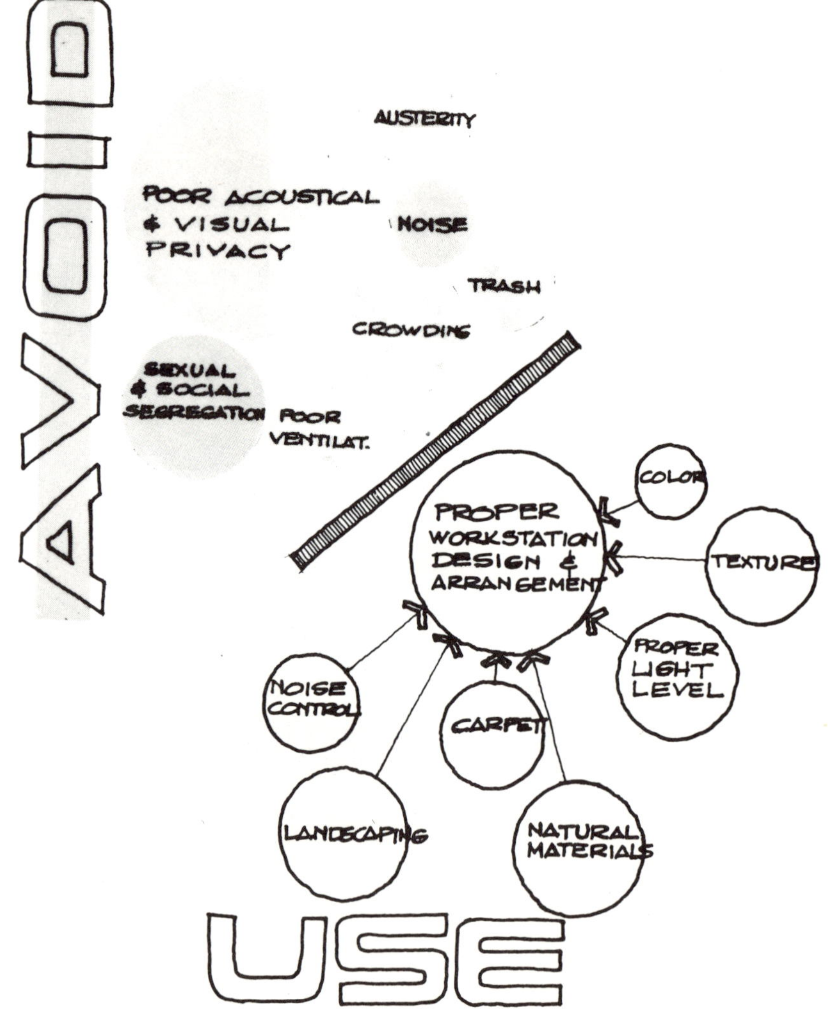

Figure 6.12

ROOM REQUIREMENTS

SCHOOL for THE BLIND
ATLANTA, GA.

GRAY & ASSOC.
MIAMI   FLORIDA

DEPT: PHYSICAL ED.
room: DIRECTOR'S OFF.

FURNISHINGS

furniture

DESK & CHAIR
CONF. TABLE
CONF. CHAIRS (6)
STOR. UNIT-

BOOKSHELVES 20 LF.

equipment

FILE CABS:
2 FULL
2 HALF.

ENGINEERING

hvac
STANDARD

plumbing
NONE

electrical
STANDARD.

communication
TELEPHONE
DICTAPHONE
PAGING SYST.

ARCHITECTURE

floor
CARPET

ceiling
ACOUST. TILE

walls
PAINT/DRYWALL

doors/windows
(2) / NONE.

acoustics
SOUNDPROOF

RELATIONSHIPS

primary

DIRECTOR'S OFF.
ADMN DPT.
GYM & ATHLETIC
FIELDS & LOCKERS.

secondary

CLERICAL
MEDICAL CTR.

comments

---

occupancy
1 MALE
0 FEMALE
2 AVG. VISITORS

dimensions
12 X 18
216 SF.

n° required
± 1
± 220 SF.

existing area
120 SF.

other

r e m a r k s

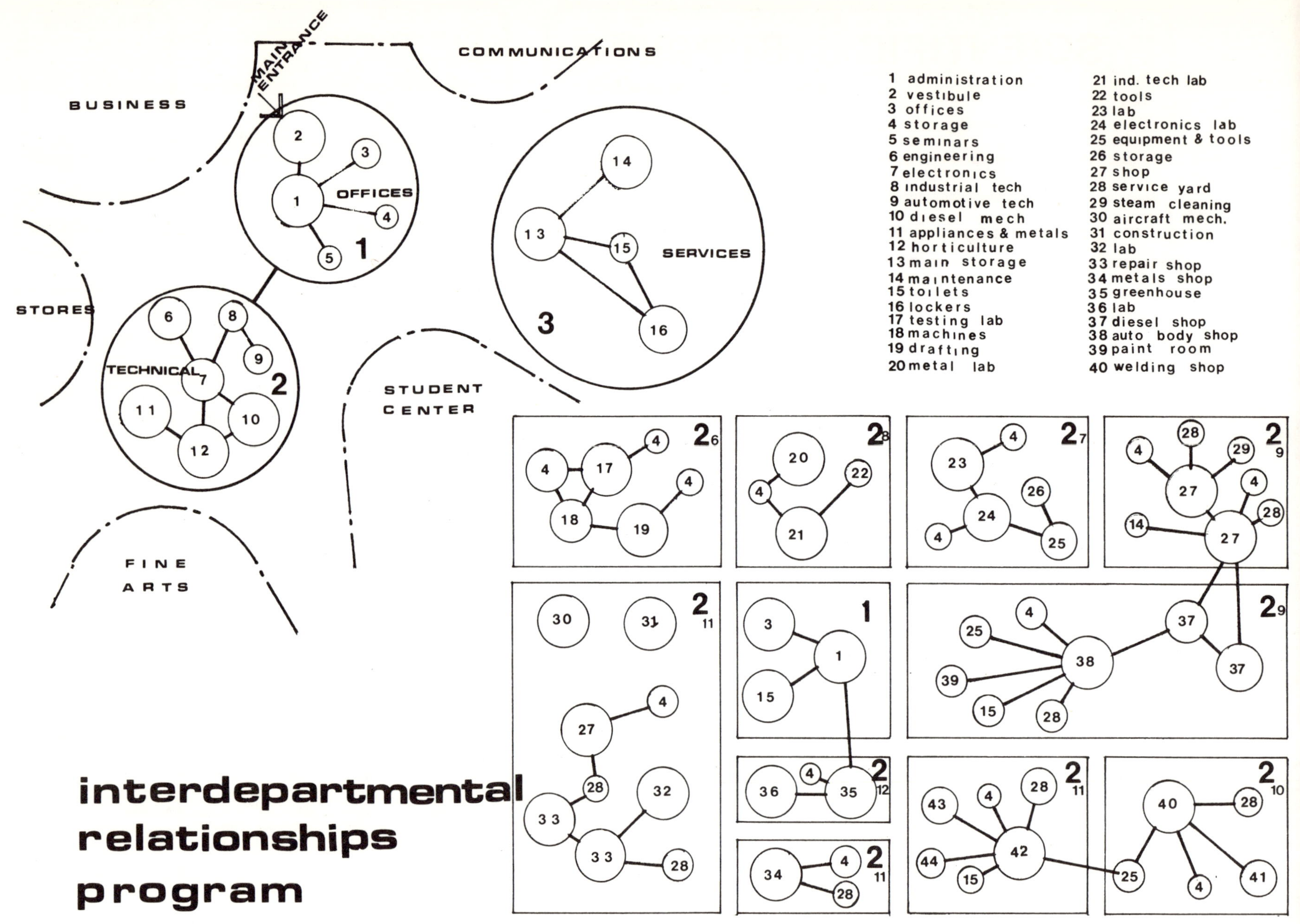

interdepartmental relationships program

Figure 6.14

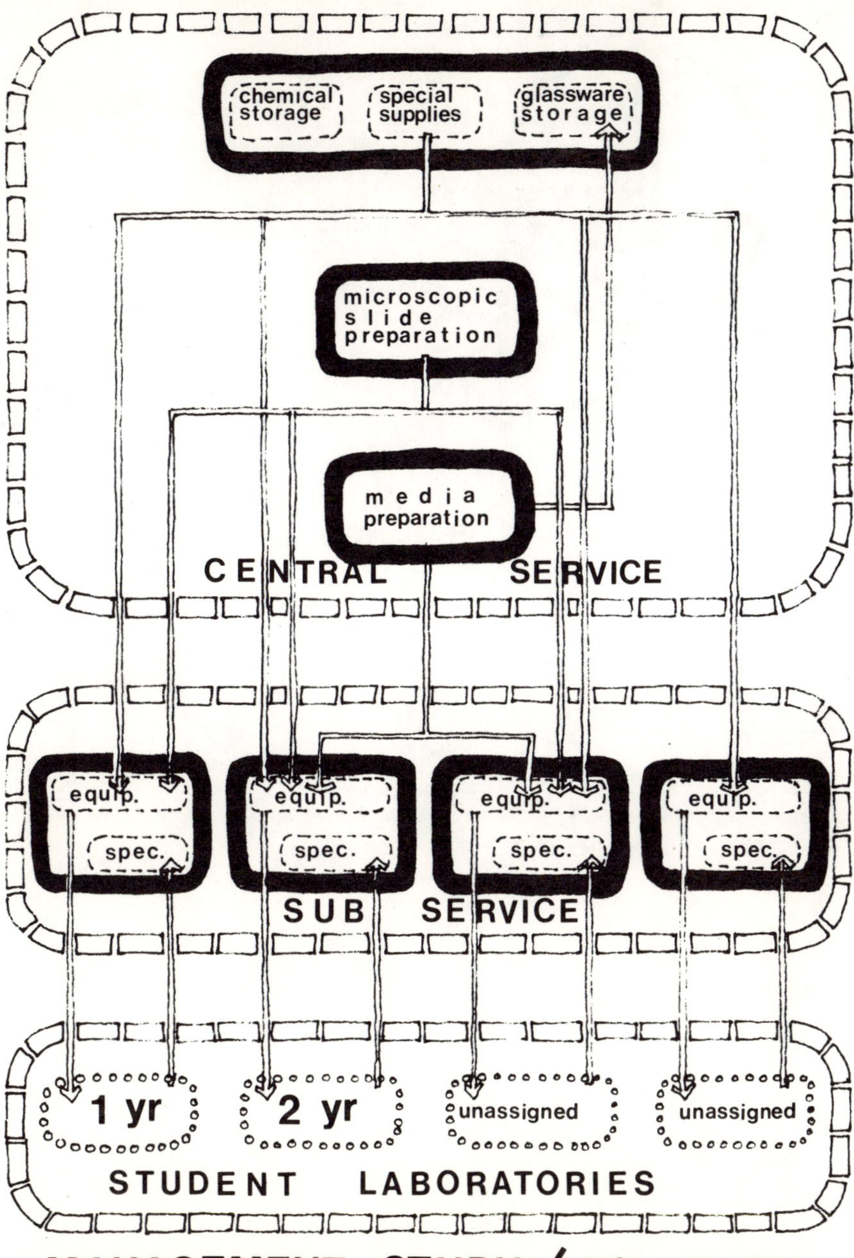

Figure 6.15

5. PROGRAMMING PROCESS. The processing and evaluation of accumulated data to determine categories of functions, priorities, values, etc. (See Figures 6.16, 6.17)

6. PROGRAM OUTLINE. A statement of varying complexity ranging from a simple problem description to a complete network of presentation techniques.

6.3 Programmatic Outline

Obviously, then, any comprehensive programming system based on these traditional approaches must contain the following basic elements:
1. Planning Procedure
2. Investigative Methodology
3. Objectives Definition System
4. Analytical System
5. Programming Process
6. Cost Analysis System and Scheduling Method
7. Presentation Techniques.

Each one of the examples previously illustrated was derived from actual systems, charts and diagrams used by professional design firms throughout the country. The primary reason for their development in the first place was the NEED for such a form or procedure. Thus their necessity is proven by the fact of their existence.

So in order to establish a comprehensive architectural programming system we must deal with each one of those basic elements and their relationships to one another.

Based on the results of the phases described previously, we can derive a general project program outline which provides the following information:

1. General Information
   1.1 Project title. Location
   1.2 Name of owner. Address
   1.3 Project number
   1.4 Participants
   1.5 Index

2. Statement of Philosophy and Objectives
   2.1 Policies. Design philosophy

3. Project Development Factors
   3.1 Marketing
   3.2 Statistical data
   3.3 Sales

4. Project Location and Site Analysis Results Requirements

5. Spatial Analysis
   5.1 Components functional description
   5.2 Spatial requirements
   5.3 Functional relationships
   5.4 Flow patterns

6. Background
   6.1 Research and references

7. Economic Considerations
   7.1 Budget
   7.2 Cost analysis

8. Project Temporality
   8.1 Phasing
   8.2 Growth/Change projections
   8.3 Time/Cost factors
   8.4 Temporality
   8.5 Time tables and schedules

9. Operational and Maintenance Requirements
   9.1 Building and site operation
   9.2 Energy conservation factors
   9.3 Building and site maintenance

10. Applicable Codes and Regulatory Agencies Surveys and Checklists

11. Construction Requirements

12. Special Characteristics

13. Appendix
    13.1 Correspondence, reports, memos and supporting data

Chapter 7 contains a more comprehensive program outline developed from the basic compliance with a systematic space operational analysis procedure. For now, this outline will serve as an illustrative layout of scope and requirements.

It is noteworthy that the analytical sequence of predictable activities must extend up to the point

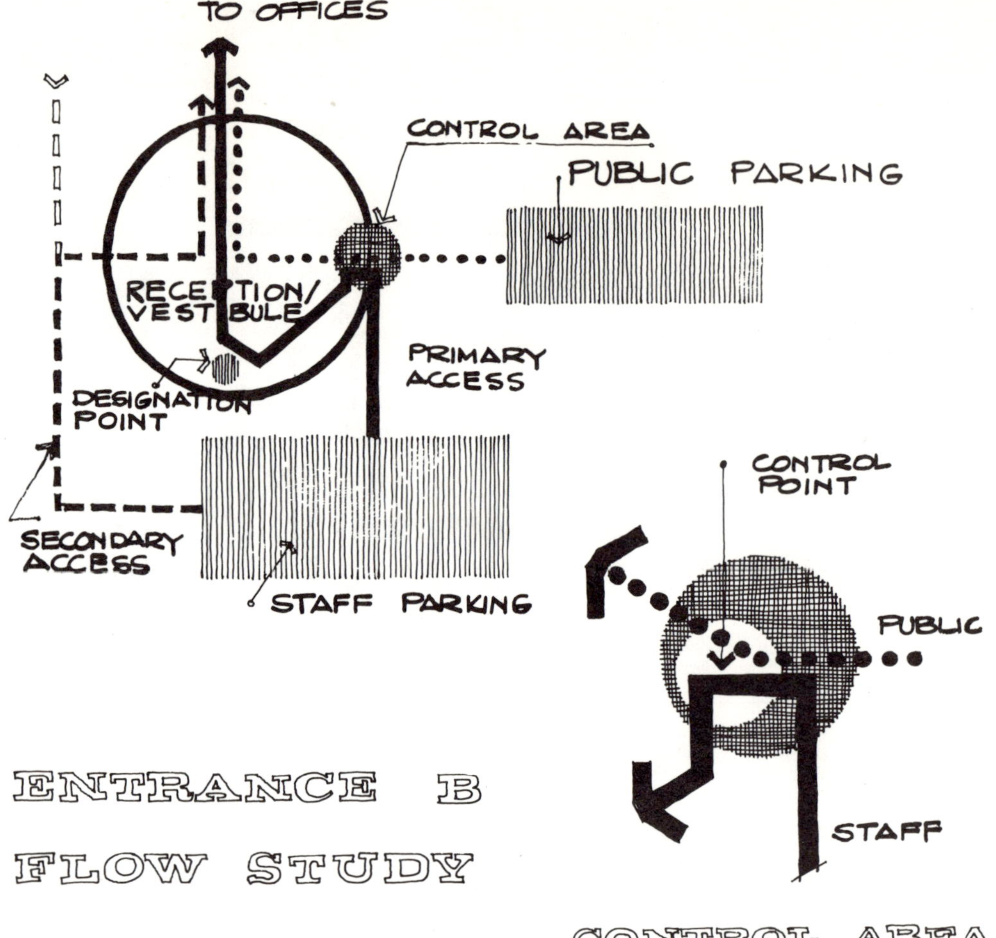

Figure 6.16

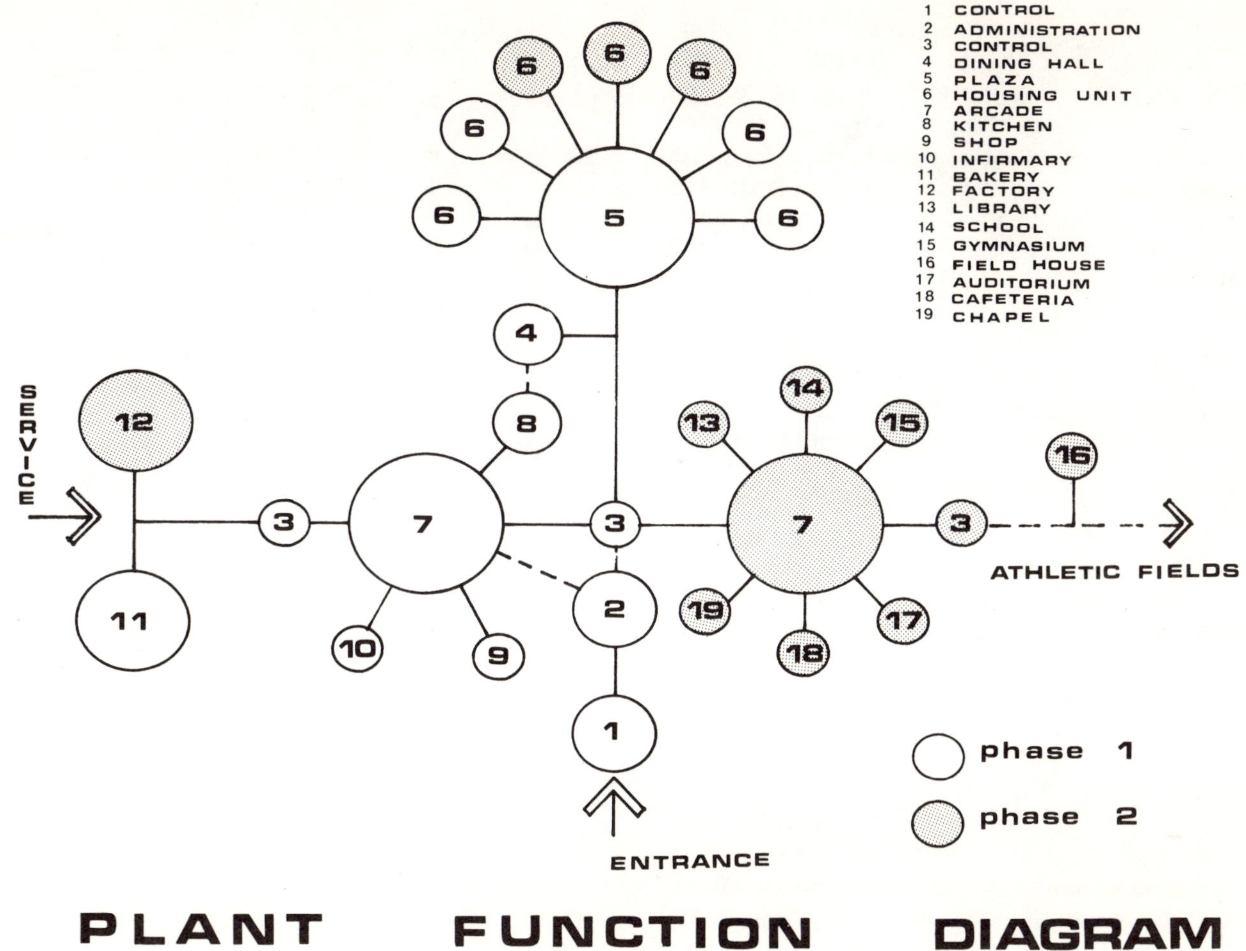

PLANT FUNCTION DIAGRAM

Figure 6.17

where the synthetical development of the design process begins; otherwise, the latter must include the missing analysis sequences within its own development, resulting in the slowdown and inefficiency to be expected from the inevitable backtracking motions of misdirected efforts and overlapping of roles.

By the same token, too much analysis and programming at the outset can be extremely damaging, because the first steps of the design process can become confusing and overwhelmed by an excess of information which becomes detrimental to the successful execution of schematics. The amount and the kind of data to be included in an architectural program at a given point must be determined in consideration of the needs of the design activity which immediately follows it. This specific factor, which we will refer to as the program-design factor, actually determines to what extent a programming activity must be executed and when, and its relation to the design activity can be plotted in a simple diagram as shown in Figure 6.18. It is obvious therefore, that the programming activity can continue throughout the schematics stages of professional design services and up to the design development phase at which point the majority of its functions must be entirely fulfilled.

This does not mean that lack of essential information can be justified for the sake of simplification. Not only can the lack of information act as a formidable setback in the effective realization of schematic designs, but the "logical" conclusions which might be derived from that inadequate data can become a very dangerous corollary. Chapter 10 describes actual developments and simplifications of programming methodologies in more detail. For now it is enough to alert everyone to the dangers of overdevelopments as well as oversimplifications.

The worst approach to architectural programming is the subsition of logical assumptions for what should be actual findings of analysis procedures or the reduction of what should be conclusions and directives of the programming process to common sense statements, since, in most cases, logical assumptions become illogical after the fact, and common sense, being the least common of all senses, is an extremely dangerous tool to employ at will during the execution of programming procedures because its innate subjectiveness can reshape the programmatic parameters in the wrong manner and mistakenly redirect their trends towards erroneous goals.

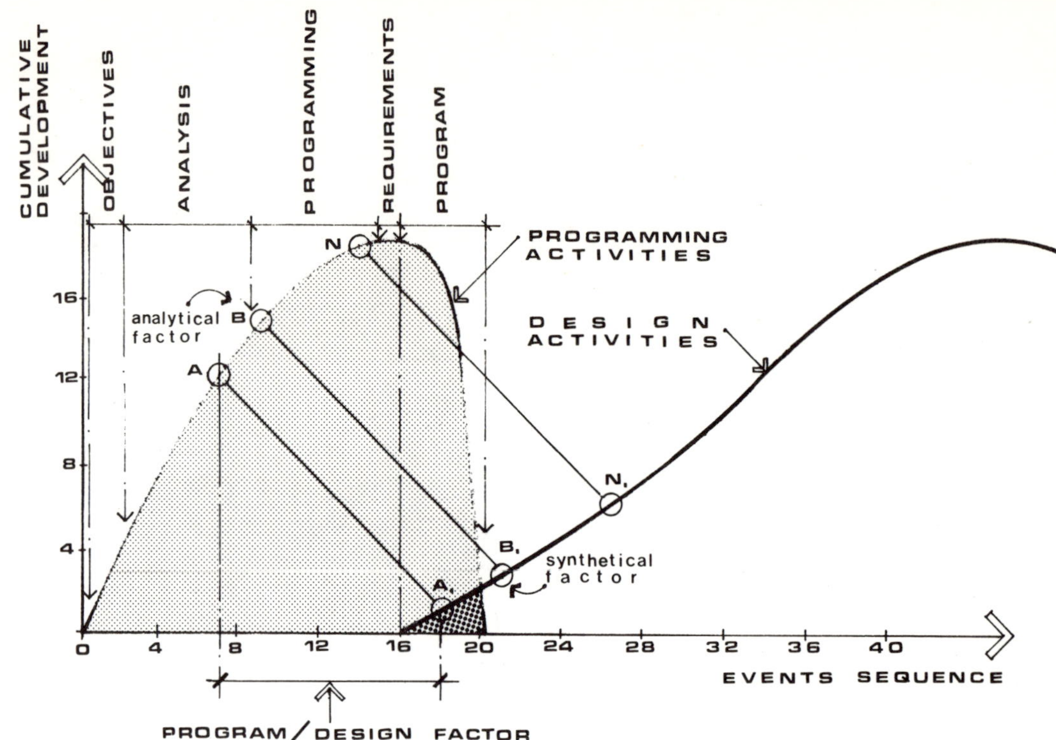

Figure 6.18

---

[13]Obviously, program structure, developments, outline and general characteristics vary extensively whether developed for use by designers or computer-aided design procedures.

ARTMENT/DIVISION FUNCTIONAL DESCRIPTION: ADMN. DEPT.-PERSONNEL DIV.
ON: TO OFFER EQUAL EMPLOYMENT OPPORTUNITIES TO ALL APPLICANTS, SUBSEQUENTLY PROVIDING A QUALIFIED OF EMPLOYEES FOR COUNTY AGENCIES. PROGRAMS OPERATED BY THIS DIVISION INVOLVE EMPLOYEE RECRUITMENT, SALARY ADMN., MANPOWER NEEDS ANALYSES, COLLECTIVE BARGAIN., UNION CONTRACT ADMN.& EMPLOY. TRAIN.

OMPONENT DESIGNATION	FUNCTION	CONSTITUTIVE ELEMENTS	FUNCTION	ACTIVITY DESCRIPTION	ROOM N° / AREA	PUBLIC ACCESS	M 0.5	F 0.5	VIS/HR	EQUIP.	A	B	C	D	E
NAGEMENT	E	MANAGER	E	GENERAL ADMINISTRATION / UNION CONTRACTS NEGOTIATIONS	104 / 160	NO	1 / 0	0 / 0	6 / 2	①	4(3)	3(4)	NA	5(5)	3(4)
		CONFERENCE	C	ORAL BOARDS, CONFERENCES & INTERVIEWS	104A / 200	NO	0 / 0	0 / 0	20 / 10	④	2	4(11)	NA	4(13)	4
		ASSISTANT MANAGER	E	PERSONNEL ANALYSIS, MGMT ASSIST. EXAMS DEVELOP. RECRUIT	105 / 154	NO	1 / 0	0 / 0	6 / 1	②	4(3)	2	2	4(5)	2
		PERSONNEL TECHNICIAN	E	NON-PROFESSIONAL RECRUIT. & CLASSIFICATION	107 / 152	NO	1 / 0	0 / 0	6 / 1	③	4(6)	2	2	4(6)	2
		PERSONNEL ANALYST	C	HELP ASSISTANT MANAGER WITH PERSONNEL ANALYSIS FUNCTIONS	NA / NA	NO	– / –	– / –	NA	NA	NA →				
ERICAL	E	RECEPTION	E	RECEPTION AND APPLICATION FORMS	107 / 250	YES	0 / 0	0 / 0	80 / 15	⑤	5(10)	5(11)	2	5(10)	4(12)
		APPLICA. FILL. (20)	E	EMPLOYMENT APPLICATION FORMS FILLING	NA / NA	YES	– / –	– / –	60 / 8	NA	NA →				
		GENERAL OFFICE	E	CLERICAL FUNCTIONS. INCLUDES SUPER. WORKSTATION & FILES	106 / 370	NO	0 / 0	4 / 1(1)	6(2) / 1	⑤	2(8)	5(9)	2	5(8)	3(7)
ORAGE	C	RECORDS STORAGE	E	REMOTE/PRESENTLY (19) RECRUIT. FILES, REGISTERS, APPLICA. PRINT	G1-8 / 50	NO	0 / 0	0 / 0	NA	NA	5(15)	5(11)	NA	NA	5(16)

DIVISION SPATIAL EVALUATION — A: 2(18) | B: 4(11) | C: 2 | D: 5(17) | E: 3

II

Space Operational Analysis

7

Basis

7.1 Structure

No matter what the goal, no matter what the objective, the analysis scope, the concepts or requirements, ALL phases involved in architectural programming fall within the parameters of three universal factors:

SPACE
MAN
TIME

These universal factors determine the basic conditions which will cause, shape, modify, and define every portion of any programming activity. For the purposes of this text we will refer to each condition as:

SPATIAL
HUMAN
TEMPORAL

The goals and objectives of any human endeavor are a direct product of these conditions. Analysis studies do nothing but describe, define and evaluate them, and any programmatic concept is a simple definition of guidelines based on their interrelationships.

The essential objective of human operations is basically compliance, in the greatest possible degree, with each of the conditions established by these determining factors, by means of analytical studies and programmatic concepts.

In a graphic manner, the sequential steps of architectural programming could be represented as shown in Figure 7.1

The diagram in Figure 7.2 is an actual illustration of their respective interactions.

By nature of their structure in the function of mankind, each one of these respective compliance conditions is subdivided in the following categories:

Spatial
Natural
Environmental
Human
Functional
Physical
Psychological
Sociological
Regulatory
Economic

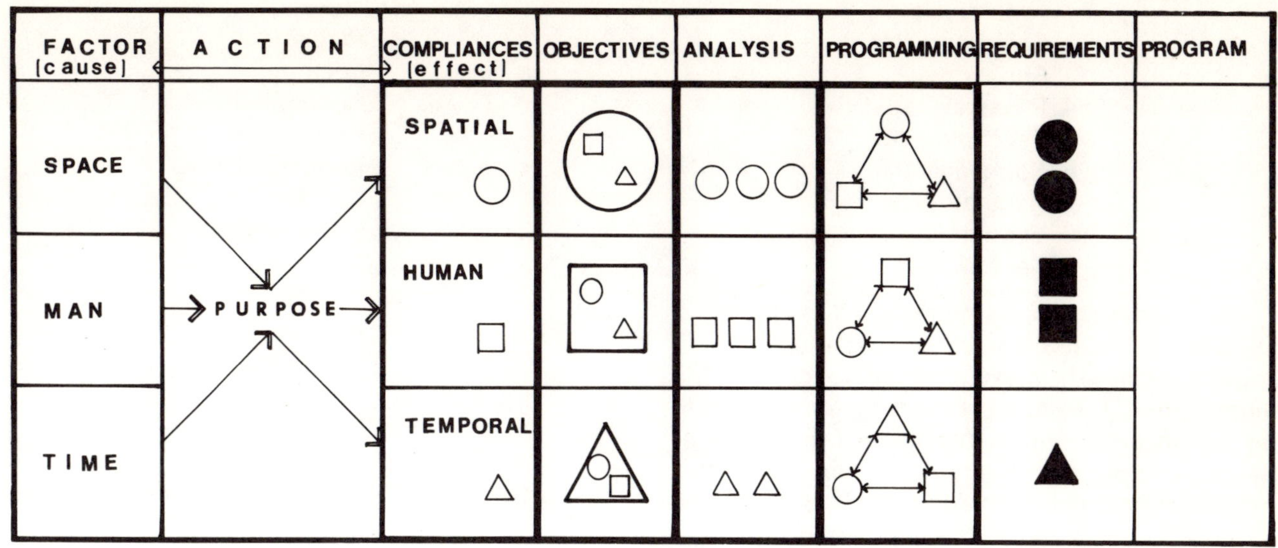

Figure 7.1

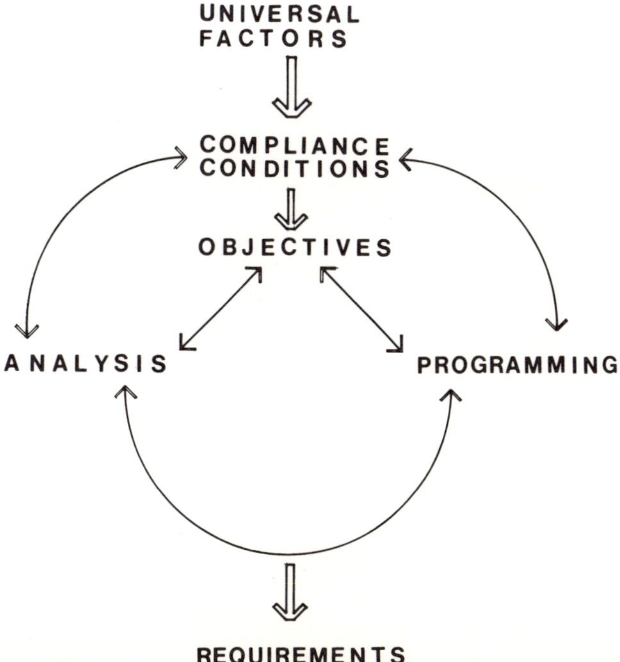

Figure 7.2

Temporal
Past
Present
Future

Natural conditions are those pertaining to Nature as such, while environmental conditions represent natural phenomena in function of man.

By the same token, functional conditions are those which define man's activities; physical conditions pertain to man as a living organism on the planet; psychological conditions deal essentially with the psyche; sociological with man within man and regulatory with human attitudes within society; and finally, economic conditions measure man's achievements within society.

The temporal conditions are, of course, past, present and future, although for reasons of simplification we will keep on referring to them as temporal, since past, present, and future subdivisions are obvious phases of every time-related consideration and compliance objective.

So in establishing the compliance conditions of any programmatic activity we can list 9 basic areas:

1. Natural
2. Environmental
3. Functional
4. Physical
5. Psychological
6. Sociological
7. Regulatory
8. Economic
9. Temporal

The analysis process represented in Figure 7.1 will be described in detail in our next chapter. Here we will determine how each compliance condition is studied since, for practical reasons, some of them can be grouped within the same framework of analysis and programming procedures.

The method described in this book for the particular study of each compliance condition is:

Site Analysis
    Natural
    Environmental

Spatial Analysis
    Functional
    Physical
    Psychological
    Sociological

Regulatory Surveys
    Regulatory

Economic Analysis
    Economic

Temporal Analysis
    Temporal

Programmatic concepts have three major subdivisions each, which encompass their essential constituents:

Structure
Direction
Value

As in the case of programmatic concepts applied to the spatial factors these constituents could be defined as follows:

Structure
    Placement
    Location
    Accessibility
    Protection
    Performance
    Comfort
    Maintainability

Direction
    Erosion
    Permanence
    Variability
    Controls
    Predictability
    Flow
    Sequence

Value
    Modification
    Prevention
    Improvement
    Efficiency
    Cost

Or as applied to the human factors and their subsequent compliance conditions:

Structure
    Grouping
        Integration
        Fragmentation
        Segregation
        Distribution
    Alternatives
    Flexibilities

Direction
    Approach
    System
    Policy
    Adjustment
    Control
    Prevention
    Accentuation
    Planned Encounters
    Exceptions

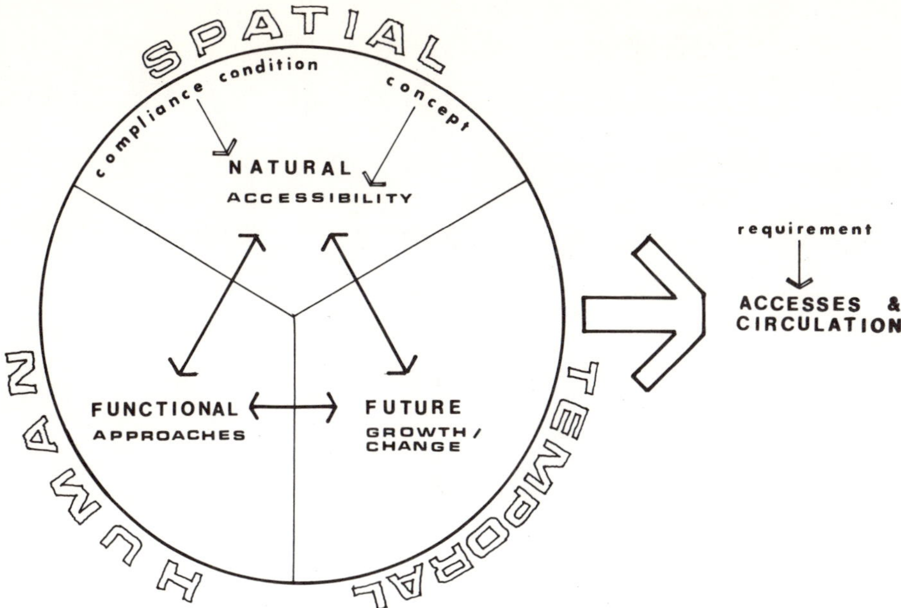

Figure 7.3

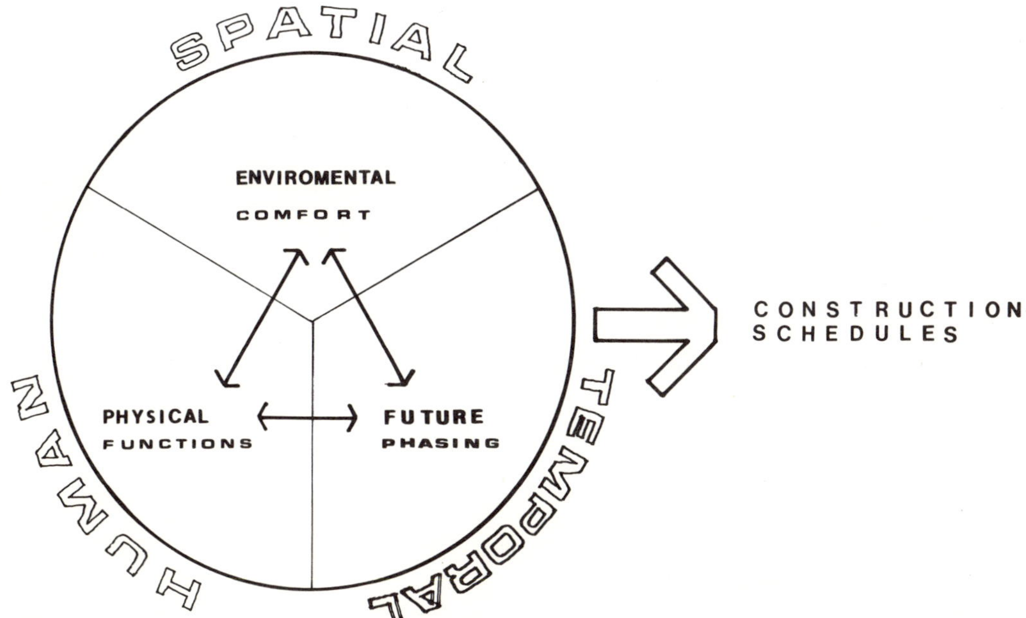

Figure 7.4

Value
- Function
- Priority
- Efficiency
- Hierarchy
- Worth
- Remedy
- Cost
- Improvement
- Allocation

Or in the case of programmatic concepts as applied to the temporal compliance conditions:

Structure
- Past
- Present
- Future

Direction
- Growth projections
- Change projections
- Phasing
- Escalation

Value
- Permanence
- Historic
- Preservation

Once these concepts, together with the corresponding compliance conditions analysis, interact among each other, the programmatic requirements occur. (See Figures 7.3, 7.4, 7.5, 7.6)

Summarizing, we can encompass the architectural programming procedures in the diagram illustrated in Figure 7.7.

7.2 Compliances

Just as the three basic universal factors can be subdivided into nine compliance conditions, these subdivisions have, by reason of their functional application, several constituents which determine the direction, scope and purpose of their analysis.

A general checklist of these basic compliance constituents can be summarized as follows:

1. Natural Compliances
   1.1 Site
   1.2 Surroundings
   1.3 Region
   1.4 Urban location
   1.5 Functional placement
   1.6 Accessibility
   1.7 Natural conditions
   1.8 Elements
   1.9 Weather
   1.10 Seasons
   1.11 Energy and Resources

2. Environmental Compliances
   2.1 Temperature
   2.2 Light
   2.3 Sonic conditions
   2.4 Shelter
   2.5 Environmental impact
   2.6 Preservation
   2.7 Pollution

3. Functional Compliances
   3.1 Purpose
   3.2 Activities
   3.3 Movement
   3.4 Flexibility
   3.5 Scale
   3.6 Use
   3.7 Manpower

4. Physical Compliances
   4.1 Measures and Scale
   4.2 Sex
   4.3 Age
   4.4 Health
   4.5 Hygiene
   4.6 Security

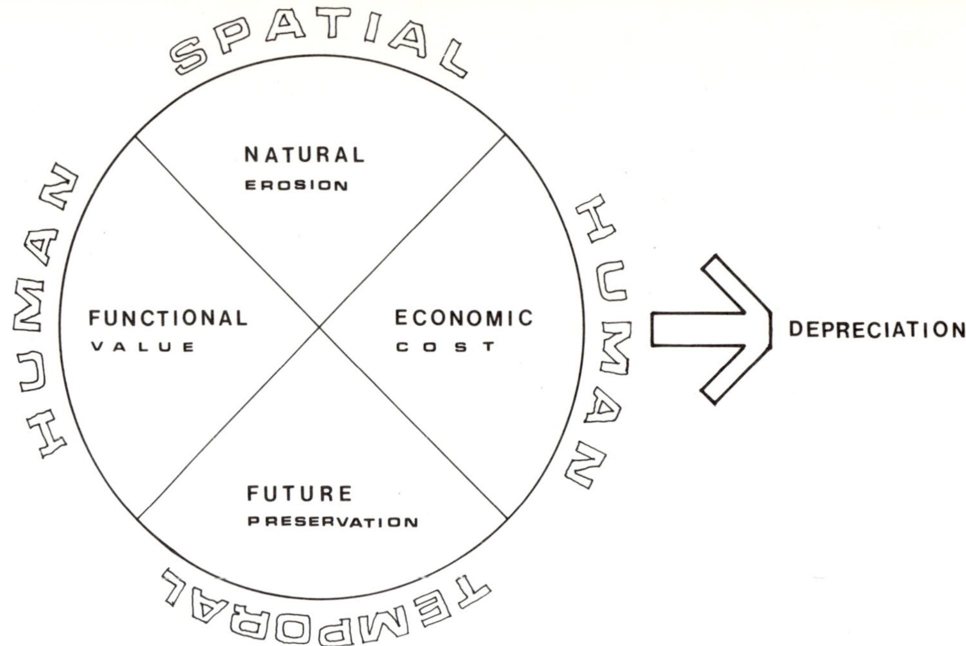

Figure 7.5

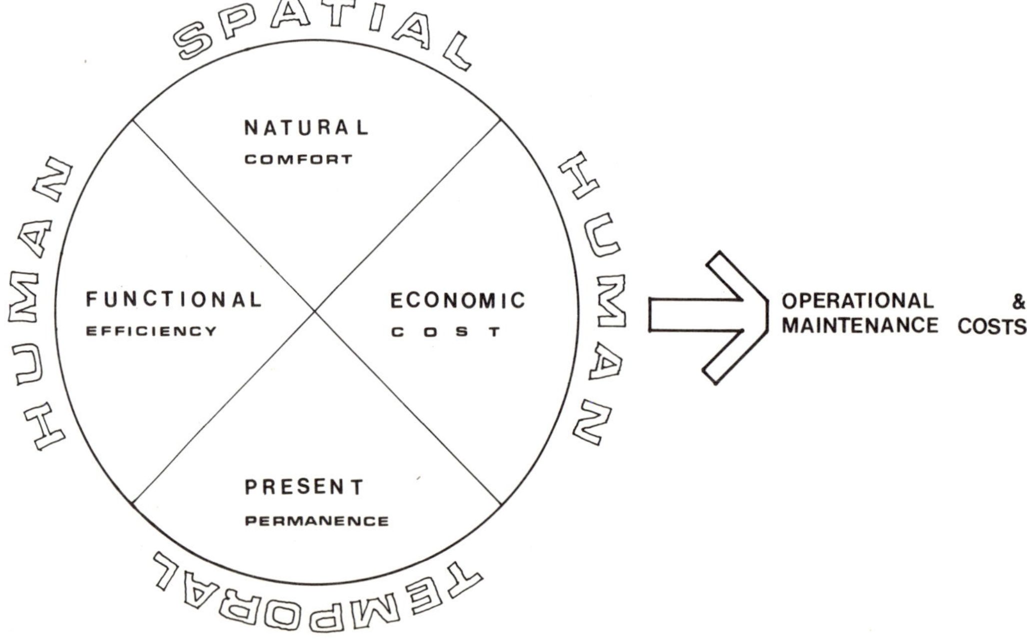

Figure 7.6

ARCHITECTURAL PROGRAMMING

FACTORS (cause)	PURPOSE	COMPLIANCE CONDITIONS (effect)		OBJECTIVES	ANALYSIS	PROGRAMMATIC CONCEPTS	REQUIREMENTS	PROGRAM
SPACE	PURPOSE	NATURE	NATURAL	DEFINITION	SITE ANALYSIS	STRUCTURE	DEFINITION	
			ENVIRONMENTAL			DIRECTION		
						VALUE		
MAN		HUMAN	FUNCTIONAL	DEFINITION	SPATIAL ANALYSIS	STRUCTURE	DEFINITION	
			PHYSICAL			DIRECTION		
			PSYCHOLOGICAL			VALUE		
			SOCIOLOGICAL					
			REGULATORY		REGULATORY SURVEY			
			ECONOMIC		ECONOMIC ANALYSIS			
TIME		TEMPORAL	PAST	DEFINITION	TIME ANALYSIS	STRUCTURE	DEFINITION	
			PRESENT			DIRECTION		
			FUTURE			VALUE		

Figure 7.7

The importance of the previous checklist lies in the fact that programmatic objectives derive from their study. Such as in the case of a specific occupancy date deriving from the schedules constituent of temporal compliances, or energy conservation and pollution control objectives being a result of the energy constituent of natural compliances and the pollution constituent of environmental compliances.

Throughout this study we will find that the interdependency of these constituents also determines to a great extent the conceptual directrixes of programming activities. Furthermore, this interdependency determines programmatic requirements without the influence of concepts if taken only in the light of essential conpliances' needs as in the case of most physical compliances and, to a great extent, in the cases of natural, environmental, regulatory, and temporal compliances.

7.3 Objectives

As we just mentioned, the basic structure of programmatic objectives is directly related to the compliance conditions with the three universal factors: SPACE, MAN and TIME. The major subdivisions or classifications of these programmatic objectives, however, would not be based on strict derivatives, but rather on pertinence.

First, not all objectives are in direct compliance with predetermined conditions. Often, objectives comply only partially or not at all. Thus we can establish a primary subdivision as follows:

Full Compliance
Partial Compliance
Non-compliance

The basic structure of these objectives is the result of their functional balance compared with the other compliance conditions considered during a specific programming activity (see diagram in Figure 7.1).

These subdivisions, which we could very well call Objectives' Degrees of Compliance, depend entirely on a value system established by a scale of priorities in relation to the overall compliance constituents list. That is, a set of priorities is established as governing them and, this scale of priorities will determine the position of a particular objective in the compliance subdivisions.

Full compliance objectives are those which completely satisfy the specific requirements outlined by the analyzed situations which determine the compliance conditions. This might seem to be the primary task of all objectives, but such is not the case; for example, in the requirement for compliance with the natural conditions of a determined site, where the word "compliance" might not imply "full compliance" but only a partial conformity with a specific portion of land. It is very unlikely the "perfect" site that would permit a full objective compliance will ever be found if we consider as variable elements all the particular characteristics contained within its boundaries.

By the same token, we will often find several objectives which outline a non-compliance requirement when evaluated against pre-established conditions. This will probably occur when a determined compliance intent is in direct conflict with a compliance requirement of higher priority; that is, accessibility vs. security, functional placement vs. environmental impact, shelter vs. preservation, codes vs. urban location, or even aesthetics vs. character.

In the definition of these objectives' degrees of compliance (or objective trade-offs), programmatic concepts play a key role in the final determination of priorities and hierarchies; in the case of full compliance by determining HOW it is to be achieved; in the case of partial compliance by determining HOW FAR such degree of compliance is to be extended; and finally, in the case of non-compliance, by measuring the applicable priorities and outlining HOW a specific lack of compliance will affect the project, what can be expected of it and what can be done about it, or simply up to what point it may or may not be acceptable.

Chapter 8 contains several basic rules to follow in the determination of these objectives. For now, let us concentrate on the basic structure of the analysis steps which will study these objectives and their relationship to the essential compliance conditions.

7.4 Analysis

If analysis is the separation of any intellectual or substantial whole into constituents for individual study, then the primary step in any analytical procedure is the clear establishment of such separation. This phase is as important in the process as any other.

The separation process cannot be effected by merely fragmenting an operation into its smallest possible constituents since, even in the simplest of cases, every portion of a whole has its own place and function within a total organized pattern and cannot be studied individually without reference to its immediate complements. In analyzing an operation — NO MATTER WHICH — the first step must always be a logical and systematic separation of constituents.

This separation accomplishes two very important things. First, the analysis can be executed within a framework of organized groupings; and second, the basic structure of the operation can be understood more clearly since, logically, such groupings will enable us to understand the primary functions they represent and the relationships they maintain among one another and with the whole.

To illustrate this point let us assume that the analyzed subject is an automobile. If our first step were the study of a manual containing a detailed description of each part of the vehicle, it would probably be an impossible task to a layman unless each description presented not only basic information of the parts, but also a statement of function. This function would most likely refer to another function, depending on a third function, and so on. If, on the other hand, we had in front of us the actual parts, without any description at all other than their apparent physical characteristics and if we were to establish an analysis of each one based on such observations, we would, in the best of cases, end up with a more or less clear description of parts, but at no point would there be any reference to their operational function, logical placement, or even purpose. Thus we would end up describing odds and ends, but certainly not the constituents of an automobile since there would probably be no reference to the automobile as a whole to begin with.

The difference between analyzing an operation with a systematic fragmentation approach and making an erratic list of elements is the same as that between disassembling a radio in order to repair it, and beating it with a rubber mallet in order to see what its contents are.

The best way of effecting this fragmentation is by considering the major subdivisions of any operational task and moving from there to their essential components, from which the detailed study of constitutive elements can be made without risk of misplacement, functional inadequacy, etc. Thus the first steps are:

1. Establishment of Divisions. As in the case of a manufacturing company: Management, Sales, Promotion, Accounting, etc.
2. Establishment of Divisional Components. Assuming Management from above: Presidency, Vice-presidency, Clerical, Records and Filing, Storage, etc.
3. Establishment of Constitutive Elements. Assuming Clerical from above: 12 Secretaries, 1 Office Manager, 1 Office Boy, 1 Chief of Personnel, 1 Storage Room, 1 Women's Restroom, 1 Men's Restroom, etc.

The second part of the analytical procedure is that which describes the interrelationship of these components and their constituents. This is a description not only of basic functions, but also of their interdependencies, and it can be achieved in each one of the three levels listed above. Depending on the degree of detailed information needed, these relationships can be more or less useful in each particular case. For example; in site planning a basic relationship study of divisional functions can be much more useful than a detailed description of ALL elements and their interdependencies. We must not forget, just as too little information can be extremely detrimental to a design task, there is also the danger of "data clog" in cases where too much information is fed too soon into the design process. On the other hand, for an actual building design a simple divisional relationship chart will not suffice, since it will not provide a detailed description of components, their activities' description and functional relationships.

In establishing this primary separation of elements we must proceed with caution and always follow a systematic approach. The divisional analysis, then, will contain two major steps:

1. Establishment of constitutive components characteristics
2. Establishment of constitutive components relationships

In analyzing the characteristics and relationships of these components in terms of each one of the compliance conditions, we will find, among others, the determining analytical factors as shown in Table 7.1

In actuality, most of these factors do not apply every time, but only a few of them as the case may require. Basically, the intent of each step is descriptive in nature and evaluative in form. (See Figure 7.7) A method for the establishment of these characteristics' relationships is explained in detail in Chapter 8.

Once the analytical procedure has achieved this, it is up to the programmatic concepts to set the guidelines by which a building program will be determined.

Table 7.1

Compliances	ANALYSIS	
	Divisional Analysis	Establishment of Divisions, Divisional Characteristics, Divisional Relationships.
	Components Analysis	
	Components Characteristics	Components Relationships
Natural	Regional Data	Interactions
	National Data	Frequency
	Site Analysis	Projections
	Descriptions	Evaluations
	Energy	Flexibilities
	Conservation	Temporality
	Demands	Type
		Class
		Effects
Natural (cont'd).		Trends
		Grouping
		Spatial Needs
Environmental	Thermal Data	Levels
	Day/Night	Interactions
	Noise Levels	Class
	Statistical Data	Projections
	Luminous Data	Temporality
	Type	Trends
	Comfort Zones	Effects
	Needs	Grouping
		Values
		Spatial Needs
Functional	Description	Balance
	Intent	Distribution
	Constitutive Elements	Grouping
		Trends
	Value	Effectiveness
	Loss	Values
	Statistical Data	Description
	Areas	Dynamics
	Volumes	Temporality
	Type	Frequency
	Direction	Volume
	Flow	Type
	Traffic	Class
	Variations	Patterns
	Structure	Interactions
	Extent	Effects
	Manpower	Flexibilities
	Adequacy	Evaluations
	Demands	Spatial Needs
	Permanence	
Physical	Description	Balance
	Statistical Data	Distribution
	Scale	Grouping
	Area Parameters	Trends
	Volumes	Values
	Preservation	Temporality
	Illnesses	Volume
	Disability	Maintenance

Compliances	ANALYSIS	
Physical (cont'd.)	Handicapped	Services
	Security	Treatments
	Hazards	Interactions
	Needs	Effects
		Systems
		Location
		Evaluations
		Spatial Needs
Psychological	Definitions	Patterns
	Levels	Interactions
	Behavioral	Levels
	Characteristics	Groupings
	Needs	Effects
		Projections
		Spatial Needs
		Implications
		Temporality
		Values
Sociological	Statistical Data	Value
	Levels	Levels
	Grouping	Interactions
	Descriptions	Grouping
	Areas	Frequency
	Volumes	Projections
	Scale	Temporality
	Extent	Flexibilities
	Adequacy	Effects
	Demands	Trends
	Special	Spatial Needs
	Characteristics	
Regulatory	Bureaucratic	Interactions
	Components	Inter-
	Statistical Data	dependences
	Planning	Ramifications
	Zoning	Levels
	Fire	Significance
	Health	Flexibilities
	Highway &	Effects
	Traffic	Evaluations
	OSHA	Projections
	Affirmative	Temporality
	Action	

Compliances	ANALYSIS	
Regulatory (cont'd.)	Contracts	
	Building	
	Commissions	
	Neighborhood	
	Associations	
	Minority Groups	
	Descriptions	
	Demands	
	Variances	
	Procedures	
Economic	Areas/Cost	Cost Parameters
	Volumes/Cost	Effects
	Improvements	Projections
	Sales Program	Characteristics
	Marketing	Areas/Cost
	Analysis	Volumes/Cost
	Levels	Temporality
	Quality	Ramifications
	Implications	Marketing
	Characteristics	Projections
	Value	Sales Projections
		Cost/Efficiency
		Ratings
Temporal	Historic Value	Significance
	Description	Evaluation
	Data	Implications
	Dynamics	Projections
	Occupancy	Interactions
	Growth	Characteristics
	Change	Temporality
		Phasing

7.5 Programmatic Concepts

As we established before, all programmatic concepts have three major subdivisions — Structure, Direction and Value — which create value judgments on grouping, alternatives, systems, policy, encounters, projections, etc., when applied to compliance conditions (see Figures 7.3, 7.4, 7.5, 7.6), and when, in their interaction with the analysis of a specific component for a compliance condition they produce the program requirements for that

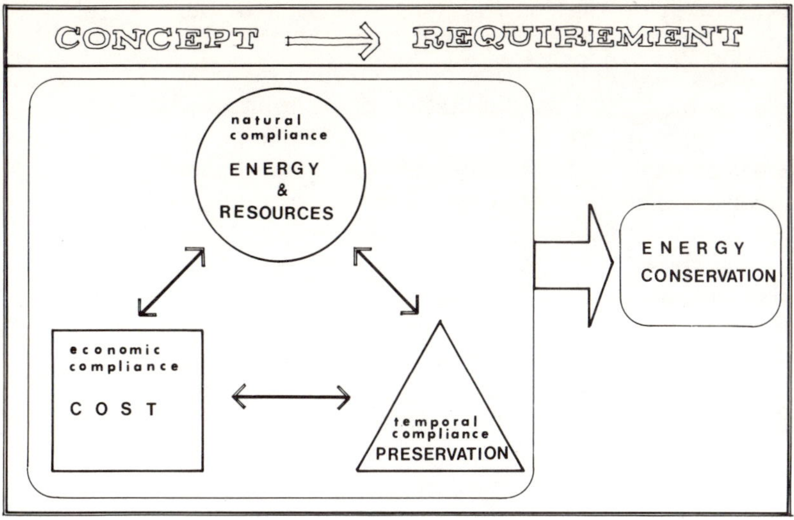

Figure 7.8

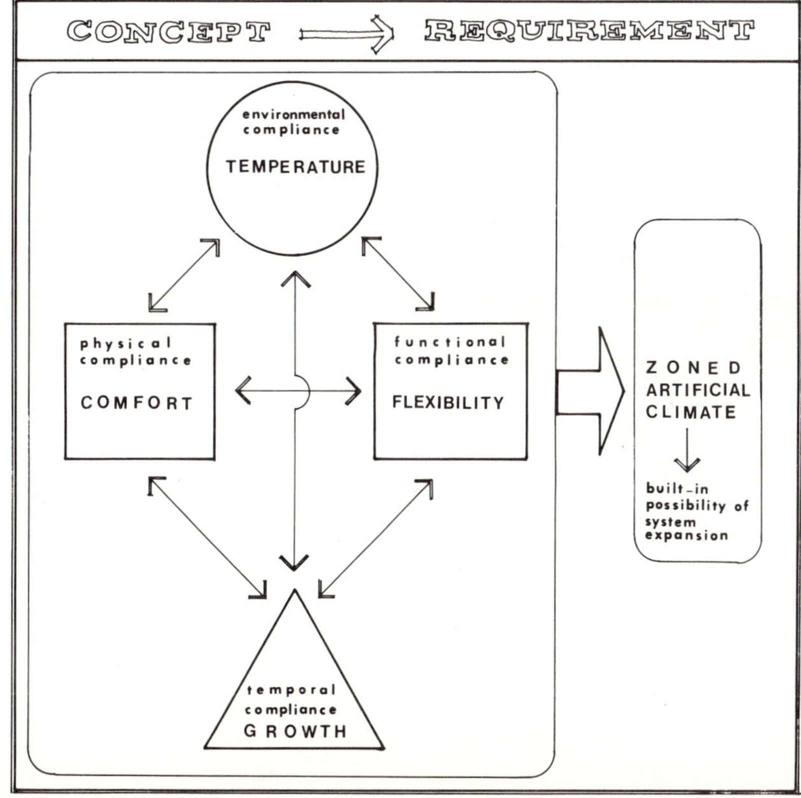

Figure 7.9

definite area of study. But there are certain characteristics of these program concepts which should be considered major determinants of any programming procedure and they should not be overlooked, no matter what the circumstance.

For example, during the conceptual phase of programming most controllable variables (wants) appear. If most needs are a direct product of the Objectives-Analysis-Concept interaction, most *wants* will result from a direct conceptual approach to an analysis evaluation, in which the concept governs the fact far beyond the parameter of logical adjustment. Words like *purpose, intent, desire,* or *preference* can often disguise a simple *want* which may or may not fit the overall assemblage of programming requirements. This should not be taken to mean that wants cannot (or should not) be considered actual requirements. The fact is that a great majority of program requirements are simple wants, but in effect, the identification of these controllable variables is important since without it the determination of flexibilities, and sometimes alternatives, becomes impossible.

Another characteristic of programmatic concepts is their application to the primary phases of analysis. During the formulation of the objectives, the programmatic concept can help tremendously in the determination of priorities which will in turn determine the compliance degrees of these objectives, become a key factor in partial compliances, and be a major determinate in non-compliance conditions (where a conflict of priorities will most likely appear).

Elaborating on the previous diagrams which represent the conceptual interactions of elements, we can see how, during the concepts stage of programming, each compliance condition is identified, measured, and weighed against the others in order to produce a final statement of needs or requirements. (See Figures 7.8, 7.9)

Program concepts are essential factors of functional performance in operations. They shape, alter or modify the basic structure of performance and, in most cases, determine its degree of effectiveness. These concepts, however, should not be confused with design concepts because, although they have points in common, a design concept represents a synthesis approach towards an architectural solution, while a programmatic concept is an abstract guideline for courses of action to be taken when faced with conflicting or incompatible compliance conditions.

The final result of this Objectives-Analysis-Concepts' interaction will be the program requirements. However, the order in which these steps are carried out is not necessarily the one set forth in Figure 7.1. Although the ideal situation would be to maintain that order, no matter which sequence is followed, the final step will always be the determination of these program requirements. Without them there is no program, and without a program there can be no functional design whatsoever. It is as simple as that.

7.6 Program Requirements

Certain considerations involving program requirements should be analyzed before even trying to define what programmatic requirements really are:

1. A list of requirements is not a program nor a program statement.
2. Program requirements are not always factual considerations.
3. *Needs* are not the only constituents of program requirements.
4. A *want* does not automatically constitute a requirement, but it most certainly can determine one.
5. Program requirements represent the basis for the final program statements.
6. Program requirements are the basis for design approaches, solutions and evaluations.

As the result of the Objectives-Analysis-Concepts' interaction, the programmatic requirements represent the next-to-final step in architectural programming. This is the phase where the conceptual basis determines the path a specific problem solution must follow in order to effect its task, and this broadly defined, constitutes the essence of this phase.

Based on their nature, program requirements can be classified as qualitative or quantitative. Based on their structure, they can be classified as *needs* or *wants* and, based on their function, we can subdivide them in direct relation to the compliance objective they define as shown in Figure 7.10.[14]

It is here during the requirements determination that the distinction between *needs* and *wants* becomes clear, and for the first time decisions must be faced regarding the point at which certain wants must be made part of the requirements chart and thereby become part of the problem outline.

Needs and wants are an integral part of programming. To fight this duality is to fight the conceptual basis of any programmatic activity. Any design professional who has been involved in the programming phase of a residence will know exactly the meaning of these words. On the other hand, although all building types follow essentially the same behavioral outline in regard to conceptual needs or wants, the determination of this outline becomes much more complicated in operations of greater complexity and specialization. It is fairly easy to distinguish between needs and wants in a single family dwelling simply because the operations involved are common knowledge to all of us; but in the case of a research laboratory, for example, the distinction between them can become a very serious matter.

Because of this the analysis phase should always precede the establishment of needs, if only to evaluate their standing.

At this stage we realize how valuable it is to inform the client of the importance of this distinction at the very outset of programming since his/her help can be the most valuable asset in differentiating between the two.

Normally, wants appear for the first time during the establishment of objectives. The problem lies in the fact that at first the programmer is not aware of their structure. Only after the analytical procedure has been completed do the first signs of wants (or controllable variables) appear, and we begin to find

<table>
<tr><th rowspan="2">COMPLIANCE
CONDITION</th><th colspan="2">R E Q U I R E M E N T S</th></tr>
<tr><th>QUANTITATIVE
(NEEDS OR WANTS)</th><th>QUALITATIVE
(NEEDS OR WANTS)</th></tr>
<tr><td>NATURAL</td><td>acreage, f.a.r., dimensions, etc.</td><td>proximity, orientation, views, etc.</td></tr>
<tr><td>ENVIRONMENTAL</td><td>sizes, environmental zones, cycles, etc.</td><td>comfort zoning, shelter, etc.</td></tr>
<tr><td>FUNCTIONAL</td><td>areas, volumes, dimensions, etc.</td><td>groupings, encounters, flexibilities, etc.</td></tr>
<tr><td>PHYSICAL</td><td>area parameters, distances, accesses, openings, etc.</td><td>quality, form, finishes, etc.</td></tr>
<tr><td>PSYCHOLOGICAL</td><td>relationships, encounters, modules, etc.</td><td>ego, hierarchies, privacy, image, character, etc.</td></tr>
<tr><td>SOCIOLOGICAL</td><td>location, race, creed, demography, etc.</td><td>class, type, impact, etc.</td></tr>
<tr><td>REGULATORY</td><td>building, fire, highway, etc.</td><td>zoning & landscaping ordinances, planning boards, etc.</td></tr>
<tr><td>ECONOMIC</td><td>budgets, allocations, costs, etc.</td><td>values, quality, etc</td></tr>
<tr><td>TEMPORAL</td><td>allocations, completions, occupancy, etc.</td><td>phasing, execution, permanence, historic value, etc.</td></tr>
</table>

Figure 7.10

words like convenient, desirable, preferable, etc., affect components' requirements and interrelationships.

In Chapter 8, we will classify the components of operations and their relationships in different categories with the sole purpose of facilitating this distinction. For now, a few basic rules could further help this determination:

1. The basis for most wants come to us in the form of ADJECTIVES or ADVERBS applied to functional requirements.
2. Almost invariably there will exist a need for JUSTIFICATION when analyzing their functional performance.
3. Very seldom will they decrease cost.
4. Often, a clear disparity will be observed between the analytical results and the wants request, that is, over-sized areas, arbitrary relationships, etc.
5. No precedence will be found in reference materials during the investigative stages.

Some of the rules of thumb mentioned above apply also to innovative concepts which, while they might be considered wants by definition, can result in advantageous approaches to the resolution of functional problems; but we must not forget if an item is classified as a controllable variable this does not mean that it must be excluded from the requirements list, but simply that it should be classified as such. Only through a careful evaluation of its weight and significance as a factual constituent of the whole can it be included or discarded from the list.

Some wants are easily identifiable; some are not. The most dangerous ones are those which appeal not only to the client, but also to the design professional, in spite of the fact that they might be unrealistic, unaffordable, or simply illogical.

Not too long ago an architectural firm was given a commission to design a small office building for an insurance company branch office on a mountain site with extremely appealing surroundings. One of the requests of the corporation president was to have his own office overlook the canyon below from a cantilevered terrace. This particular requirement appealed so much to the programming architect that most of the design was worked around this concept. The problem began when the president realized that he was going to face serious objections from his board of directors because of this extravagant feature; and although the facilities were eventually redesigned to soften this emphasis, a tremendous amount of time and effort were simply wasted because there was no clear identification of this requirement as a *want,* and no verification of its possible consequences.

In dealing with these situations, it is advisable to submit every identifiable want to comprehensive scrutiny. A requirement based on a mere desire can only be accepted as part of a program through the logical justification of its balanced position within the whole constituted by the programmatic requirements.

7.7 Program Outline

There are three basic rules to follow in developing a program outline:

1. Keep it simple.
2. Keep it clear.
3. Keep it brief.

Just as in programmatic requirements, the two basic divisions of every compliance statement are qualitative and quantitative, but there is still one other unique characteristic to be noted: priority.

So far we have listed the major subdivisions of all compliance conditions as:

> Natural
> Environmental
> Functional
> Physical
> Psychological
> Sociological
> Regulatory
> Economic;
> Temporal

Still, this might not be the actual order of importance in which a specific project's priorities could be defined. Because of such priority levels it is sometimes convenient to establish a different order in the listing of compliances. This must not be construed as a suggestion that only the first three or four should be considered. Not at all. As a matter of fact, compliance requirements of a sixth or seventh order, while not placed at the top of the priorities list, might still become essential design determinants; but the fact does remain that the above order helps to clarify accentuations and hierarchies.

Moreover, whenever a compliance requirement does not apply to a specific project this MUST BE NOTED, since the remark "not applicable" can have as great a significance in the development of a design solution as any other requirement statement.

For example, a project might not have any applicable sociological requirements; in that case, when the programmer is establishing the order of priorities, the sociological implications should be listed at the end with a clear NOT APPLICABLE statement next to it. This will not only enable the designer to formulate a clearer picture of the project's scope, but will also state very definitely that whatever architectural solution is to be pursued will not take into account these implications. This becomes very important in cases where further reviews are necessary. We should not forget that the FIRST APPROVAL a design professional obtains from the client is that concerning the project's program.

In this regard, it is extremely important for the design professional to obtain written approval of the program outline before beginning the design phase in cases where programming and design are two definite sequential steps.

The basic portions of a comprehensive program outline[15] as defined should then be:

General Information
    Project title
    Location
    Name of owner, address
    Project number
    Participants
    Index
Statement of Objectives (in order of priority)
    Natural
    Environmental
    Functional
    Physical
    Psychological
    Sociological
    Regulatory
    Economic
    Temporal
Background, Research & References
Project Requirements
    Site analysis
    Spatial analysis
        Components functional description
        Spatial requirements
        Functional relationships
        Flow patterns
    Regulatory survey
    Economic analysis
    Temporal analysis
Special Characteristics
Operational & Maintenance Requirements
Construction Requirements
Appendix
    Project development factors
        Marketing analysis
        Statistical data
        Sales program
        Other
    Supporting data (when applicable)
        Notes
        Memos
        Correspondence

7.8 The Planning Procedure

The importance of a planning procedure in relationship to the architectural programming phase can be as basic as the programming phase is to the design process.

Several methods have been established for planning the actual programming phase in professional design, most of them in the form of sequential flow diagrams P.E.R.T.[16] or critical path methods (CPM). In either case, they establish the same thing:

the necessary steps which must be taken in order to arrive at a comprehensive program outline for a building project.

The major advantage of these steps lies more in their provision of a listing and description of the elements involved than in their actual scheduling system, since, in spite of all arguments to the contrary, it is very difficult to hold every participant to an inflexible schedule of actions during the programming phase. It must be noted, however, that it is during this procedure that most if not all of the human elements involved in a project come together in work sessions, conferences, etc., for the first time.

To actually assume that a CPM schedule can be worked out on every project is more than inaccurate, it is unrealistic. Yet, it is still desirable.

An organizational chart which facilitates this type of schedule can be illustrated in Figure 7.11.

In this particular example, the actual responsibilities and actions are underlined to show who is or will be responsible for each step, and by just assigning time values to each operation, a flow diagram can be established as shown in Figure 7.12.

The most important constituents in each case will always be the clients, users, consultants and programmers, since as we have repeatedly stated, they are the key elements in any programming phase. Still it must not be forgotten that all other elements (when applicable), can play a major role in the development and accuracy of these fact-finding operations, as well as in the establishment of programmatic concepts and project requirements.

7.9 Investigative Methods

Any investigative method, when used properly, can be extremely helpful in preparing a programmer for the tasks s/he must perform during his/her analytical activities. The primary division of research approaches must inevitably be:

1. Existing references research — where by means of previously executed projects one can obtain pertinent information.
2. Designed research — where either by unavailability or non-applicable circumstances the investigative method must be designed to fit the particular problem.

If the first type can be considered useful and adequate in certain cases, the second one is definitely ideal in each instance.

However, there are some basic rules which we must never forget when initiating any research investigation.

1. REFERENCES RESEARCH. First, it is imperative to recognize that REFERENCES ARE NOT DETAILED, BUT GENERAL GUIDELINES. The importance of this cannot be overrated. No single reference material will apply EXACTLY to any specific problem other than the particular one it illustrates. As a matter of fact, assuming detailed guidelines because of reference material can be the most damaging single factor a programmer faces when preparing for a specific project analysis, since s/he will probably end with preconceptions and ideas which might result in a biased attitude towards his/her investigative approaches and, logically, their results.

We can now find reference material in almost every discipline covered by architectural programming. This reference material comes to us in the form of publications, documentaries, brochures, previously executed projects and even verbal advice. But we must not forget that in each case any reference statement represents simply a general statement of direction and not an invariable course of action.

For the purposes of this segment, we will divide reference materials into two major categories based on their structure:

DIRECT

INDIRECT

By direct, we mean those which offer us a physical example of operations, characteristics, etc., and by indirect those which offer a more or less clear description of their elements by audio-visual means.

Both are essential. If a direct investigative method can offer a living example of what we are trying to understand, it is very unlikely that a single case study will suffice in establishing cardinal trajectories in programming procedures.

	CONTRACTS	PRELIMINARY INTERVIEWS	OBJECTIVES DEFINITION	SUPPORTING DATA / SITE	SITE ANALYSIS	SUPPORTING DATA / SPACE	SPATIAL ANALYSIS	REGULATORY SURVEYS	ENGINEERING SYSTEMS / CRITERIA	ECONOMIC ANALYSIS	TIME ANALYSIS	SPECIAL CHARACTERISTICS	OPERAT./ MAINT. REQUIREMENTS	CONSTRUCTION REQUIREMENTS	PROGRAMMATIC CONCEPTS	PROGRAM REQUIREMENTS	COORDINATION & REVIEW	PROGRAM
CLIENT	●	●	●					○	●	●	●	●	●		●	●	●	
CLIENT CONSULTANTS	●	●		●		●			●	●		●	●	○			●	
MANAGING ARCHITECT	●	●	●							○							●	
PROGRAMMER		●	●	●	●	●	●	●	●	●	●	●	●	●	●	●	●	●
PROJECT ARCHITECT			●					●	●	○	●	○		●			●	
ENG. STRUCT.								○	●			○	●	●			●	
ENG. H.V.A.C.								○	●			○	●				●	
ENG. PLUMB.								○	●			○	●				●	
ENG. ELEC.								○	●			○	●				●	
INTERIOR DESIGNERS							●					○					●	
SPECIAL CONSULTANTS									●			●	●					
OTHER/SPECIFY																		
REGULATORY AGENCIES								●										
CONTRACTOR									●			○		●				
SUBCONTRACTORS									●			○		●				
SALES REPRESENTATIVES									●			○	○	●				

● INDICATES DIRECT INVOLVEMENT

○ POSSIBLE INVOLVEMENT

Figure 7.11

ARCHITECTURAL PROGRAMMING SCHEDULE
DESIGN
1 FINALIZE CONTRACTS
2 PRELIMINARY INTERVIEWS
3 OBJECTIVES DEFINITION
4 SPATIAL ANALYSIS DATA
5 SITE ANALYSIS DATA
6 SPATIAL ANALYSIS
7 SITE ANALYSIS
8 BUDGET REVIEW
9 CHARTS
10 DIAGRAMS
11 ECONOMIC DATA
12 SPECIAL CHARACTERISTICS
13 REGULATORY SURVEY
14 ENGINEERING RESEARCH
15 ECONOMIC ANALYSIS
16 STRUCTURAL EVALUATION
17 PLUMBING
18 H.V.A.C.
19 ELECTRICAL
20 REPORT
21 ENGINEERING REPORTS
22 OPER. & MAINT. RQMTS.
23 CONSTRUCTION RQMTS.
24 PROGAMMATIC GUIDELINES
25 SPATIAL REQUIREMENTS
26 SPECIAL REQUIREMENTS
27 COORD./REVIEW & SCHEDULES
28 PROGRAM DRAFT/REVIEW
29 PROGRAM BOOKLET
30 CLIENT REVIEW
31 MODIFICATIONS & ADJUSTMENTS
32 CLIENT ASSISTANCE
33 CONSULTANTS REVIEW
34 FINAL PROGRAM

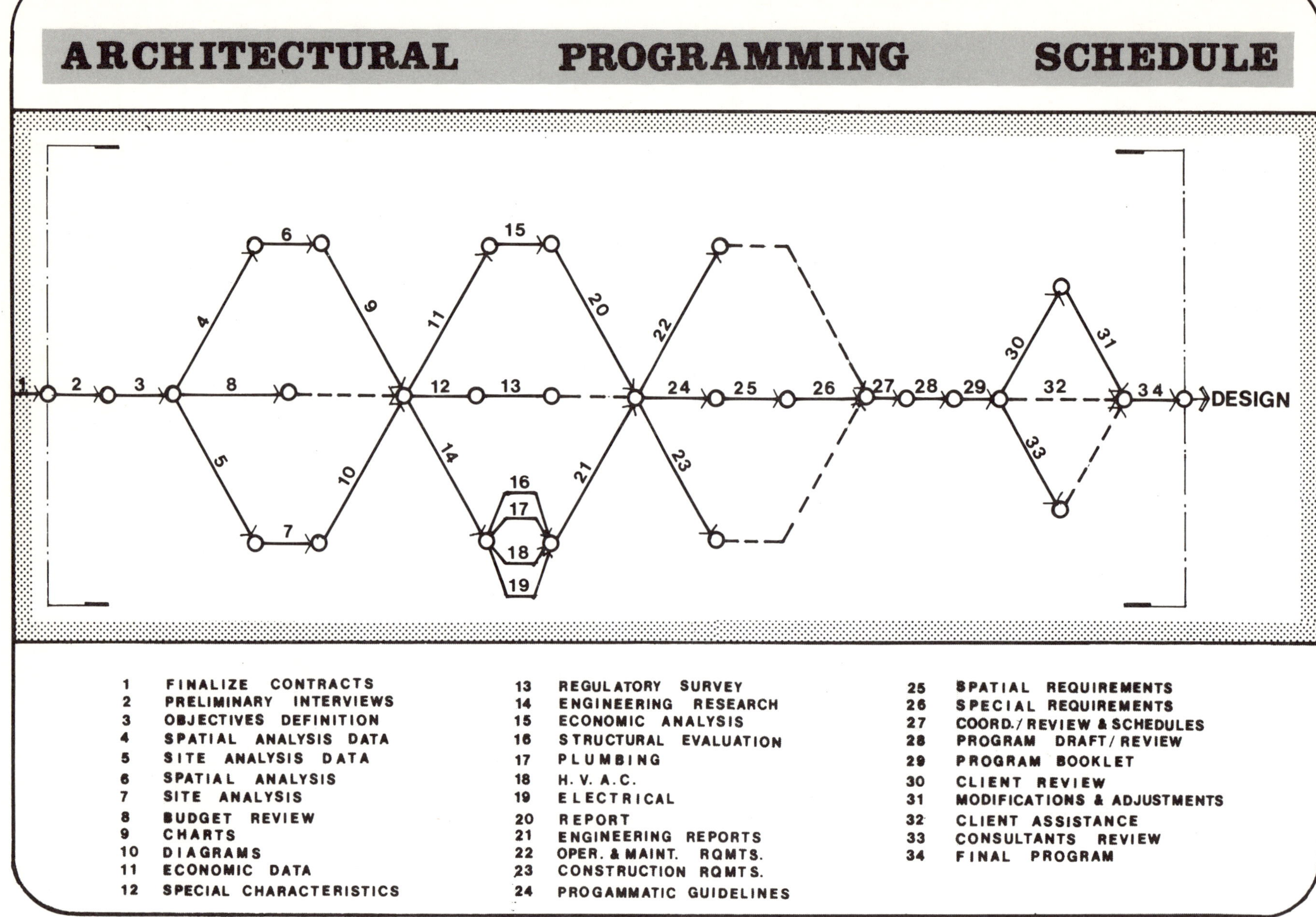

Figure 7.12

That is why the importance of audio-visual reference material has always been so great in every programming task. It not only offers generalized and even specialized descriptions and information, but it does so in a much more pluralistic manner.

By the same token, it must be noted that changes in times, location, and even urban placement can affect a building type study so vastly that any assumption regarding the validity of its statements must always be questioned.

References are no substitute for experience. As a matter of fact, very few things are. But this does not mean that experience can justify the lack of investigation. It only facilitates and simplifies research, but only research. And research conclusions should never be based on anything else because any assumptions based on experience are as dangerous as inflexible guidelines obtained from references.

Direct references should only be construed as individual case studies. Indirect references should be considered research material only and not programmatic concepts or requirements. Furthermore, no single portion of the analysis studies should be undermined by the handy reference statement, but only complemented by it.

It is convenient to form an actual reference library for programming, since not all professional design reference material applies to this particular phase of our practice. Detailed technical information regarding building products, contract documents, or even design philosophies do not apply directly to programming and should not be included in its reference shelf. They belong somewhere else. Chapter 10 contains more information regarding the programming library.

2. DESIGNED RESEARCH. Designed research is specifically created in a series of particular circumstances. While this is true of almost any investigative system in general, in many cases a specialized methodology must be developed to fit the need. Insofar as a general organizational matrix is concerned, Figure 7.13 might further illustrate this point.

In this matrix the degrees of impact and usefulness of the research techniques (listed horizontally), are tabulated when referred to each of the major decision areas (items) listed in the vertical column.

By doing this we do not only clarify the importance of certain techniques and their relationships to decision areas and team workshops, but also aid their direction, degree of importance and scheduling.

Just as any programming procedure must inevitably be designed and adjusted to each individual project, these research techniques vary immensely in their layout and content with each particular set of conditions. It is impossible, therefore, to offer a rigid methodology by which to arrive at these research techniques' layout, much less their particular content, since factors which vary from location to building types or operational goals can affect them so drastically that any attempt in that direction would not only be fruitless but also misleading.

It is important, nevertheless. to outline the possible effects of investigative methods in the subsequent stages of architectural programming because, no matter what system is employed, the majority of all decision-making processes that are to follow will essentially depend on the result of this primary phase:

1. Not all clients are aware of the necessary background information essential or complementary to programming activities. Therefore, research can help both the programmer and the client fully understand their own problems.
2. Research can expose the need for specialized consultants in given fields, which need could otherwise remain unnoticed through stages in which their intervention was advisable or even essential.
3. Through familiarization with the pertinent building type data, the programmer can begin to reap the benefits of the preliminary encounters with a client from the very first work session, since the analytical course of

action will most likely be established in a much more rational manner than a mere preliminary discussion of disorganized statements and facts.

4. Research acts as an incentive for clients to express and discuss their points of view on key matters much more readily, since they are in the presence of someone who knows what they are talking about and understands their position.

5. Research is the only way known to compensate for lack of knowledge and experience.

---

[14] For clarification purposes, the distinction between needs and wants is an arbitrary one in this example.

[15] Refer to Chapter 9 for actual case studies illustrating this outline.

[16] P.E.R.T. — For Program Evaluation and Review Technique, a form of network planning of process flow, time organization and progress monitoring.

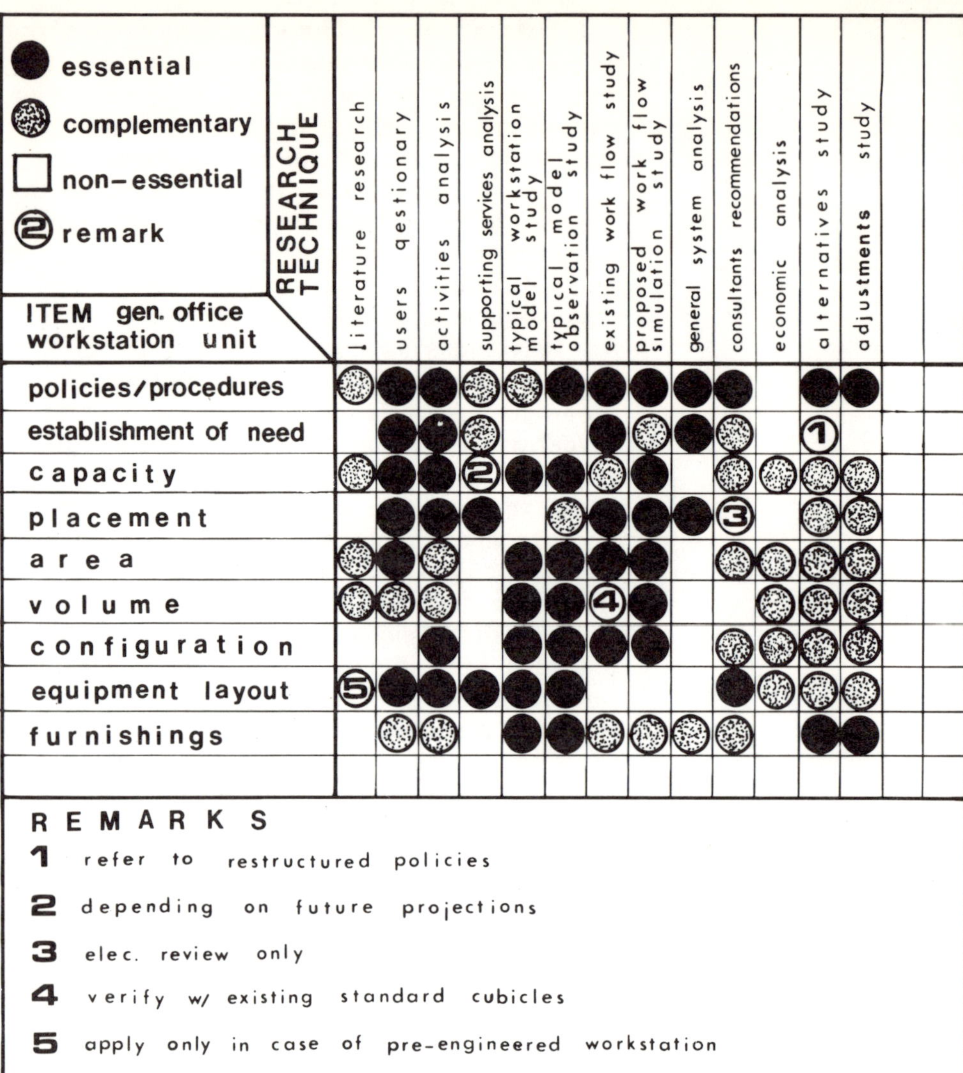

Figure 7.13

8
Techniques

8.1 Definition of Objectives

The greatest disadvantages of this primary stage of architectural programming are also, perhaps, its most valuable assets: those of generalization and indetermination. Any client's objective is nothing more than a purpose of compliance with universal factors. Needs, wants, concepts and solutions are all originated there, through balanced or unbalanced relationships to given situations. If it is true that generalized objectives are of little value to the design process, it is also true that this generalization opens the door to further analytical and conceptual discussions which will hopefully clarify those objectives. Furthermore, conclusions will be reached by means of the logical establishment and balanced distribution of facts from two expert elements: the design professional and the client.

It is unrealistic to expect a client to arrive at our offices with ALL the answers. More than that, it is undesirable. A client is not (in almost all instances) a design professional and will probably have little conception of what the construction industry is all about. For the client, the professional design experience is not only one of education and discovery, but also one of disciplined understanding and self-evaluation.

Programmatic objectives are the starting point of any construction project. Their clear definition is, therefore, essential; however an objective's definition is not a programmatic requirement in itself, but simply a statement of intent from which to begin the analysis and programming procedures that will follow.

As in the case of program requirements, fundamental program objectives can be classified in two major categories based on their essential structure:

QUALITATIVE

QUANTITATIVE

Qualitative objectives are those of an intellectual nature, while quantitative objectives are those resulting from factual compliances.

As dictated by the "analysis-programming" balance, programmatic concepts will play a major role in the definition of qualitative objectives, while analytical procedures will indicate the best course of action for the final transformation of quantitative objectives into actual program requirements.

A clear example of a quantitative objective would be "The need to build a facility which will accommodate twice the personnel now employed by our company"; while a qualitative objective would be: "To provide a suitable environment for the physically disabled."

It is fairly easy to distinguish and differentiate them. Quantitative objectives will most likely offer facts while qualitative objectives will generally offer concepts as starting points for discussion of subsequent programmatic activities.

Nevertheless, in each case the analysis and programming processes will inevitably interact in spite of their respective primary or secondary roles. For example, in building a facility to accommodate a certain number of employees, it will become necessary to establish area parameters and circulation patterns and to determine (through a definite conceptual process) HOW this will be done. In the second case, it will become necessary to study (through a carefully executed analysis procedure) the required sizes and clearances for wheel chairs, ramps, etc.

One simple rule clarifies the different course of action in each case: When faced with a qualitative objective, try to determine HOW and by which FACTS it is to be achieved. This approach will draw attention to the otherwise obscured analysis procedure. When faced with a quantitative objective, try to determine WHY it ever became an objective in the first place and how logical its primary establishment was. The programmatic concepts behind the statement will inevitably surface if we take this approach.

What we are doing in each case is arriving at the missing link through the obvious presence of the existing programming element, either determining the analysis steps when the programming concepts are our only clue, or vice versa since analysis and programming are not isolated actions that take place independently of one another, but very closely related processes which interact through every phase of the architectural programming activities.

In dealing with conceptual or qualitative objectives it becomes obvious that a clear understanding of ideas must be established. The key words here are "communication" and "understanding" at every level.

In dealing with quantitative objectives a clear definition of facts is the essential task. The key words here are "general study" and "assimilation."

Assuming as a program postulate the objective of "providing a suitable environment for the physically disabled," we will find this statement of intent too theoretical to actually become a design determinant; and although theoretical objectives are useful in preparing a program booklet, since they fill pages with beautiful words, they are of little use to a designer.

Once a purpose has been checked against a compliance objective, a programmatic objective is determined by simply applying a concept or defining a course of action which will determine HOW this purpose is going to be achieved; in this case: HOW this livable environment for the physically disabled is going to be achieved, through site selection, functional objectives, physical needs, psychological determinants, etc. From there we will find a clear analytical procedure leading us on our way: site analysis, physical analysis of wheel chairs, ramps, toilets, accesses, circulations, turning radiuses, parking, etc. Once this study is completed, the simple application of programmatic concepts such as priority, hierarchy, integration, segregation, fragmentation, etc., will enable us to arrive at the actual program requirements for this particular objective.

The process is fairly simple as long as it is kept within the scope of architectural requirements and possibilities, excluding unnecessary intellectual complexities that have no bearing on the actual project. Such complexities are for the most part a definite waste of time and are only justified if they clarify ideas and thus keep the doors of communication open.

Any classification of program objectives other than the one established above would be extremely inaccurate and undesirable. The same programmatic objective can have two different meanings when stated by two different clients, or simply when established in two separate regions of the same country. For example, the design objective of a "warm and inviting" building cannot be interpreted in the same way in Minnesota as in Florida, because the term "warm" varies in each case from an actual insulation and heat loss definition in the first instance to an appeal of earthiness in the second; and while "inviting" in Minnesota might indicate a sunlit garden, in Florida it would probably call for a shaded porch.

By the same token, words like "shelter," "image," "character," and "appeal" can have such a tremendous variety of interpretations by reasons of their own conceptual nature, that a list of all possible variations would become unrealistic and useless. That is why generalized objectives must be translated into directrixes and guidelines for definite programmatic and design procedures and developments. Statements of intent are useful, but they are never enough.

Generally, there are several common denominators for these definitions when applied to architectural materials and components, and these common denominators can be extremely helpful in determining which course a specific design philosophy must follow in order to achieve the tasks established by a program objective. Brick and wood, because of their direct earth origin, will usually be selected as warm materials as opposed to concrete or metals, since the latter are more readily identified with the efficiency concepts of the Industrial Revolution. Curtain walls or steel shapes are usually labeled cold materials since they represent the latest technologies of aesthetic achievement in construction developments and, therefore, architectural "etiquette."

Still, it must not be assumed that these classifications are absolute. The physical and psychological effects of colors, which are basic architectural constituents, vary from region to region. Thus, while in Wisconsin a white building could be considered impractical or unappealing because of its heat reflectance, in Arizona the same color building would be desirable and appealing for the very same reason.

This is where the regionality factor plays a key role in the actual definition of programmatic objectives as well as in the clear understanding of the conceptual basis behind them. Factors like urban placement, socio-economic characteristics and even building types have a tremendous effect on the basic meaning behind every statement of intent.

In Value Engineering Techniques, the system employed for the determination of functional objectives is the "two word definition" method, in which ANY function must be described by a verb and a noun combination ONLY: this phrase is known as the "two word abridgment."

In this method, the verb answers the question, "What does it do?", focusing the analysis on the function rather than on the item and thus defining a work function. The noun in the abridgment answers the question, "What does it do it to?", thus stating what the work function acts upon. This two word abridgment of a verb-noun combination (as in, convey waste, execute contracts, deliver products, etc.), when the noun portion of the abridgment describes a measurable item, is helpful in the determination of quantitative statements since the noun used can always be measured. However, when the two word system contains a non-measurable noun (as in, provide comfort, assure security, create image, improve character) we find ourselves facing a qualitative statement.

When properly applied to the architectural programming phases, this system is without a doubt one of the most formidable weapons a programmer can employ in the definition and determination of compliance objectives.

Another system that can be implemented during this stage is the "Purposes-Target-Results" (PTR) method, in which an ideal target serves as module

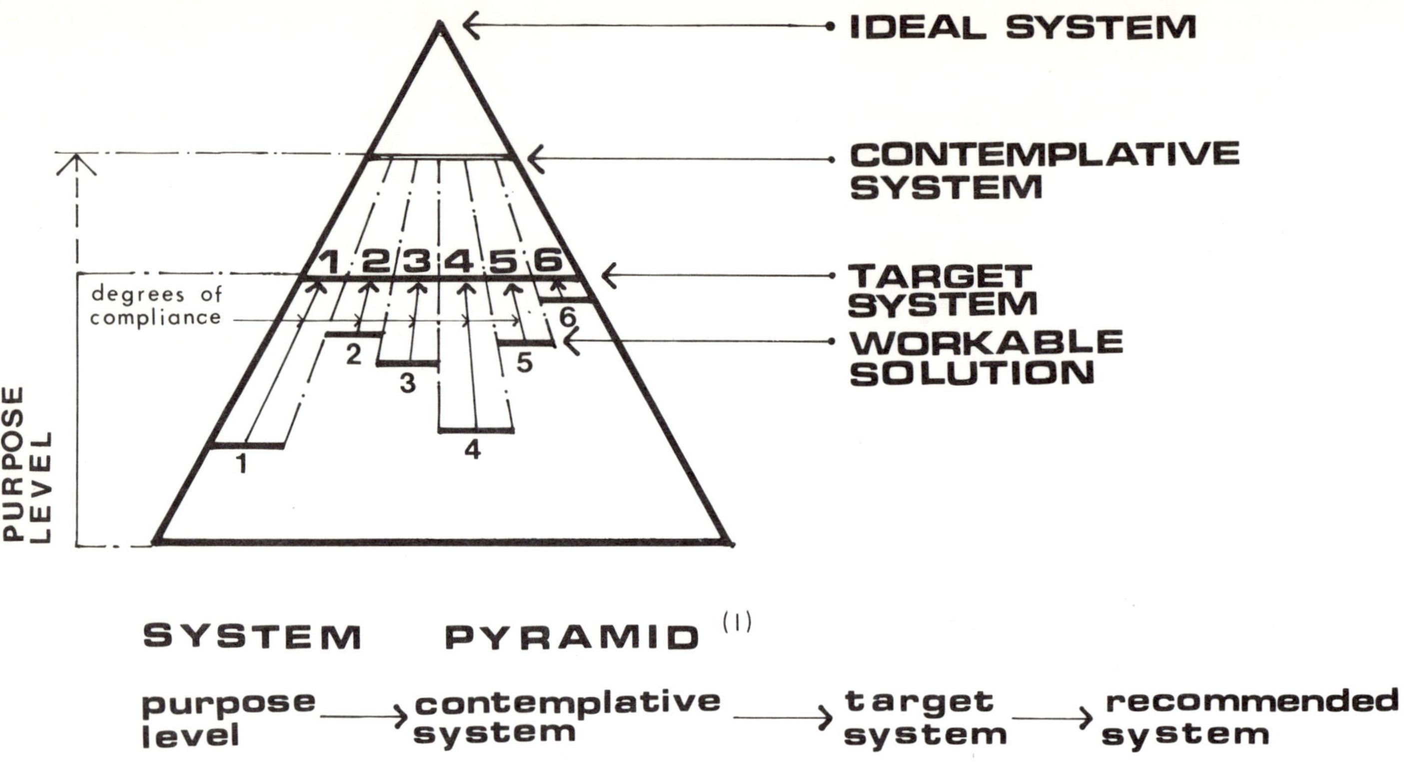

Figure 8.1

for the definition of objectives and goals, through a viable degree of compliance with a contemplative system as established by a logical definition of future possibilities.

Although this system has been used primarily for planning procedures where excessive data collection is not recommendable, it works well during the objectives definition stage precisely because of its limited data collection characteristics. One noteworthy fact is that during the objectives definition phase, where the procedures are closely related to project planning, the excessive analysis of existing conditions and data collection can be extremely detrimental to the proper definition of

goals and tasks, because such analysis not only envelopes the thinking process in pre-established modules, but also restricts creativity and innovation. In other words, when faced with a planning procedure or an objectives definition process, the question should be "What will be right for tomorrow?", rather than "What is wrong today?"

The PTR system basically follows the outline illustrated in Figure 8.1.

Through the establishment of an objectives degree of compliance, the viable possibilities and future developments are outlined and directed toward an ideal target. That target has been pre-established as the factual result of a contemplative process projected too far into the future to serve as a possible pattern of achievement within the limitations of present conditions.

Summarizing, we could say that, whatever the system, the basic approach to be taken when confronting any programmatic objective is that which will indicate the one procedure (analysis or programming) that is missing in the primary statement of intent. This approach will ensure that the actual architectural programming process always begins with all elements operating in a balanced fashion, with its analytical and conceptual phases working hand in hand towards the definition of their common goal: the program requirements.

There must never be an excess of one element nor lack of another but a balanced distribution of their achievements and interactions. By supporting and complementing one another throughout the development of the entire architectural programming procedure, these two elements (analysis and programming) will not only structure that procedure but also delineate and clarify the general outline and essential tasks.

8.2 Site Analysis

The major difference between site analysis and spatial analysis is the same as the one between structural steel and reinforced concrete: Structural steel is REVISED, while reinforced concrete is DESIGNED. Site analysis consists basically of a revision of compliance conditions while spatial analysis consists more of a creative attitude which operates within a framework of logistics and facts.

In order to evaluate a site properly, we must first determine which of its components affect the evaluation and then derive our analysis checklist from those components. To do this, it is important to first establish a list of essential components and then to determine their constituents, just as in any analytical procedure.

Site analysis is divided into two major divisions:
1. Natural Characteristics — basically meaning components characteristics;
2. Artificial Conditions — basically meaning components relationships.

Each of these major divisions has the following components and constitutive elements:

Natural Characteristics

    Structural
        Soil Conditions
        Geological Considerations
        Subsurface Water

    Physical
        Natural Drainage
        Slopes, Contours, Fractures
        Views
        Orientation

    Environmental
        Temperature
        Snow/Frost
        Precipitation
        Surface Water
        Natural Surroundings
        Flora
        Fauna
        Conservation
        Pollution

Artificial Conditions

    Technical
        Functional Location
        Functional Compliance
        Growth and Change Projections
        Historical Value
        Accessibility
        Circulations

    Physical
        Site Utilities and Services
        Existing Structures
        Neighboring Structures
        Operational Factors
        Maintenance and Taxes
        Sound Conditions
        Improvements

    Regulatory
        Planning Regulations
        Zoning
        Building
        Fire

Health
Highway and Traffic
OSHA
Legal and Contractual Conditions

The methods for executing a detailed analytical study of each one of these components vary according to a specific project's scope and extent. For obvious reasons, it would be impossible to offer a detailed description and illustrative examples of each, but a brief outline can define the corresponding analysis methods applied to each component as follows:

Natural Characteristics
  Structural Characteristics
    Soil Conditions
    Geological
      Considerations   Soil Borings and Geological Analysis
    Subsurface Water
  Physical Characteristics
    Natural Drainage
    Slopes, Contours
    Fractures   Land Surveys
    Views
    Orientation

  Environmental
    Sun   Solar and
    Wind   Wind Analysis
    Temperature
    Snow/Frost   Temperature/Snow and Rain Analysis
    Precipitation
    Surface Water
    Natural Surroundings   Site Survey
    Flora/Fauna

Artificial Conditions
  Technical
    Functional Location
    Functional Compliance
    Growth and Change
      Projections   Programmatic Concepts applied to survey data
    Historical Value
    Accessibility
    Circulation

Physical
  Site Utilities and Services
  Existing Structures
  Neighboring Structures
    Survey   On site Survey
  Operational Factors
    Economic Analysis
  Sound Conditions
  Improvements
Regulatory
  Planning Regulations
  Zoning
  Building
  Fire
  Health   Regulatory Survey (see Sub-Chapter 8.5)
  Highway and Traffic
  OSHA
  Legal and Contractual
    Conditions

SOIL BORINGS. This is the physical analysis of soils as sampled by core drilling sections of the subsoil in different locations of a given site and generally done by specialized laboratories. A typical example of the resulting report can be found in Appendix A.

SURVEYS. Generally, surveys are also executed by specialists and may or may not contain all the information that would be required during the programmatic process. Therefore, design professionals should insist that only pertinent information be included in such surveys, since the cost of these services depends mostly on the extent of the information contained in the document, and any useless data will only serve to increase the price with no positive results. However, a survey often lacks some vital data which could affect the viable means of outlining a project program and, consequently, could affect the end result. In such cases we must insist that such information be included within the scope of the site or building survey. A survey is as legal as any other construction document, and projects based on the information contained in them are based on stated facts, not on assumptions.

Any comprehensive survey checklist must include the following items:

SURVEY CHECKLIST

1. Descriptive Data
   1.1   Name of surveyor, registration number, address, telephone
   1.2   Job number and date of survey
   1.3   North indication
   1.4   Drawing scale
   1.5   Legal description

2. Site Boundaries
   2.1   Property lines
   2.2   Public walks
   2.3   Planting strips
   2.4   Surface water

3. Site Dimensions
   3.1   Property boundaries, dimensions, radii, etc.
   3.2   Corner angles and direction of boundaries
   3.3   References to center lines of streets, adjacent or nearby structures, etc.
   3.4   Site area

4. Elevations and Topography
   4.1   Reference to datum points
   4.2   Grades and contours
   4.3   Top of curbs
   4.4   Elevations at crown of roads
   4.5   Manhole covers' elevations
   4.6   Mounds, hills, abrupt differences in level, etc.

5. Existing Conditions and Surroundings
   5.1   Roads and drives
   5.2   Right-of-ways
   5.3   Easements and easement types
   5.4   Walks
   5.5   Power poles
   5.6   Light poles
   5.7   Telephone poles
   5.8   Manhole covers
   5.9   Fire hydrants
   5.10 Traffic signals' standards
   5.11 Miscellaneous signs
   5.12 Surface water
   5.13 Existing Structures
   5.14 Rock formations, caves, etc.

6. Drainage
   6.1   Culverts
   6.2   Ditches
   6.3   Catch basins
   6.4   Inverts' elevations

7. Landscaping
   7.1   Trees
   7.2   Ground cover
   7.3   Shrubs

8. Utilities
   8.1   Water mains and taps
   8.2   Gas mains and taps
   8.3   Electrical
   8.4   Telephone
   8.5   Sewerage

9. Construction Aids
   9.1   Bench marks and monuments
   9.2   Property corners, stakes, etc.

All surveys may not contain the information outlined above, but a general review of this checklist, will help identify pertinent information for specific cases.

SOLAR ANALYSIS. Solar analysis is generally executed through a solar analysis chart as illustrated in Figure 8.2.

The advantage of these charts is that they illustrate the relative angle of the sun for a given location and the duration of sunlight within a given day. Many available references explain in detail the procedures to follow to execute these solar charts. For more detailed information on this subject, refer to Appendix B.

WIND ANALYSIS. This is done by compiling the corresponding meteorological data in charts similar to the one illustrated in Figure 8.3.

Another wind analysis technique now widely used is the development of "wind roses." For more details on this method refer to Appendix C.

TEMPERATURE, SNOW AND PRECIPITATION. In each case, the data is provided with great accuracy by the local weather bureaus and collected in charts like the ones illustrated in Figures 8.4, 8.5, and 8.6.

Consequently, the tabulated results can be represented graphically as illustrated in Figures 8.7, 8.8 and 8.9 respectively.

SITE EVALUATION. The descriptive analysis of a site is only the first step in any comprehensive analytical procedure. As we established earlier, analysis is not only the detailed study of the constituents of a whole, but also the result of such study, a statement which automatically assumes an evaluation of the collected data.

A method for effecting this evaluation is provided by the evaluation chart in Figure 8.10.

Each natural characteristic or artificial condition is assigned a value (ranking independently from 4 to 0 — excellent to bad) based on its degree of compliance with the site requirements set forth and is also assigned a Project Requirement Factor value (from the PRF values).

The PRF's can only be distributed ONE to each natural characteristic or artificial condition. They represent the degree of importance of the characteristic or condition when applied to a specific project. Where two or three characteristics have the

SOLAR ANALYSIS DATA SHEET

PROJECT: _______________ LATITUDE: ______ prepared: ______

LOCATION: _______________ ______ checked: ______

PREPARED BY: _______________ LONGITUDE: ______ date: ______

WINTER dec 22	FALL sept 23 / mar 21 SPRING	june 22 SUMMER
SUNRISE (SR) / SUNSET (SS)	SUNRISE (SR) / SUNSET (SS)	SUNRISE (SR) / SUNSET (SS)

TIME (AM PM)	AZIMUTH Z	ALTITUDE H	TIME (AM PM)	AZIMUTH Z	ALTITUDE H	TIME (AM PM)	AZIMUTH Z	ALTITUDE H
NOON			NOON			NOON		
SR SS			SR SS			SR SS		

REMARKS (×3)

Figure 8.2

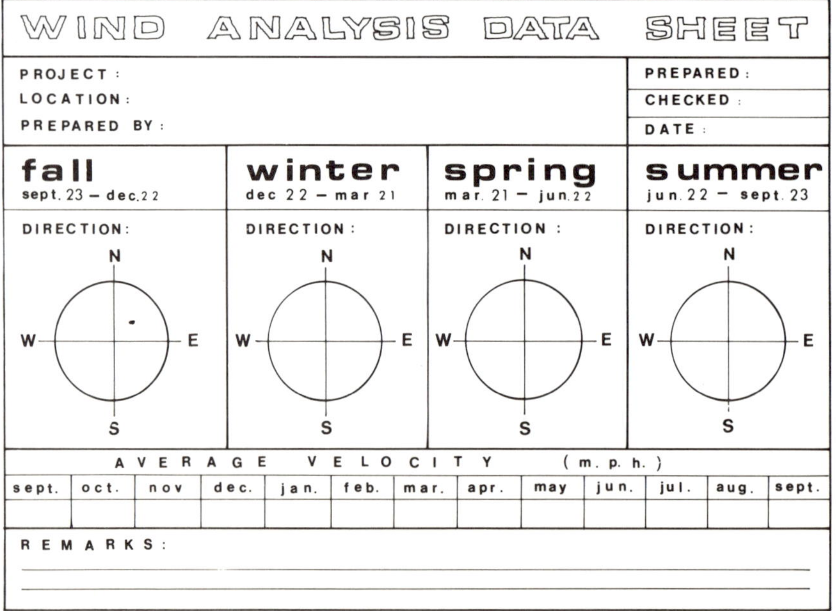

Figure 8.3

TEMPERATURE DATA SHEET

	sept.	oct.	nov.	dec.	jan.	feb.	mar.	apr.	may	jun.	jul.	aug.
min. monthly total												
max. monthly total												
average monthly												

REMARKS:

Figure 8.4

SNOW/FROST DATA SHEET

		sept.	oct.	nov.	dec.	jan.	feb.	mar.	apr.	may	jun.	jul.	aug.
snow	monthly average												
snow	monthly maximum												
	frost												

REMARKS:

Figure 8.5

PRECIPITATION DATA SHEET

	sept.	oct.	nov.	dec.	jan.	feb.	mar.	apr.	may	jun.	jul.	aug.
min. monthly total												
max. monthly total												
average monthly												

REMARKS:

Figure 8.6

SITE ANALYSIS : TEMPERATURES
PROJECT N°.
DATE
SHEET   OF
REVIEWED
SPRING
SUMMER
AUTUMN
WINTER
TEMP. °
APR
MAY
JUN
JUL
AUG
SEP
OCT
NOV
DEC
JAN
FEB
MAR
REMARKS
100°
90°
80°
70°
60°
50°
40°
30°
20°
10°
0°
-10°
-20°
-30°
-40°
-50°
RECORD HIGHS
AVERAGE MAX.
AVERAGE
AVERAGE MIN.
RECORD LOWS
NOTES

   Figure 8.7

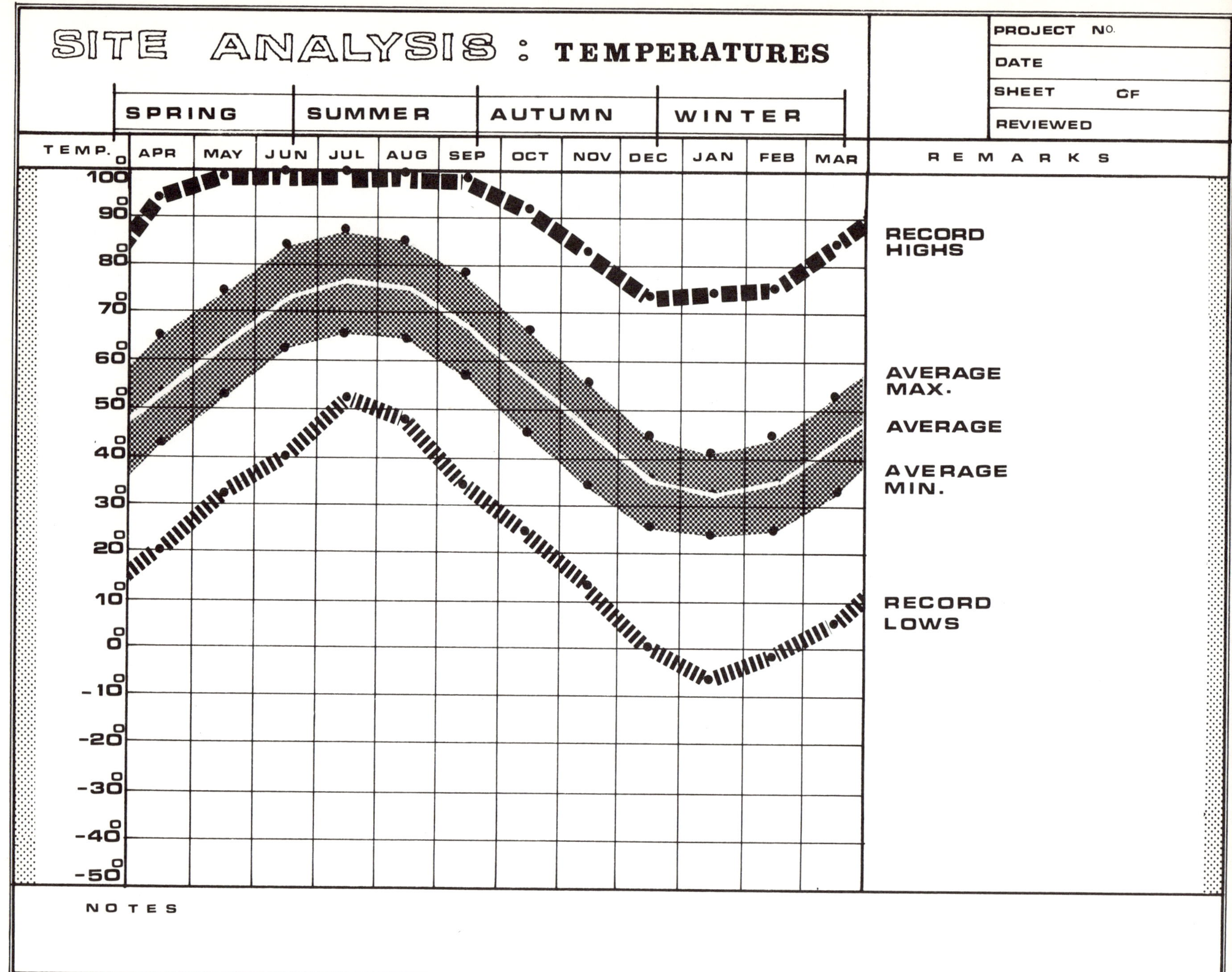

SITE ANALYSIS: SNOW
PROJECT No.
DATE
SHEET OF
REVIEWED
SPRING
SUMMER
AUTUMN
WINTER
INCHES
APR MAY JUN JUL AUG SEP OCT NOV DEC JAN FEB MAR
REMARKS
20
19
18
17
16
15
14
13
12
11
10
9
8
7
6
5
4
3
2
1
0
RECORD HIGH
MAX. IN 24 HRS — 15.5"
NORMAL
RECORD LOW
NOTES

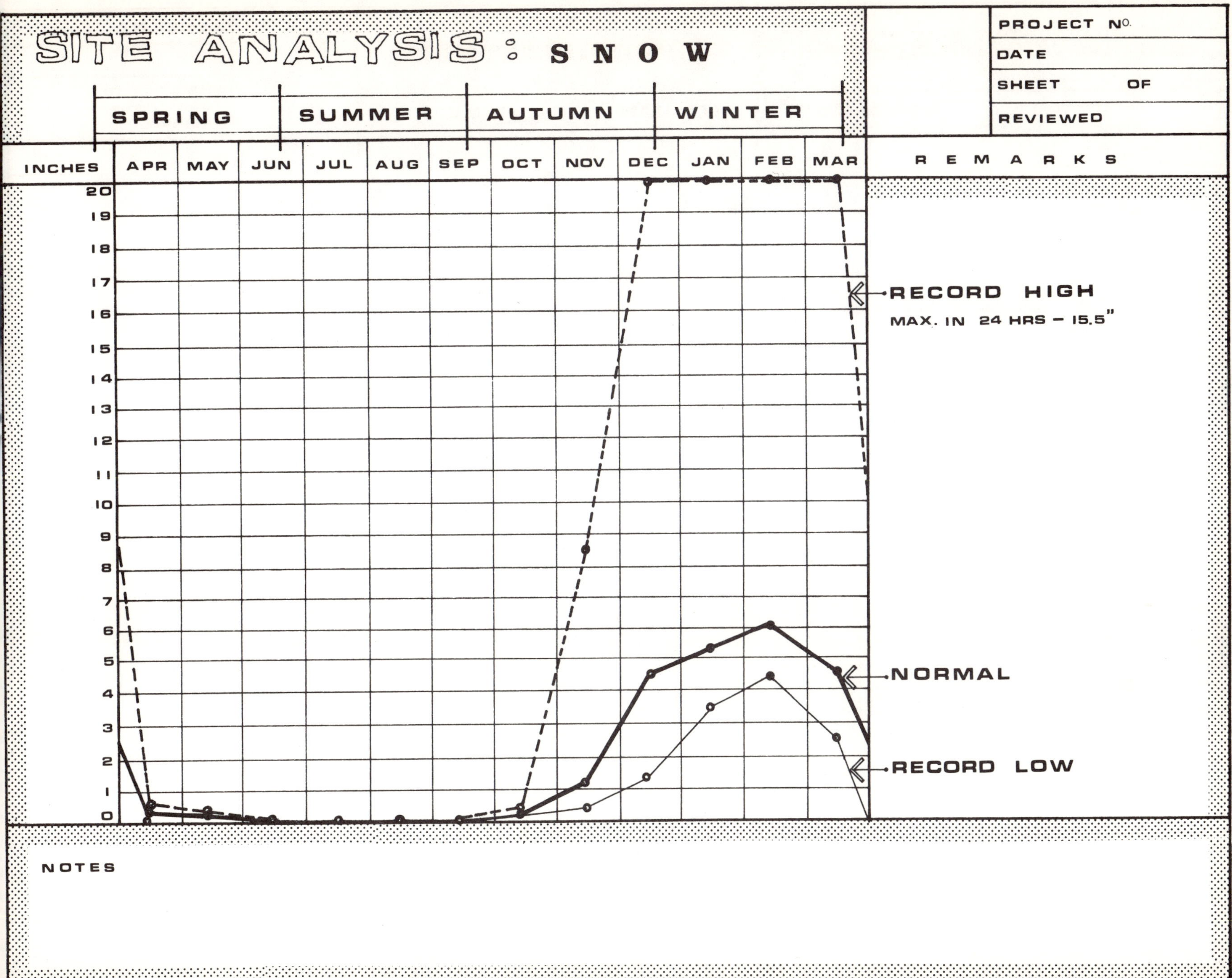

Figure 8.8

SITE ANALYSIS : PRECIPITATION
PROJECT No.
DATE
SHEET   OF
REVIEWED
SPRING   SUMMER   AUTUMN   WINTER
INCHES   APR   MAY   JUN   JUL   AUG   SEP   OCT   NOV   DEC   JAN   FEB   MAR   REMARKS
18.35"
RECORD MAX.
MAX MONTHLY — 18.35"
MAX IN 24 HRS — 7.82"
NORMAL
TOTAL YEARLY — 40.46"
RECORD MIN.
T
TOT: 108   8   8   8   9   10   11   12   9   8   9   9   7   CLEAR DAYS
TOT: 109   9   10   12   12   10   8   8   8   7   9   7   9   PARTLY CLOUDY DAYS
TOT: 148   13   13   10   10   11   11   11   13   16   13   12   15   CLOUDY DAYS
NOTES

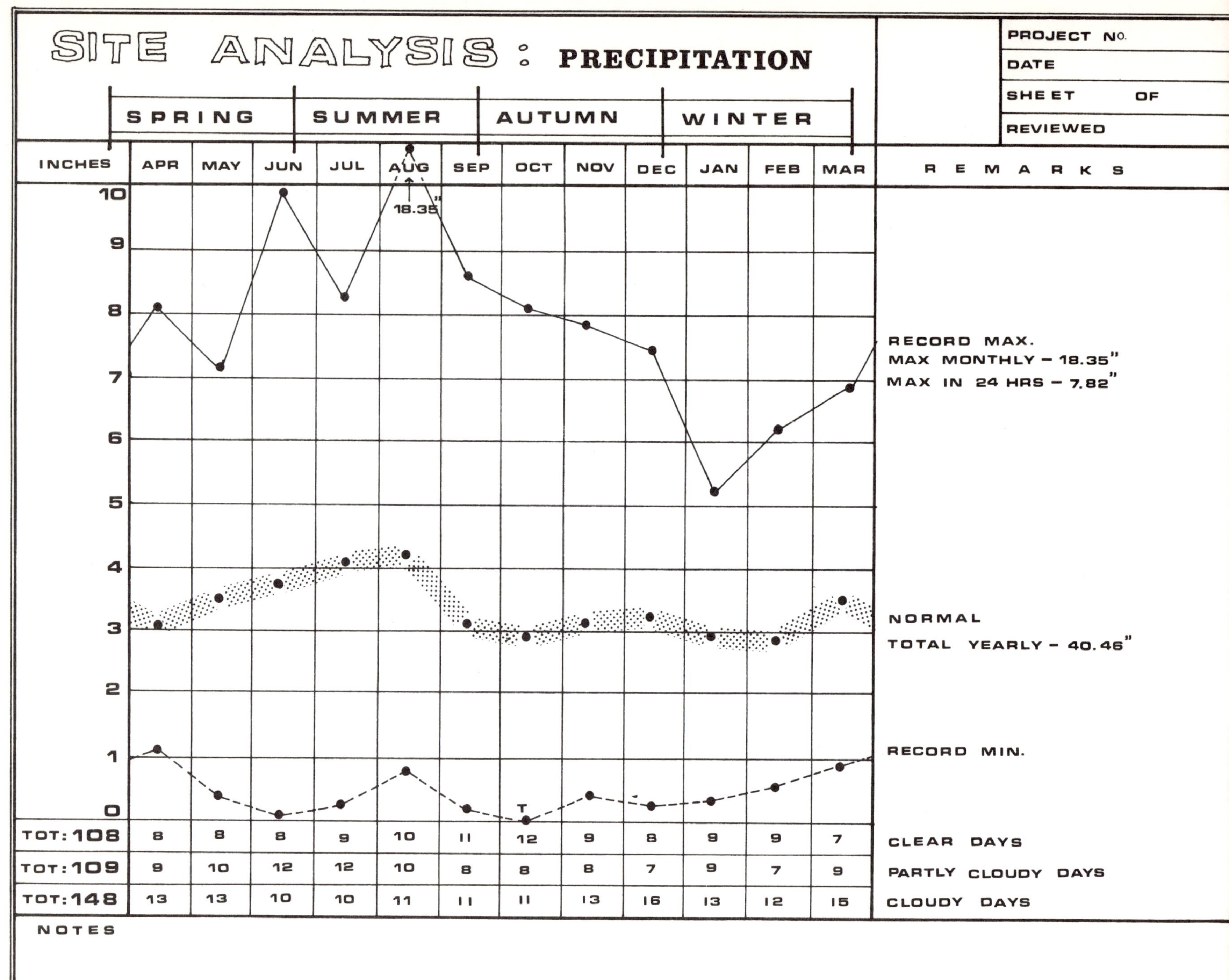

94   *Figure 8.9*

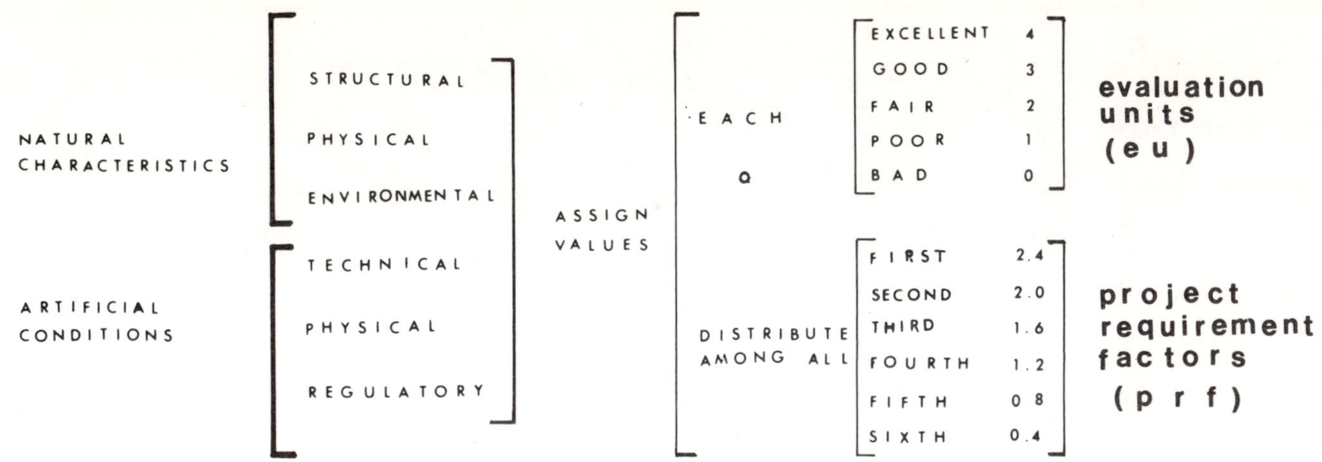

SITE EVALUATION CHART									
DESCRIPTION		prf	site designation						
			1		2		3		4
			eu	VALUE: prf x eu	eu	VALUE	eu	VALUE	eu	VALUE
natural characteristic	S									
	P									
	E									
artificial condition	T									
	Ph									
	R									
TOTAL VALUE										
COST										
S.E.R. = cost / t.value										

Figure 8.10

SITE EVALUATION CHART									
DESCRIPTION	**prf**	**site designation**							
		1		**2**		**3**		**4**	
		e u	VALUE: prf x eu	e u	VALUE	e u	VALUE	e u	VALUE
natural characteristic — **S**	1.2	2	2.4	4	4.8	2	2.4	3	3.6
natural characteristic — **P**	2.4	3	7.2	2	7.2	4	9.6	3	7.2
natural characteristic — **E**	1.8	4	7.2	3	5.4	4	7.2	2	3.6
artificial condition — **T**	0.8	3	2.4	3	2.4	2	1.6	2	1.6
artificial condition — **Ph**	0.4	2	0.8	3	1.2	3	1.2	4	1.6
artificial condition — **R**	1.8	3	5.4	3	5.4	1	1.8	2	3.6
TOTAL VALUE			25.4		26.4		23.8		21.2
COST		$505,000**		$432,000**		$510,000**		$484,000**	
S.E.R. $= \dfrac{\text{cost}}{\text{t.value}}$		19882		16363		21429		22830	

Figure 8.11

same degree of importance, an average should be assigned to each by adding two or three PRF's and dividing them by the number of elements involved. For example, if the priorities lie in the physical characteristics of a site and in its technical and regulatory conditions, then we must add 2.4 + 2.0 + 1.6 totaling 6, which divided by 3 = 2.0.

Therefore, assign a value of 2.0 to each physical, technical, and regulatory condition, and distribute the remaining values of 1.2, 0.8, and 0.4 to the remaining structural and environmental characteristics and physical conditions, following the ORDER OF IMPORTANCE of each one in relation to the project.

For example, assuming 4 different sites and applying the factors explained above, a site evaluation chart would be as shown in Figure 8.11.

The Site Evaluation Ratio (S.E.R.) is obtained by dividing the actual cost of the site by the total value of the site (obtained by multiplying the project requirement factors by the evaluation units of each site and adding all the results in the value column). This Site Evaluation Ratio (S.E.R.) represents a revised cost in function of a site's compliance with specific project requirements. The smallest ratio appears as an adjusted cost/function factor which determines the best-priced site from all possible selections. The total value of the site cannot exceed 31.2 points (the minimum possible value is 0); and a site with fewer than 20 points (or 64% of the maximum value) should not be considered feasible, since it does not meet the minimum requirements of compliance no matter what the cost factor might indicate.

8.3 Spatial Analysis

INTRODUCTION. In all of architectural programming, no single procedure is of greater importance than the actual spatial analysis of operations. There

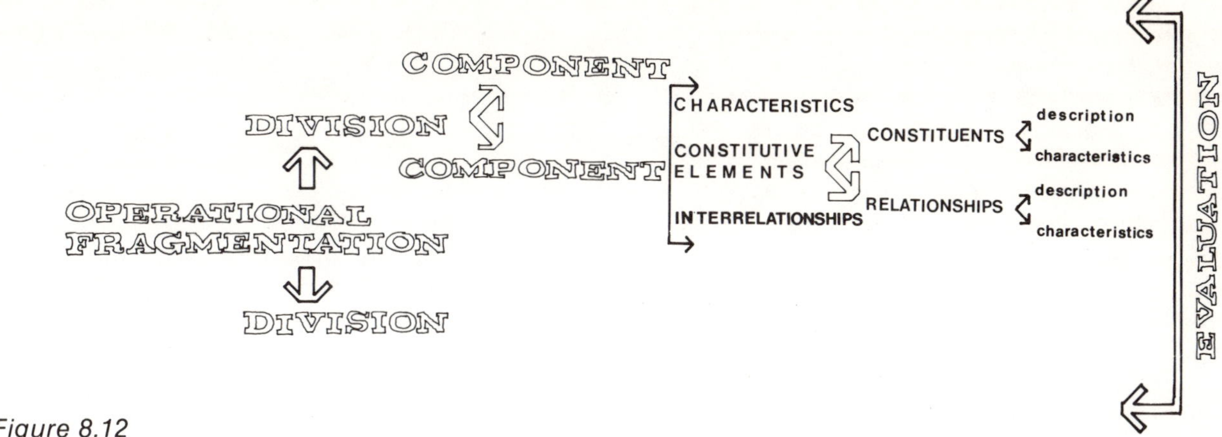

Figure 8.12

are several reasons for this, each of which includes the very basic postulates on which this process is based.

1. No matter what the case may be, the spatial analysis is the one procedure which must always be effected. Any of the other phases of architectural programming could possibly be circumvented or become unnecessary, depending on the case; but WITHOUT SPATIAL ANALYSIS THERE IS NO ARCHITECTURAL PROGRAMMING AT ALL.

2. All other procedures (site analysis, temporal analysis, economic analysis, or regulatory surveys) can be done more or less accurately by special consultants such as planners, operational consultants, economic consultants, etc.; but the actual spatial analysis of operations can only be executed properly by the basic elements who (by dealing with the users' needs) create Architecture: The design professionals.

3. Spatial analysis is the one single process on which every other professional design phase will depend. In addition other parts of architectural programming, from site analysis to regulatory surveys, will be based on it. For these reasons any spatial analysis procedure must be considered the central core of any architectural programming method. Therefore, in the following pages, we will try to establish a method of spatial analysis which can be executed independently of any other programming phase but which will also have the flexibility of being incorporated within the total package of all programming activities.

For this purpose we have divided the spatial analysis method into two basic sections:

Spatial Components' Characteristics

Spatial Components' Relationships

The first of these sections represents the actual analysis of the spatial components of operations; the second, the study, description and evaluation of their interaction.

As stated in Sub-chapter 7.4, the first step in any analysis process must be the logical and systematic separation of constituents. One way of doing this is to logically divide operational tasks into:

1. Divisions

2. Components

3. Constitutive Elements

These three levels can serve to simplify or detail, as the case requires, any or all portions of a spatial analysis process. It is, therefore, of extreme importance to clearly establish how this fragmentation will take place. Thus, the actual establishment of these basic elements will become our first step in any spatial analysis method.

After this initial step, the two basic sections — Characteristics and Relationships — must be defined. Finally, there must be an evaluation of the results (See Figure 8.12)

OPERATIONAL FRAGMENTATION. Every operation contains major subdivisions which represent, in varying degrees, the essential description of its structure. For example, in a hospital complex we find these divisions clearly indicated by separate wards, stories or even buildings; while in an industrial complex we find clearly defined functions of offices, warehouse, etc., each one of which contains groupings of components such as Management, Sales, Purchasing, etc., in the first division, and Receiving, Manufacturing, Storage, Shipping, etc., in the second.

An operational division is a defined hierarchical grouping of operational components. Their functions are general in nature, and can probably be defined in the same manner. By the same token, they can be very readily understood, since almost all human operations contain a definite series of divisions with clearly defined functions.

Still, this primary fragmentation, as simple as it may seem, can be of tremendous help in understanding the general activities upon which a specific operation is based and hence in illuminating the roots of its structure.

Essentially, an operational fragmentation must provide an understanding of functions on a divisional level. Almost every client (assuming his/her task takes place at the executive level of a specific operation) can probably provide this information readily, since the daily dealing with the operation qualifies the client more than anyone else to determine its divisional fragmentation.

There can be, of course, several levels in this fragmentation process, such as in the case of extremely large corporate structures where the functions vary to such an extent that the actual divisions become corporations in themselves. In such cases it is advisable to treat each major division as a project in itself. It would be impossible to assume, for architectural programming purposes, that the major divisions of General Motors are Chevrolet, Oldsmobile, Buick, etc., since each one of those subdivisions represents an enormous corporation in itself.

In such cases, the rule of thumb is DIVIDE THE GENERAL OPERATION INTO MAJOR SUBDIVISIONS ACCORDING to ACTUAL BUILDING OCCUPANCY and on that basis commence the divisional fragmentation process. After all, site planning will deal with the primary grouping or groupings of buildings as a discipline in itself, while the spatial analysis of each structure is accomplished much more simply by treating each building as a self-contained operation with its spatial relationship to other structures established at a site planning level.

Once the divisions for a specific project have been outlined, the next step is the determination of the division components. This is done by simply breaking each division down into the different functions it performs.

In small projects, we may find no divisional fragmentation at all; in the case of a small retail establishment for example, the divisional fragmentation is almost impossible to establish, for obvious reasons. In such cases we should proceed directly to the establishment of components as a preliminary study of functions.

One basic rule to remember when determining the components of a specific division is that COMPONENTS ARE NOT PEOPLE, ROOMS, OR SPACES, but FUNCTIONS — clearly defined operational functions. A component is not a vice-president, a file room, or a vestibule, but Vice-presidency, Filing, and Reception. Each human element, room, or space is actually a constitutive element of these components, and the definition of these constitutive elements becomes the third step in this general fragmentation of operational functions.

One primary characteristic of all divisions is their operational value; but not all components can be characterized this way since many operation constituents are not strictly essential to the operation itself, but only complement or support those others which, by virtue of their function, constitute the core of operational activities for a given project. For this reason, we will consider three major functional classifications in determining the operational description of each component and its constitutive elements according to its function:

1. Essential
2. Complementary
3. Supporting Services

Essential constituents would be those without which the basic function of any component (or its elements) would be disrupted or imperiled, such as humans, machinery, equipment and products, animals, plants, operational natural environment, active storage, etc.

Complementary constituents would be those which COMPLEMENT the actual function of the component or element they affect, such as parking (except for the specific case of parking garages, etc. when automobile storage becomes the essential function of a project), inactive storage, recreation, circulations, transition areas, natural environment, etc.

Finally, Supporting Services are those which maintain or support the other two, such as maintenance, mechanical and electrical equipment, health and hygiene, etc.

The classification of the structure of operational components is as follows:

```
Components
    Function
        Essential
        Complementary
        Service
    Constitutive Elements
        Activity
        Description
        Function
            Essential
            Complementary
            Service
        Structure
            Class
            Type
            Number
        Location
        Relationships
            Direction
            Value
            Description
```

Thus the constitutive elements are the individuals, spaces, rooms, and detailed functions making up a component, while the establishment of each one's function, class, type, etc. is the primary step in the description of their characteristics according to performance within the operational complex.

CONSTITUTIVE ELEMENTS' CHARACTERISTICS. This description is achieved by simply assigning areas, volumes, rooms, or spaces to each one of these elements in order to provide the necessary field for a proper operational performance and by analyzing the frequency of use, type, special requirements, direction, and volume of activities involved in each case, as well as the sex, number, class, etc., of each occupant or functional constituent involved in the operational performance.

A general classification of constitutive elements' characteristics could be organized in the following manner:

```
Activity Description
Functions
    Essential
    Complementary
    Service
Structure
    Class[17]
        Static
            Designation
            S.F./Occupant
            Area
            Volume
            Frequency of Use
            Special Requirements
        Dynamic
            Designation
            S.F./Occupant
            Area
            Volume
            Frequency of Use
```

The figure 8.13 is a reproduced form, "Spatial Components Analysis Data Sheet", transcribed below.

SPATIAL COMPONENTS ANALYSIS DATA SHEET	ARCHITECT:			PROJECT NO.
PROJECT	OWNER/REP. :	TITLE:		SHEET OF
LOCATION				DATE:
DIVISION FUNCTIONAL DESCRIPTION	PHONE : EXT. :	DEPT : SUITE:		TIME:
				INTERVIEWER:
COMPONENTS FUNCTIONAL DESCRIPTION(S)			REVIEWED:	APPROVED:

DIV.	COMPONENT DESIGNATION	FUNCT.	CONSTITUTIVE ELEMENTS.	FUNCT.	CLASS	ACTIVITY DESCRIPTION	LOCATION	ROOM NO.	OCCUPANCY PA	OCCUPANCY M 0.5	OCCUPANCY F 0.5	VIS	FQ USE	AREA	VOL.	EQUIP.	EVALUATION RATINGS A	B	C	D	E	F	OTHER (specify)	REMARKS

SPATIAL EVALUATION SURVEY.	RATINGS DESIGNATION	COMPONENT SPATIAL EVALUATION	GENERAL NOTES & COMMENTS
A – EXISTING LOCATION RATING	1 EXCELLENT	component:	
B – EXISTING AREA RATING	2 GOOD	component:	
C – EXISTING PERSONNEL AMOUNT RATING	3 FAIR	component:	
D – EXISTING ARTIFICIAL CLIMATE RATING	4 POOR	component:	
E – EXISTING ENVIRONMENT RATING	5 BAD		
F – EXISTING EQUIPMENT RATING		DIVISION SPATIAL EVALUATION	

Figure 8.13

    Direction
    Special Requirements
Transient
    Volume
    Direction
    Special Requirements

Type
    Human
        Description
        Age
        Sex
    Equipment
        Description
        Sizes
        Volume
    Animal/Plant
        Description

        Number
      Number
    Location/Room No.
    Relationships
        Direction
        Value
        Description

An example of a data collection form, which emphasizes the analysis questionnaire and user evaluation is illustrated in Figure 8.13.

Another simple form for data collection which accomplishes the investigation of components and constitutive elements' characteristics can be outlined as illustrated in Figure 8.14.

In either case, whether with the survey and user evaluation format shown in Figure 8.13, or with the systematic analytical and programmatic approach

SPATIAL COMPONENTS ANALYSIS DATA SHEET

ARCHITECT | PROJECT NO.
PROJECT | PREPARED | DATE | TIME
LOCATION | CHECKED
DESCRIPTION | APPROVED | SHEET | OF

DIV.	COMPONENT DESIGNATION	FUNCT	CONSTITUTIVE ELEMENTS	FUNCT	CLASS	ACTIVITY DESCRIPTION	LOCATION	ROOM NO.	OCCUPANCY					SPECIAL CHARACTERISTICS						REMARKS
									M 0.5	F 0.5	VIS	F/U	S.F. OCC	AREA	VOL.	EQUIP.	MECH.& ELEC.	ENVRMT	OTHER	

Figure 8.14

illustrated in Figure 8.14, the results should always be aimed at the clear and comprehensive determination of the operation's components, their constitutive elements, as well as an accurate layout of their essential characteristics.

Although it may seem pointless to classify the functional value of a component or the functional value and class of a constitutive element, since for all practical purposes they will be a part of the project in one way or another, the primary purpose of these classifications is to clarify and document functional performance and class, for reasons which will become apparent in the spatial components relationships analysis; as well as the determination of controllable variables or "wants" as referred to in Chapter 7.

Furthermore, in any analytical procedure it is best to establish as many characteristics as possible; this means not only formal characteristics but functional and operational characteristics as well, since design represents, more than just a mere arrangement of areas, the satisfactory three-dimensional layout of spaces according to the spatial requirements of a specific operation, based not only on FORM but on FUNCTION as well.

One common investigative approach to spatial analysis surveys is through the use of questionnaires which can be distributed or filled out by department heads, users, etc., and later on reviewed by the design professionals. Among the advantages of this procedure is the reduction in man-hours normally required for personal interviews. However, this procedure does not produce the clear and accurate results that personalized interviews and work sessions offer. In many cases questionnaries are generally subject to misinterpretations and erroneous conclusions. For this reason, the questions and possible replies chosen for study must be designed in the most clear and concise manner. For an illustrative example of this subject refer to the Sub-Chapter 10.1 comments on the employee survey of Case Study number 2 in this text.

COMPONENTS RELATIONSHIPS. The three single factors that regulate a relationship from a spatial analysis point of view are:
1. Direction
2. Value
3. Description

DIRECTION. One important thing to remember when analyzing spatial relationships is that RELATIONSHIPS BETWEEN SPACES ARE NOT NECESSARILY RECIPROCAL. The importance of this can not be stressed enough. The production manager of a manufacturing plant might want to maintain visual contact at will with the assembly line of the plant, but this does not mean that he wants any or all the elements of the assembly line to maintain visual contact with him. And while a doctor might want easy access to the consultation rooms from his private office, the reverse could prove detrimental to his practice.

Therefore, in the spatial analysis chart of relationships, we have modified the standard link between spaces from the traditional solid line as represented in Figure 8.15 to the one illustrated in Figure 8.16, where an arrow indicates the manner and direction of one component's link with another thus defining their relationship more precisely.

VALUE. Probably the most important single characteristic of any relationship between two spaces is its value. For the purposes of this study, we have subdivided *value* into two major categories: Positive and Negative, each one having different degrees of functionalism:

Positive
    Essential
    Complementary
    Non-essential

Negative
    Non-desirable
    Non-acceptable

In the spatial relationships' diagrams, these values will be represented by a circle aligned with the arrow it describes in the manner shown in Figure 8.17.

Thus, in figure 8.18 the relationship A to B is essential, while the relationship B to A is only complementary; and in Figure 8.19 the relationship of C to D is non-essential, while the link from D to C is simply non-acceptable.

DESCRIPTION. The actual description of the relationships is done by analyzing all of the possible factors which regulate it, such as those shown in Figure 8.20.

These factors are graphically represented, as illustrated, by either solid lines or broken solid lines right after the value symbol in the case of Dynamics; solid lines or dash/dot lines throughout the link in the case of Temporality; or sub-indexes in the case of Type, Class, Frequency and Volume.

An example of this detailed description would be represented as shown in Figure 8.21.

Reading from left to right (clockwise) in the direction of the relationship: The spatial link from A to B would be complementary ◒ if indirect auditive (1.3), of humans (2.1) with occasional frequency (3.3), low volume (4.3), and permanent (————); and reading from right to left in the direction of the B to A relationship, we find that this link is non-desirable ⊘ if direct (————), physical (1.1), human (2.1), repetitive (3.2) with medium volume (4.2), and permanent (————).

Modifications of this diagram for cases having special conditions can be easily effected by means of "loops" in each necessary characteristic index. A loop is nothing but a footnote which, by changing a

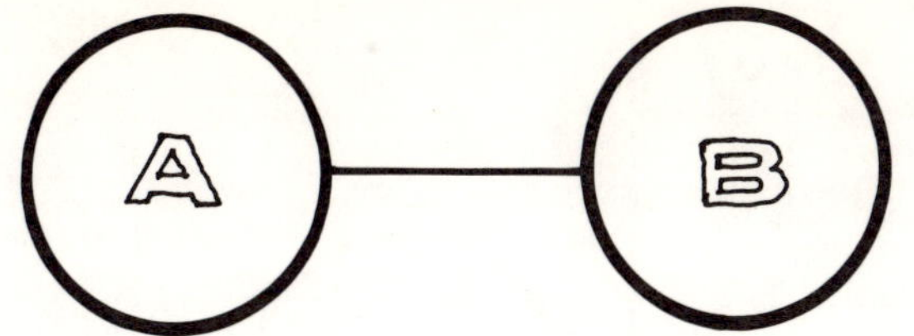

Figure 8.15

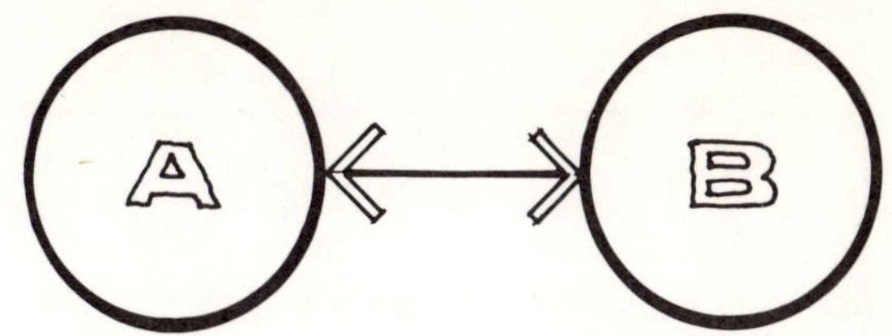

Figure 8.16

POSITIVE

NEGATIVE

value

ESSENTIAL ●

COMPLEMENTARY ◐

NON·ESSENTIAL ○

NON·DESIRABLE ⊘

NON·ACCEPTABLE ⊗

Figure 8.17

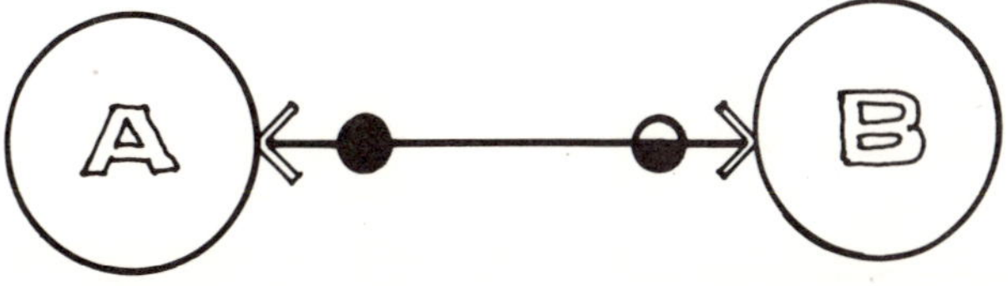

Figure 8.18

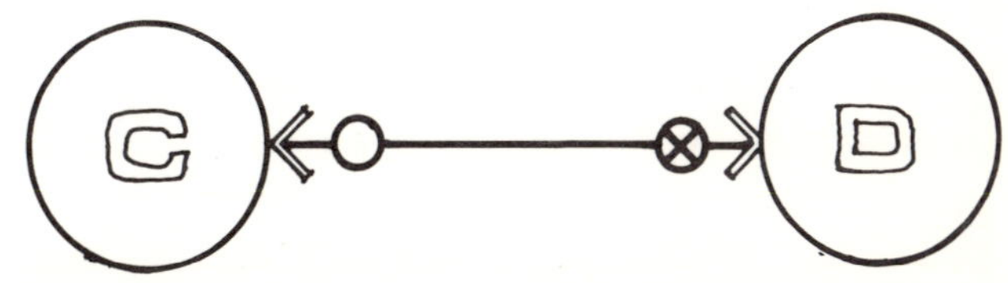

Figure 8.19

SPATIAL RELATIONSHIPS DESCRIPTION		
DESIGNATION	DESCRIPTION	GRAPHIC REPRESENTATION
DYNAMICS	DIRECT INDIRECT	
TYPE	PHYSICAL AUDIOVISUAL AUDITIVE VISUAL	1.1 1.2 1.3 1.4
CLASS	HUMAN / HUMAN EQUIPMENT / EQUIPMENT ANIMALS, PLANTS / REC. HUMAN / EQUIPMENT HUMAN / ANIMAL, PLANTS ANIMAL / EQUIPMENT OTHER (DEFINE)	2.1 2.2 2.3 2.4 2.5 2.6 2.7
FREQUENCY	CONSTANT REPETITIVE OCCASIONAL SCARCE	3.1 3.2 3.3 3.4
VOLUME	HIGH MEDIUM LOW	4.1 4.2 4.3
TEMPORALITY	PERMANENT TEMPORARY	

Figure 8.20

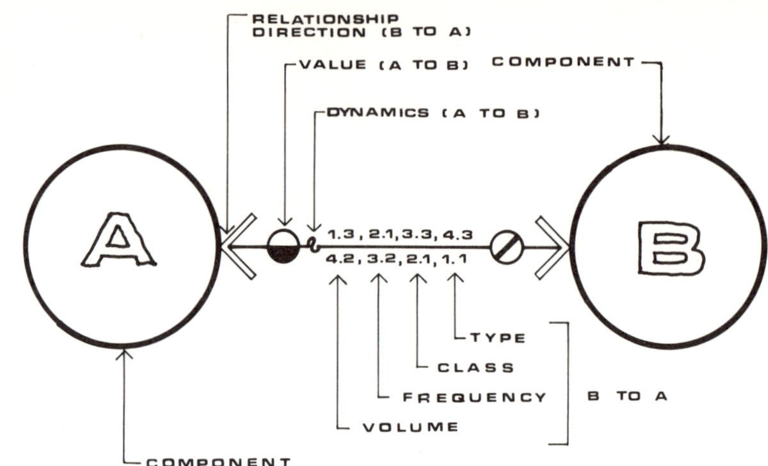

Figure 8.21

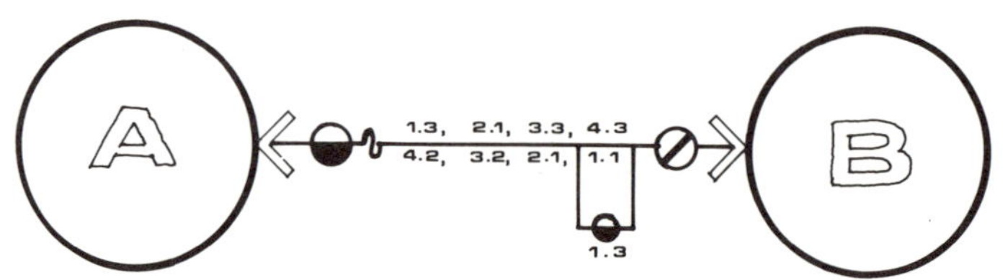

Figure 8.22

particular characteristic, will affect the value of the particular relationship described.

For example, in Figure 8.22, the relationship between B and A is non-desirable if of physical type, human class, repetitive frequency and medium volume; but when a change in this value is possible (such as in a case where an auditive relationship could prove useful enough to make the value change from "non-desirable" to "complementary") the modification is noted by means of a loop around the type index 1.1 (physical) to indicate that if this type is changed to 1.3 (auditive), the relationship becomes complementary.

Thus, the final diagram can be modified to clarify the fact that a physical link between B and A can affect the relationship enough to make it non-desirable, while an auditive link would make it complementary without changing any other characteristic. It is important to note the permanence or temporality of a relationship because flexibility, phasing, or scheduling will often cause changes in some links, and these factors CAN DIRECTLY AFFECT A DESIGN SOLUTION.

Although some combinations might appear improbable (such as indirect-physical, or indirect-audio-visual), let us not forget that "improbable" does not necessarily mean "impossible." An indirect-physical link is very common in large office complexes where messengers carry messages, memos, correspondence, between departments; and any closed-circuit television system is a typical example of an indirect-audio-visual link between two spaces.

In Sub-chapter 10.1 (Simplification and Developments), we will explain how many of these different characteristics might not have to be listed under specific conditions. The basic idea is to describe the relationships among spaces in the most detailed manner possible without going to extremes. For example, it might not be necessary to define ALL the characteristics of links at the preliminary design stages; but they might prove useful during the design development phase, since keying and door swings will most likely become important issues at that time. In such cases, it is convenient to select those characteristics to be determined at each stage, but always using the same index numbers — that is: type (1), class (2), frequency (3), volume (4) — because to alter these would most likely cause confusion and misunderstandings. It is better to leave a blank space between two index numbers than to start shifting values around.

Even if the actual relationship is not defined in terms of type, class, frequency and volume, the most important factors SHOULD NEVER BE OMITTED, and those are: direction, value and dynamics. No spatial relationships diagram can be considered complete without them.

As far as the actual data collection form for the establishment of these charts is concerned, the use of the matrix in Figure 8.23 is probably best.

The relationships are read in the direction of the arrow formed by splitting in half the corresponding rhombus that joins two spaces. Referring once more to Figure 8.22, the actual value, direction and dynamics, as well as characteristics of the relationships, are established through questionnaires, interviews, or work sessions and noted as illustrated above.

B to A:

Value	= Complementary	(◐)
Dynamics	= Direct	(————)
Type	= Physical	(1.1)
Class	= Human	(2.1)
Frequency	= Occasional	(3.3)
Volume	= Low	(4.3)
Temporality	= Permanent	(————)

A to B:

Value	= Non-desirable	(⊘)
Dynamics	= Indirect	(—ᆺ—)
Type	= Physical	(1.1)
	or Auditive	(1.3) with Value change to Complementary (◐)
Class	= Human	(2.1)
Frequency	= Repetitive	(3.2)
Volume	= High	(4.1)
Temporality	Permanent	(————)

Note that each link between two spaces falls in a separate square, thus allowing a clear separation of each link for individual study. If the relationship between two spaces IS RECIPROCAL, there is no need to split the square in half; the description would simply be listed in the entire space.

In the chart itself, reciprocal relationships should be represented as shown in Figure 8.24. Where the relationship B to C is essential ● direct (———) physical (1.1), of humans and machines or equipment (2.4) constant (3.1) and with low volume (4.3), and the line between C and B is the same, that is, reciprocal.

In spite of this reciprocity, the direction, value and dynamics are still represented. The reason for this is the obvious possibility of changes in the value and reciprocity of links among spaces occurring during the programming phase.

These representations of spatial requirements assume a temporal classification of PRESENT throughout. In Sub-chapter 8.6 we will define more clearly the temporality aspects of these representations. For now it is important to note, however, that the symbols used throughout this text to indicate TIME in establishing these diagrams are shown in Figures 8.25, 8.26, and 8.27.

In Figure 8.26, "D" links to a future space "E" in the manner indicated, while "E"'s link to "D" will be as illustrated. Or in Figure 8.27, "F" related to a particular space "G" that no longer exists (due to either organizational restructuring or redistribution), and vice versa, as indicated.

These temporal indications are extremely useful in situations where phasing, growth, and change projections are taken into account in the initial spatial relationships' analysis.

In these graphic representations, the actual spaces should also contain certain information within their perimeters. Referring to Figure 8.14 (Spatial Components Analysis Data Sheet), we can include within the actual space designation circle the items shown in Figure 8.28.

It is also advisable to draw the circles to signify the actual area representation or assigned value to be occupied by the space, since this will give the

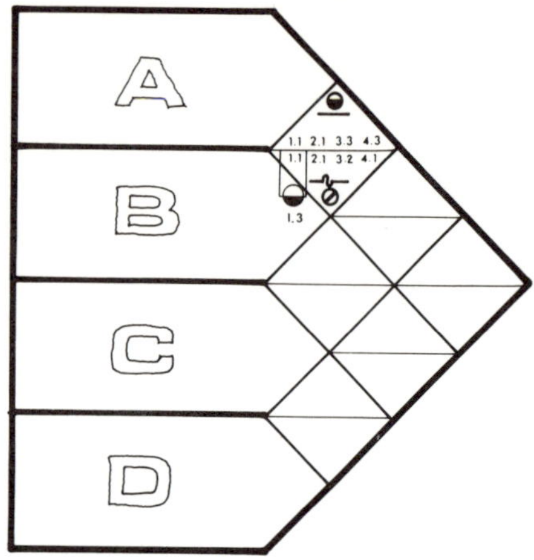

SPATIAL RELATIONSHIPS DESCRIPTION MATRIX

Figure 8.23

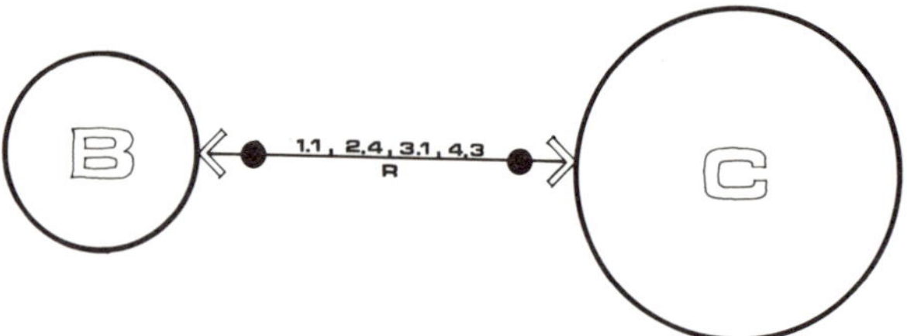

Figure 8.24

T I M E	G R A P H I C REPRESENTATION
P A S T	+ + + + +
PRESENT	—————
FUTURE	– – – – –

Figure 8.25

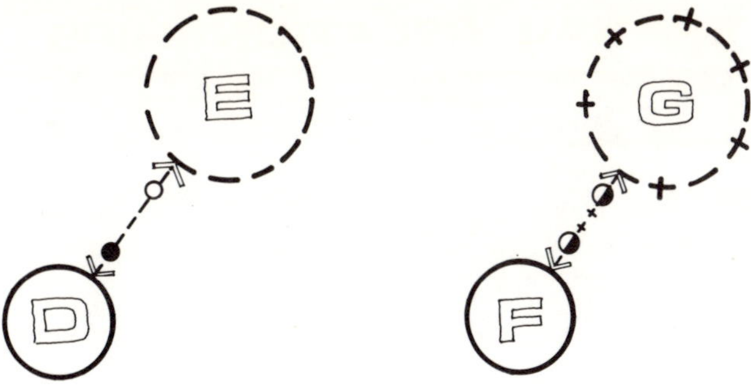

Figure 8.26   Figure 8.27

Figure 8.28

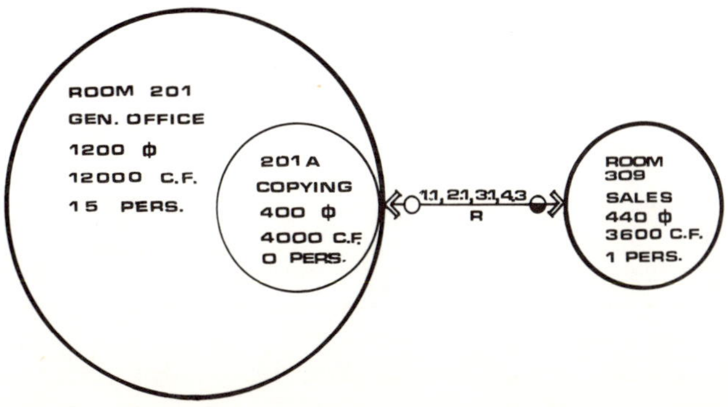

Figure 8.29

designer a much clearer idea of proportions and sizes. Whenever a space with very definite individual characteristics is contained within another, this should be represented graphically as shown in Figure 8.29. One particular copying room is located within the general offices of a company and, although a basic part of those offices must relate directly to the office of a sales secretary, there is no need for an actual line between the general offices space and this secretary's private cubicle.

One very important point is UNLESS THERE IS A DEFINITE VALUE IN ANY SPATIAL LINK, positive or negative, there SHOULD NOT BE ANY LINK INDICATION AT ALL between the two spaces. It is actually pointless to describe the possible relationship between two areas which, because of their function, have no relationship whatsoever with one another.

Summarizing, the results of Sub-chapters 8.2 and 8.3 can be encompassed in the chart in Figure 8.30.

FLOW PATTERNS. Generally, it is advisable to use these diagrams to indicate the trajectory of elements within a complex. These visual aids not only offer the advantage of outlining the movement and direction of active elements within operational systems, but can also be expanded to illustrate the volume, frequency, and structure of those movements.

One simple way of doing this is represented as shown in Figure 8.31.

For example, in the diagram illustrated in Figure 8.32 we can easily see that, while the traffic from A to D is high and repetitive, the flow pattern from B to E is low and occasional. At the same time we can also illustrate the possible flow alternatives from B to D by means of a cross-pattern through C, plus the final modified distribution of high repetitive volume from D to E, in order to exit two repetitive medium-low flow patterns at D and E.

Notice that, because of this flow diagram, the functional performance of space D must include this flow modification capability as an essential performance requirement which is, therefore, a design determinant.

SPACE OPERATIONAL ANALYSIS.

SITE ANALYSIS

NATURAL CHARACTERISTICS	structural	SOIL CONDITIONS GEOLOGICAL CONSIDERATIONS SUBSURFACE WATER	
	physical	NATURAL DRAINAGE SLOPES, CONTOURS & FRACTURES ORIENTATION & VIEWS	
	environmental	SUN, WIND & TEMPERATURE SNOW, FROST & PRECIPITATION SURFACE WATER & SURROUNDINGS FLORA/FAUNA CONSERVATION & POLLUTION	**SITE**
ARTIFICIAL CONDITIONS	technical	FUNCTIONAL LOCATION & COMPLIANCE GROWTH & CHANGE PROJECTIONS ACCESSIBILITY & CIRCULATIONS HISTORICAL VALUE	**EVALUATION**
	physical	SITE UTILITIES & SERVICES EXISTING & NEIGHBORING STRUCTURES OPERATIONAL FACTORS SOUND CONDITIONS IMPROVEMENTS	
	regulatory	PLANNING & ZONING REGULATIONS BUILDING, FIRE & HEALTH REGULATIONS HIGHWAY & TRAFFIC REGULATIONS OSHA & FEDERAL REGULATIONS LEGAL & CONTRACTUAL CONDITIONS	

SPATIAL ANALYSIS

1 divisional fragmentation — PAST / PRESENT / FUTURE

2 components analysis.

PAST → IDENTIFY BY GRAPHIC REPRESENTATION : +++++

3 — COMPONENTS CHARACTERISTICS

FUNCTION	essential complementary service		E C S
CONSTITUTIVE ELEMENTS	activity description		
	function	ESSENTIAL COMPLEMENTARY SERVICE	
	structure	CLASS TYPE NUMBER	
	location		
	relationships	DIRECTION VALUE DESCRIPTION	

4 — COMPONENTS RELATIONSHIPS

DIRECTION			←→
VALUE	positive	ESSENTIAL COMPLEMENTARY NON-ESSENTIAL	
	negative	NON-DESIRABLE NON-ACCEPTABLE	
DESCRIPTION	dynamics	DIRECT INDIRECT	
	type	PHYSICAL AUDIOVISUAL AUDITIVE VISUAL	1.1 1.2 1.3 1.4
	class	HUMAN MACH./EQUIPMENT ANIMAL/PLANT HUMAN/MACHINE/EQUIP. HUMAN/ANIMAL/PLANT ANIMAL/PLANT/MACH. OTHER	2.1 2.2 2.3 2.4 2.5 2.6 2.7
	frequency	CONSTANT REPETITIVE OCCASIONAL SCARCE	3.1 3.2 3.3 3.4
	volume	HIGH MEDIUM LOW	4.1 4.2 4.3
	temporality	PERMANENT TEMPORARY	

FUTURE → IDENTIFY BY GRAPHIC REPRESENTATION : ------

Figure 8.30

Figure 8.31

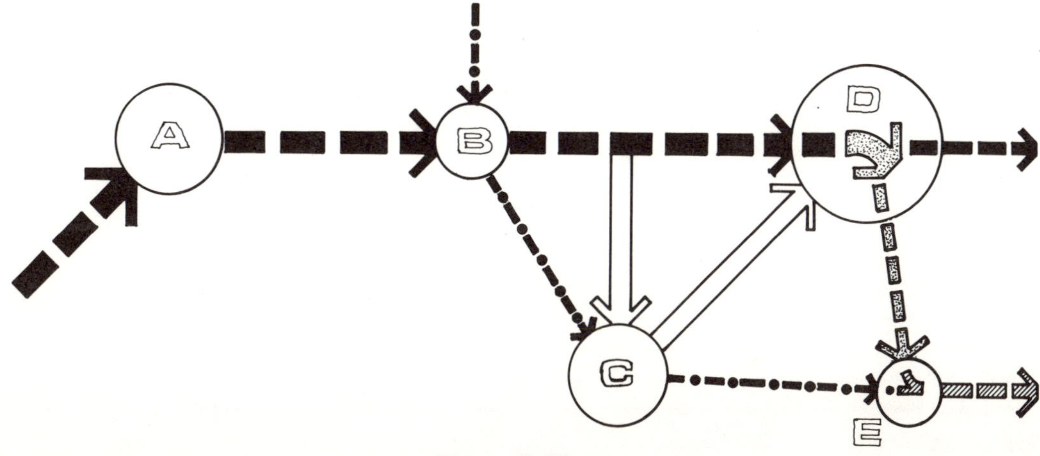

Figure 8.32

Although they are directly related, flow patterns must not be confused with spatial relationships diagrams. The major difference between these two graphic aids is that, while flow diagrams represent the pattern of movements of dynamic elements within operational systems, spatial relationships diagrams are actually patterns of functional relationships among spaces. Their complementary values are reciprocal and, in many cases, consequential but in no case should it be construed that one can supplant the other.

Most spatial relationships diagrams, when properly executed, can illustrate more or less accurately the flow of movements taking place within their scope. Still, a definite graphic representation of these movements is always recommendable in situations where flow plays a key role in the conceptual phase of architectural programming. Chapter 9 contains different case studies which will more clearly illustrate the previously outlined methods.

In cases involving extensive subdivisions, it is advisable to establish spatial relationships diagrams and, in some instances, flow diagrams on a divisional basis and refer them to a general diagram which, in turn, relates these divisions among themselves. This can not only simplify the actual analysis procedure but also allow a clearer understanding of each portion of the project in organized segments.

8.4 Economic Analysis

The economic analysis level of the architectural programming phase is the one commonly known as budgeting or budget analysis[18] and just as in any other professional design discipline, we can find different degrees of accuracy and comprehensiveness as the processes become more and more definite. Any economic analysis procedure begins with a general budget study and ends with a detailed outline of cost-breakdowns prior to the bidding phase.

As the architectural execution progresses, so does the economic study of its elements, in more or less the following order:

Architectural Programming ..Budgeting
Preliminaries ....Preliminary Cost
Analysis
Design Development ....Detailed Cost
Analysis and Cost
Control
Working Drawings ....Cost Control and
Final Cost
Estimates

Budgeting, therefore, is the only portion of the economic analysis procedure that concerns us at this stage. Unlike detailed estimates, a budget analysis is prepared at a time when detailed information on a project has not yet been determined. It is unrealistic, then, to expect detailed quantities take-offs or accurate descriptions of systems and special conditions at that time. Still, we must not forget that a budget study represents the first documentation of a project's future costs and their distribution.

Therefore the budget analyst shoulders the enormous responsibility of laying the foundation for a project's costs without sufficient, if any, pertinent data. Nor can he expect help from estimators, contractors, or specialized cost analysts, since their function is primarily to obtain costs from detailed documents and other reliable data. The budget analyst becomes then, the center of a complex and everlasting chess game involving costs, systems, design philosophies, materials and construction procedures where engineers, technicians and designers, like pawns, shift their strategic positions continually.

PROJECT COST AND CONSTRUCTION COST. A budget, if it is to serve its purpose, must include a comprehensive breakdown of all foreseeable costs which a project might incur, of which construction costs are only a portion.

To illustrate this, let us refer to the budget analysis chart represented in Figure 8.33, where the essential items which actually structure the project cost are listed and defined.

BUDGET ESTIMATE FORM	PROJECT NO.
PROJECT	DATE
LOCATION	PREPARED BY

GROSS AREA	NET AREA	BLDG TYPE	EFFICIENCY

ITEM	COST
A BUILDING COST (GROSS AREA X SQ.FT. UNIT COST)	
B SITE DEVELOPMENT	
C FIXED EQUIPMENT	
D HEATING VENTILATING & AIR CONDITIONING	
E PLUMBING & FIRE PROTECTION	
F ELECTRICAL	
G TOTAL CONSTRUCTION COSTS SUBTOTAL 1	
A + B + C + D + E + F = G	
H SITE ACQUISITION	
I UNUSUAL SITE CONDITIONS	
J OFF SITE WORK	
K GROWTH & CHANGE PROVISIONS	
L LANDSCAPING	
M FURNISHINGS	
N ALLOWANCES / CONTINGENCIES	
O MISCELLANEOUS	
P ADDITIONAL BUDGET ITEMS SUBTOTAL 2	
Q PROFESSIONAL FEES	
R SURVEYS AND INSURANCE	
S LEGAL, ACCOUNTING AND ADMINISTRATIVE COSTS	
T LEASING, ADVERTISING AND PROMOTION	
U FINANCING COSTS AND TAXES	
V OWNER'S BUDGET ITEMS SUBTOTAL 3	
W TOTAL BUDGET REQUIRED	
G + P + V = W	

NOTES:

Figure 8.33

CONSTRUCTION COSTS.

1. *Building Cost:* The approximate amount of construction costs based on a square foot cost index for that specific building type.

2. *Site Development:* All necessary site improvements including grading, excavation, backfill, walks, curbs, driveways, drainage, parking, paving, electrical wiring, plumbing, irrigation and special systems within the site boundaries up to five feet around the building perimeter.

3. *Fixed Equipment:* Items such as food service equipment, seating, special technical equipment, sports and automotive equipment, paint and spray booths, dock levelers, communication systems, flag poles, etc.

4. *HVAC, Plumbing, Sewerage and Fire Protection and Electrical Work:* All corresponding work and equipment contained within a perimeter of five feet outside the building limits.

ADDITIONAL BUDGET ITEMS.

1. *Site Acquisition:* The total of the figures indicating the direct and indirect costs of site purchasing, such as land cost, taxes, fees, etc.

2. *Unusual Site Conditions:* Demolition requirements, abnormal amount of grading and backfilling, special access roads, preservation of existing structures or natural features, unusual soil conditions, etc.

3. *Off-site Work:* Any work or improvements to be carried out beyond the actual project limits.

4. *Growth and Change Provisions:* Structural, electrical, or mechanical oversizing due to future additions, expansions, etc.

5. *Landscaping:* To include indoor planting and fountains in addition to outdoor planting, soil treatments, existing plant relocation, planters, benches and any landscape related grading.

6. *Furnishings:* If applicable, to include all furniture, drapery, artwork, carpeting and special interior design features and finishes as required.

7. *Allowances and Contingencies:* Varies anywhere from 3 to 10% of budgeted construction costs.

8. *Miscellaneous:* Art and sign programs, income producing and self-liquidating items such as vending machines, laundry equipment, etc. and tenant allowances and concessions.

OWNER'S BUDGET ITEMS. These items are self-explanatory and in many cases are not included in a budget analysis, since the client can execute these portions more accurately. Also, in many cases, the information is neither readily available to the budget analyst nor requested by the client.

Of all the items described above, the most difficult one to approximate is that of the square foot cost per gross area of construction, because the factors involved in this approximation depend greatly on assumed qualities of construction, size and building efficiency ratios.

Several publications (such as *Means & Dodge Construction Costs Data Manuals*) currently offer a fairly clear differentiation between the cost indexes for specific building types on a square and cubic foot basis, with adjustments for location, quality levels and building sizes. At this particular point, it is better to refer to those publications as the most reliable source of such data.

PLAN EFFICIENCY. Plan efficiency has been defined by some in a quantitative way as the ratio between the net assignable areas of a structure and its gross area, where net area is the sum of all areas on all floors of a building assigned to, or available for assignment to, an occupant, including every type of space functionally usable by an occupant.[19] Gross area is the sum of all floor areas included within the outside faces of exterior walls for all stories, or areas, which have floor surfaces.[20]

Plan efficiency has been represented in the form of a decimal value (which becomes a round numerical factor once multiplied by a constant 10, 100, etc.) or in the form of a percentage or percentage ratio. This type of efficiency formula is basically represented by the following relationship:

$$\frac{\text{Net Area}}{\text{Gross Area}} = \text{Efficiency}$$

Unfortunately, this formula is true for the particular

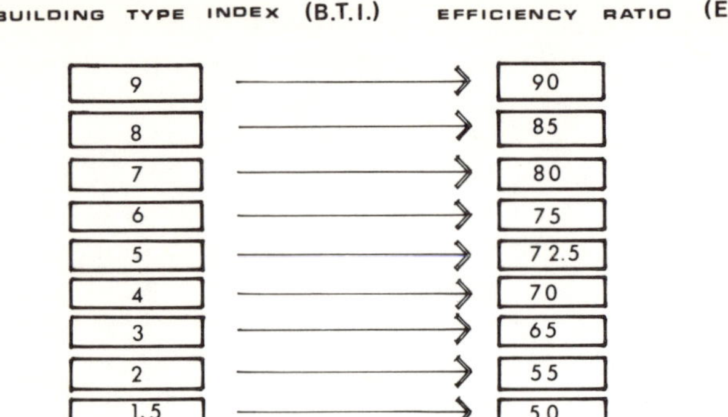

Figure 8.34

efficiency ratio of one project compared only with itself in terms of areas and usable spaces, but without adjustment factors or comparison indexes of any kind. In spite of this, plan efficiency encompasses a lot more than mere simple ratios when issues like building types, function, location, construction requirements, capacity, are taken into account.

Nevertheless, if we are to roughly establish basic plan efficiency ratio values with the means, procedures and data available to us today, we can begin by using the following system as a general guideline until more pertinent information is available:

$$\frac{\text{Net Area}}{\text{Gross Area}} \times 100 = \text{Efficiency Ratio}$$

By obtaining Building Type Indexes based on plan and volumetric fragmentation, cost indexes and average levels of quality, we can establish a general relationship as shown in Figure 8.34.

Since the square and cubic foot cost ratios represent levels of construction quality among all buildings, and serve as an approximate index of fragmentation level among building types, these Building Type Indexes have been determined by averaging the medium-low, medium and medium-high levels of construction on the basis of square and cubic foot ratios, and relating these ratios to unit costs of $100 per square foot and $10 per cubic foot

so as to include ALL building types in a common value pattern.

The results obtained are as follows:

Building Types Indexes
    Residential
        Apartments — 4.5
        Housing for the aged — 3.0
        Motels — 3.7
        Hotels — 3.0
        Public Housing — 4.0
        Dormitories — 2.8

    Educational
        Elementary Schools — 2.5
        Secondary Schools — 2.5
        Senior High Schools — 2.6
        Colleges (Adm. & Class) — 2.5
        Colleges (Science & Labs) — 2.2
        Colleges (Unions) — 2.6
        Gyms — 5.0

    Cultural
        Libraries — 3.5
        Theaters — 4.0
        Auditoriums — 4.0
        Museums — 3.5

    Health
        Hospitals — 1.8
        Medical Clinics — 2.6
        Nursing Homes — 2.5
        Veterinary Hospitals — 2.6

    Religious
        Churches — 3.5
        Convents — 3.0
        Religious Education — 3.2

    Institutional and Government
        Court Houses — 2.0
        Police Stations — 2.5
        Fire Stations — 2.7
        Jails — 2.2
        Town Halls — 3.0
        Post Offices — 3.0
        Community Centers — 3.7

Commercial
- Auto Sales — 6.0
- Banks — 2.5
- Department Stores — 7.0
- Funeral Homes — 3.7
- Commercial Garages — 5.5
- Municipal Garages — 6.0
- Parking Garages — 9.0
- Offices — 3.3
- Restaurants — 3.0
- Retail Stores — 5.8
- Supermarkets — 6.2
- Medical Offices — 3.0
- Dental Offices — 3.0
- Shopping Centers — 6.5

Transportation
- Bus and Truck Terminals — 3.5
- Airports — 2.7
- Hangars — 8.7

Industrial
- Factories — 7.0
- Warehouses — 8.8
- Research Labs — 2.4

Recreation
- Social Clubs — 3.2

It is important to notice that the lower the value of the Building Type Index, the lower the efficiency ratio will be; therefore, adjustments should be made by reducing these values as the project's quality and size levels increase, but in no case should any Building Type Index be reduced by more than 1 point, or increased by more than 0.5 points.

Average sizes for the previously mentioned building types are as shown in Table 8.1. The medium column represents the approximate average size per building type while the low and high, in turn, should be assumed low and high AVERAGES ONLY and not lowest or highest projects recorded.

Approximate levels of quality, as analyzed on a twelve-year span, produce the results shown in Table 8.2.

Table 8.1

Average Building Sizes

Building Type	Approximate Average Sizes in Square Feet		
	Low (-1)	Medium	High (+0.5)
Residential			
Apartments	18,000	30,000	100,000
Housing for aged	18,000	30,000	50,000
Motels	15,000	35,000	65,000
Hotels	18,000	50,000	80,000
Public Housing	20,000	50,000	100,000
Dormitories	20,000	40,000	90,000
Educational			
Elementary Schools	30,000	45,000	60,000
Secondary Schools	50,000	90,000	125,000
High Schools	50,000	100,000	190,000
Colleges (Classrooms)	25,000	50,000	100,000
Colleges (Science & Labs)	15,000	40,000	80,000
Colleges (Unions)	15,000	40,000	80,000
Gyms	12,000	20,000	50,000
Cultural			
Libraries	6,000	11,000	28,000
Theaters	9,000	11,000	18,000
Auditoriums	12,000	25,000	38,000
Museums	10,000	12,000	20,000
Health			
Hospitals	25,000	52,000	140,000
Nursing Homes	15,000	25,000	40,000
Medical Clinics	4,000	8,000	20,000
Religious			
Churches	2,500	9,000	35,000
Convents	7,000	10,000	16,000
Religious Education	6,000	8,000	14,000
Institutional			
Court Houses	15,000	25,000	70,000
Police Stations	4,000	10,000	20,000
Fire Stations	4,000	6,000	8,000
Jails	7,000	14,000	85,000
Town Halls	5,000	10,000	24,000

Table 8.1 (continued)

Average Building Sizes

Building Type	Approximate Average Sizes in Square Feet		
	Low (–1)	Medium	High (+0.5)
Institutional (cont'd.)			
Post Offices	6,000	12,000	24,000
Community Centers	4,000	10,000	20,000
Commercial			
Auto Sales	10,000	20,000	30,000
Banks	3,000	5,000	10,000
Department Stores	50,000	100,000	140,000
Funeral Homes	5,000	8,000	12,000
Commercial Garages	6,000	10,000	18,000
Municipal Garages	5,000	8,000	16,000
Parking Garages	120,000	180,000	340,000
Offices	5,000	10,000	25,000
Restaurants	3,000	5,000	7,000
Retail Stores	4,000	8,000	22,000
Supermarkets	15,000	20,000	30,000
Medical Offices	4,000	6,000	15,000
Shopping Centers	20,000	50,000	200,000
Transportation			
Bus & Truck Terminals	7,000	12,000	20,000
Airports	30,000	60,000	150,000
Industrial			
Factories	12,000	30,000	55,000
Warehouses	10,000	20,000	50,000
Research Labs.	5,000	20,000	40,000
Recreation			
Social Clubs	15,000	35,000	45,000

Table 8.2

Levels of Quality

Building Type	Low % of A	M.L. % of A	Med. A = average costs/s.f.	M.H. % of A	High % of A
Residential					
Apartments	60	80	100	150	300+
Housing for aged	50	90	100	130	210+

Table 8.2 (continued)

Levels of Quality

Building Type	Low % of A	M.L. % of A	Med. A = average costs/s.f.	M.H. % of A	High % of A
Residential (cont'd.)					
Motels	50	85	100	125	220+
Hotels	50	90	100	130	250+
Public Housing	60	90	100	145	230+
Dormitories	50	90	100	130	250+
Educational					
Elementary Schools	50	90	100	135	250+
Secondary Schools	55	90	100	130	200+
Senior High Schools	55	90	100	135	230+
Colleges (Classes)	50	80	100	130	210+
Colleges (Science & Labs)	50	80	100	140	220+
Colleges (Unions)	60	85	100	130	180+
Gymnasiums	60	85	100	140	280+
Cultural					
Libraries	60	85	100	135	280+
Theaters	50	70	100	130	200+
Auditoriums	50	70	100	130	200+
Museums	60	85	100	140	300+
Health					
Hospitals	60	90	100	130	180+
Medical Clinics	60	90	100	135	240+
Nursing Homes	50	85	100	130	250+
Veterinary Hospitals	50	80	100	120	160+
Religious					
Churches	60	86	100	125	180+
Convents	60	90	100	130	190+
Religious Education	65	85	100	125	200+
Institutional					
Courthouses	60	90	100	150	180+
Police Stations	60	85	100	130	200+

Table 8.2 (continued)

Levels of Quality

Building Type	Low % of A	M.L. % of A	Med. A = average costs/s.f.	M.H. % of A	High % of A
Institutional (cont'd.)					
Fire Stations	50	85	100	135	230+
Jails	65	90	100	120	170+
Town Halls	60	90	100	140	200+
Community Centers	50	90	100	125	220+
Commercial					
Auto Sales	70	85	100	130	200+
Banks	50	80	100	140	200+
Department Stores	70	90	100	130	200+
Commercial Garages	50	80	100	135	290+
Municipal Garages	50	80	100	140	250+
Parking Garages	50	85	100	135	300+
Offices	50	85	100	135	260+
Restaurants	50	80	100	130	250+
Retail Stores	70	85	100	130	220+
Supermarkets	70	90	100	130	200+
Medical Offices	50	90	100	120	160+
Dental Offices	50	90	100	120	160+
Shopping Centers	70	85	100	135	250+
Transportation					
Bus/Truck Terminals	50	80	100	140	280+
Hangars	50	80	100	130	250+
Industrial					
Factories	40	80	100	150	300+
Warehouses	40	80	100	180	300+
Research Labs	45	80	100	150	210+
Recreation					
Social Clubs	55	85	100	125	180+

Unfortunately, it is impossible at this point to develop a fool-proof formula for precisely determining the building plan efficiency ratio for any specific project. Furthermore, different portions of different buildings have different efficiency levels depending on their functions; for example, mechanical areas of a school are definitely more efficient on a net to gross ratio than those dedicated to academic functions. For this reason a general breakdown of functional spaces should be carried out whenever possible when determining the overall relationship of net to gross area for a project.

Up to what point a simple rearrangement of spaces can affect the efficiency ratio of a specific plan can be easily illustrated in the diagrams shown in Figure 8.35.

The importance of plan efficiency ratios is further stressed by the fact that they inherently represent "What it will cost" (gross area), in terms of "What it is needed for" (net area); and, therefore, clarify the relative cost of those spatial requirements.

To illustrate this point, let us use the previously mentioned charts to determine the approximate gross area of a bank building with 5,000 Net Sq. Ft. Area and a medium-low quality level of construction.

> Building Type: — Bank
> Building Type Index: — 2.5
> Net Area: — 5,000 Sq. Ft.
> Building Size Adjustment: — None
> Medium-Low Quality Adjustment: — +0.2
>   (approx. adjustment)
> Efficiency Ratio: — 60 (from figure 8.34)
> Revised Efficiency Ratio: — Approx. 62 (from
>   BTI = 2.7)

So from:

$$\frac{\text{Net Area}}{\text{Gross Area}} \times 100 = \text{Efficiency Ratio}$$

Building Gross area =

$$\frac{5,000}{\text{Gross Area}} \times 100 = 62 = 8064 \pm 8,000 \text{ Sq. Ft.}$$

Although the process can be reversed to obtain the net area from the gross area, the net area should be established by means of a detailed spatial analysis as outlined in Sub-chapter 8.3, and then checked against the results of the previous charts,

PLAN EFFICIENCY STUDY

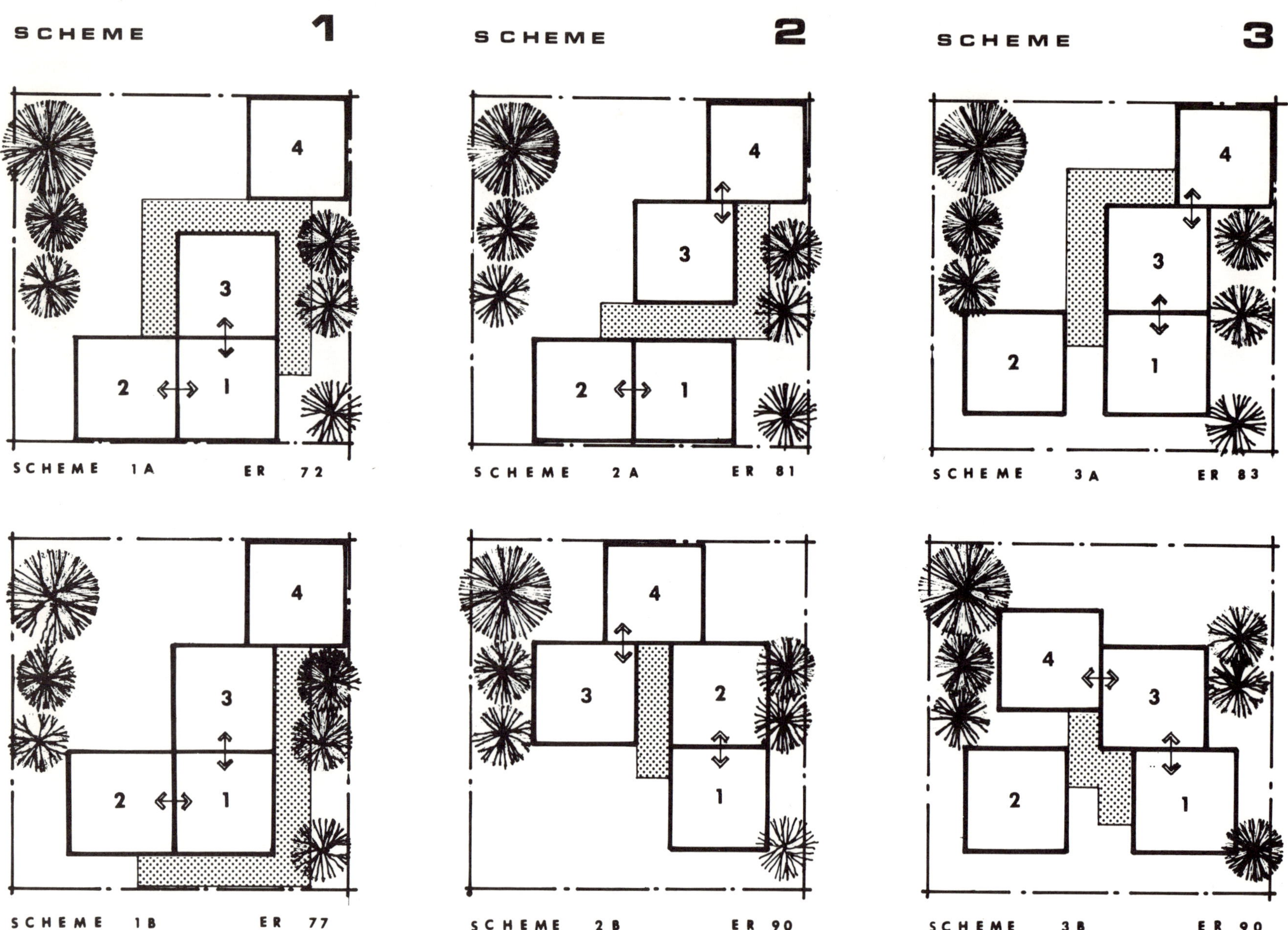

 Figure 8.35

since all spatial considerations can lead to enormous variations in the design layout and, therefore, in the efficiency ratios of a project.

In spite of these guidelines, approximations of net to gross ratios, budgeting and costs assumptions at this stage are just that — approximations, and in no way can they be expected to produce very accurate results.

No matter which approximations are obtained in applying these (or other) formulas, as any building exceeds its average size, it becomes more advantageous to establish a functional fragmentation of its components and revise the efficiency ratios by working with smaller segments, thereby reducing the margin for errors in large square footages.

Special site and building configurations, as well as special requirements and characteristics, can affect the results so drastically, those cases should be analyzed separately.

To summarize, we can say that in spite of these approximations, logical limitations and inaccuracies, and despite the fact that inexact formulas can lead to inevitable misjudgments, ANY economic analysis approach during the budgeting phase is better than none. It is essential to stress the importance of a conscientiously prepared budget, since carelessness and omissions at this stage can lead to serious complications as the project development progresses. "In the beginning, there was the budget...." We will always be reminded of that.

8.5 Regulatory Surveys

Although most regulatory surveys are done while a project is in the schematics phase, the logical procedure would be to make a preliminary evaluation of these code requirements which might affect the project's design prior to any schematics development. The reason for this is obvious: Any regulatory requirement is as much a part of the design problem as any functional need, economic constraint, or spatial determinant.

Variances and regulatory waivers, although inevitable in some cases, should be avoided whenever possible if only because their end result is inherently unpredictable.

If a program is to provide a designer with the full scope of a project, including all foreseeable determinants, characteristics and limitations, then a regulatory study of applicable codes and regulations should be an integral part of the program's general outline. Nothing is more damaging to the development of a design concept than the stumbling blocks which appear during its schematic phase as a designer tries to verify the compliances of his ideas with codes and regulations.

We must also realize that ALL regulations affecting a project are not always verified at the outset, and this causes many embarrassing moments — which should never occur — in the relationship between design professionals and their clients.

Because of the inevitable fragmentation of codes and regulations, it is impossible in this limited space to establish a comprehensive regulatory checklist covering all possible constraints and pertinent data. Yet, if a detailed outline of these types of checklists is beyond our scope in this text, we can very well establish a general description of the functions of these surveys and thus determine the essential data to be collected in order to arrive at a clear definition of the role they will play within a program's layout.

The basic regulatory surveys in this Sub-chapter are:

1. Zoning regulations
2. Building and Fire regulations
3. Health regulations
4. Highway and Traffic regulations

It has been assumed that regulations covering plumbing, heating, ventilation and air-conditioning, as well as electrical work, legal and contractual conditions, OSHA, pollution, and energy conservation will be investigated by specialized consultants.

ZONING REGULATIONS. These regulate the development of a project in regard to its urban placement, specific location and functional characteristics. On a general checklist, the basic data to be determined can be summarized as follows:

ZONING REGULATIONS CHECKLIST

Department _______________________________
Division _________________________________
Inspector ________________________________
Telephone number ____________ Extension ___
Code section index _______________________

1. Zoning classification ___________________

2. Permitted uses
   Primary _______________________________
   Accessory _____________________________
   Conditional ___________________________

3. Minimum lot dimensions
   Area __________________________________
   Width _________________________________
   Depth _________________________________

4. Setback requirements
   Front _________________________________
   Side(s) _______________________________
   Rear __________________________________
   Other _________________________________

5. Distance between buildings
   On one lot _____________________________
   On adjacent lots _______________________

6. Height requirements
   (note elevation datum point)
   Max. allowable height _________________
   Max. number of stories ________________

7. Floor Area Ratio (F.A.R.) ______________

8. Open Space Ratio (O.S.R.) & Lot coverage ___

9. Signs ________________________________

10. Off-street parking
    Width ________________________________
    Length _______________________________
    Ratio/Number _________________________
    Aisles _______________________________
    Entrances ____________________________
    Exits ________________________________
    Drives _______________________________
    Stripping ____________________________
    Handicapped
        Number ___________________________
        Width ____________________________

Handicapped (continued)

        Length ___________________________
        Location _________________________
    Lighting _____________________________
    Screening ____________________________
    Off-street loading
        Number of spaces _________________
        Area _____________________________
        Width ____________________________
        Length ___________________________
        Clearances _______________________

11. Landscaping requirements
    Landscaping __________________________
    Planters _____________________________
    Fencing ______________________________

12. Additional regulations ________________
    ____
    ____

BUILDING AND FIRE REGULATIONS. These regulate the building and construction aspects of a project as well as their fire and life safety constraints.

BUILDING AND FIRE REGULATIONS CHECKLIST[21]

Department _______________________________
Division _________________________________
Inspector ________________________________
Telephone number ____________ Extension ___
Code section index _______________________

1. Applicable codes
   Building ______________________________
   Fire __________________________________

2. Fire zones ___________________________

3. Building occupancy
   Occupancy classification
       Primary ___________________________
       Secondary _________________________
       Other _____________________________
   Fire classification
       Primary ___________________________
       Secondary _________________________
       Other _____________________________

BUILDING AND FIRE REGULATIONS
CHECKLIST[21] (continued)

4. Building type of construction ______________
   Structural framework ______________
      Number of stories ______________
      Max. height ______________
      Area/Floor ______________
   Basement(s) classification
      Type ______________
      Max. height ______________
      Area/Floor ______________

5. Exit requirements
   No. of occupants/floor ______________
   No. of exits/floor ______________
      Unit/Exit width ______________
      Corridors ______________
   Width of exits
      Stairs ______________
      Doors ______________
      Ramps' width ______________
      Ramps' slope ______________
   Distance to exits ______________
   Dead ends ______________
   Barrier-free requirements ______________
   Door swings ______________
   Special requirements ______________

6. Fire protection
   Structural frame
      Columns ______________
      Floors ______________
      Ceilings ______________
      Walls and partitions ______________
         Exterior walls ______________
         Fire walls ______________
         Party walls ______________
         Exit enclosures ______________
         Rated enclosures ______________
         Shafts ______________
         Parapets ______________
         Partitions ______________
            Load bearing ______________
            Non-bearing ______________
            Corridors ______________

BUILDING AND FIRE REGULATIONS
CHECKLIST[21] (continued)

   Roof structure ______________
   Roof covering ______________
   Vertical openings ______________
   Stairways and smoke-
      proof enclosures ______________
   Doors, windows, glass ______________
   Skylights ______________
   Attic separations ______________
   Cantilevers ______________
   Fire extinguishers ______________
   Standpipes ______________

7. Plumbing
   Fixture count ______________

TOILET FIXTURE COUNT							
OCCUPANTS	WC		UR	LAV		SH	DF
EMPLOYEES	M	F		M	F		
TOTAL							

   Barrier-free requirements ______________

8. Remarks ______________
   ____

HEALTH REGULATIONS CHECKLIST

Department ______________
Division ______________
Inspector ______________
Telephone number ______________ Extension ___
Code section index ______________
1. Building type ______________
2. Heating, ventilating,
   and air conditioning ______________
3. Sewerage, pollution control,
   environmental impact ______________
4. Openings ______________
5. Trash collection ______________

HEALTH REGULATIONS CHECKLIST
(continued)

6. Room finishes
   Floor ___________________________
   Base ___________________________
   Walls ___________________________
   Ceilings ___________________________
   Fixed equipment ___________________________

7. Special requirements ___________________________

8. Remarks ___________________________
   ____

HIGHWAY AND TRAFFIC REGULATIONS
CHECKLIST

Department ___________________________
Division ___________________________
Inspector ___________________________
Telephone number ___________ Extension ___
Code section index ___________________________

1. Site location ___________________________

2. Storm sewer ___________________________

3. Curbs, walks and sidewalks ___________

4. Accesses (ingress and egress)
   Police ___________________________
   Fire ___________________________
   Emergency ___________________________
   Security/Controls ___________________________

5. Curb cuts ___________________________

6. Landscaping, planters and fencing _______

7. Median strips and cuts ___________________

8. Signs
   Directional ___________________________
   Traffic ___________________________

9. Off-street parking ___________________________

10. Remarks ___________________________
    ____

VARIANCES AND WAIVERS. Although this text advocates a clear definition of the professional design problem prior to commencing any portion of the design process, in cases involving variances and regulatory waivers it is advisable to present schematic drawings to illustrate plans for development in non-conforming situations, because in many cases it is then easier to determine the extent of a variance and in most hearings these schematic drawings are specifically required as documentation, since a variance is generally granted for a specific project and is valid only for the definite life of that structure.

In addition, cases involving zoning appeals or reclassifications usually require a master plan before a hearing is scheduled, because the waiver of regulations covering a project's existing zoning classification might be partially (or totally) modified only when documented by a proper layout of conditional uses, thus creating the need for a physical master plan as a physical statement of intent.

In these cases, the programmer's presence at hearings themselves can be extremely helpful to the client's legal representatives (with very few exceptions it is recommended that a legal representative always be retained for these hearings), since most technical questions will be directed to him/her as the only consultant qualified to answer them.

Because of the political and regional characteristics of these hearings, it is impossible to establish a definite procedure that would encompass all the necessary functions and tasks to be performed in these situations on a national basis. The main responsibility for these tasks lies with the clients, and with the programmer serving in the capacity of technical consultant. Still, we must not forget that these procedures essentially concern themselves with the scope and function of a project, and thus with the scope and functions outlined in that project's architectural program.

8.6 Temporal Analysis

GROWTH AND CHANGE PROJECTIONS. Growth and change projections, as their name indicates, are simply approximations based on existing data as related to the future development of operations.

Although these approximations represent only the most likely growth or change possibilities of specific functions, their establishment during the programming phase constitutes an essential part of the program's structure.

In order to arrive at a viable system of projections, we must first define what is to be encompassed in these determinations, and this can only be accomplished by defining growth and change as professional design concepts. What is growth from a professional design point of view? To answer this is to determine not only the different types of growth, but also their functional development.

Growth can be defined in a generic sense as "a gradual development towards an end." In each of the possible definitions of "growth" we will find the word "development" as an essential constituent. Growth, therefore, is development "towards" or "with" purpose and direction.

Unfortunately, growth cannot be construed as a multiplying factor which will invariably affect some or all portions of a building. A computer room might grow by simply altering the computer models contained within a constant spatial perimeter, while this growth in capacity would DECREASE the human element necessary to effect the operations. Within the growth concept, changes and reductions could also be contemplated. By the same token, the necessary growth projections for a specific project can very well determine future spatial subdivisions of functions, which can lead to programmatic changes and the need for flexibility or alterations.

Spatial growth might or might not mean operational growth. Increasing the size of a warehouse does not necessarily mean that all the other operations of an industrial complex will increase in the same proportion or in any proportion at all, since the capacity, volume, costs, etc. of all elements contained within the new facilities could simply represent a larger figure in the same set of accounting records.

Still, spatial growth alone can be considered the major determinant in architectural programming, since all other kinds of growth will most likely fall within the change projections concept. Therefore, we will only concern ourselves with spatial growth in determining the different categories assigned to this projections procedure. Spatial growth can be classified according to the following:

Type
    Arithmetic
    Geometric

Location
    Site, Region, etc.
    Building, Structures,
    Services, Parking, etc.

Placement & Direction
    Adjacent
        Horizontal
        Vertical
    Remote

Functional Performance
    Increases
    Reductions
    Allocations
    Distributions

Most of the data involved in determining these approximations can be readily provided by the client, since the factors determining the growth pattern have been witnessed, and generally recorded, throughout the operation's history.

In spite of this, it is often necessary to refer to other sources for statistical growth and change data which will clarify, justify or facilitate the development of proper POPULATION projections, since this is a specialized issue which may not be made readily available by the client. In these instances, several techniques and methodologies now widely used in statistical research can be applied. For more detailed descriptions of these techniques refer to Appendix E.

No matter which approach is employed, the type of progression the growth pattern has followed should be established at the outset; that is, whether the ratio of increase has been arithmetic (also linear) or geometric (also exponential). The difference between the two will determine whether elements should be added or multiplied as projections for a

SPATIAL PROJECTIONS DATA SHEET

PROJECT NAME	PREPARED
LOCATION	CHECKED
ARCHITECT	APPROVED

PROJECT Nº

DATE

PAGE

DIV.	COMPONENT	F	CONSTITUTIVE ELEMENTS	F	LOCATION	MARK/ ROOM*	YEAR			YEAR			YEAR			YEAR			YEAR			GROWTH			CHANGES	
							# occ.	s.f. occ.	AREA	# occ	s f occ	AREA	# occ	s f occ	AREA	# occ	s f occ	AREA	# occ	s f occ	AREA	TYPE	PLACEMENT	L/D	E	VOLUME
TOTALS																										

REMARKS

Figure 8.37

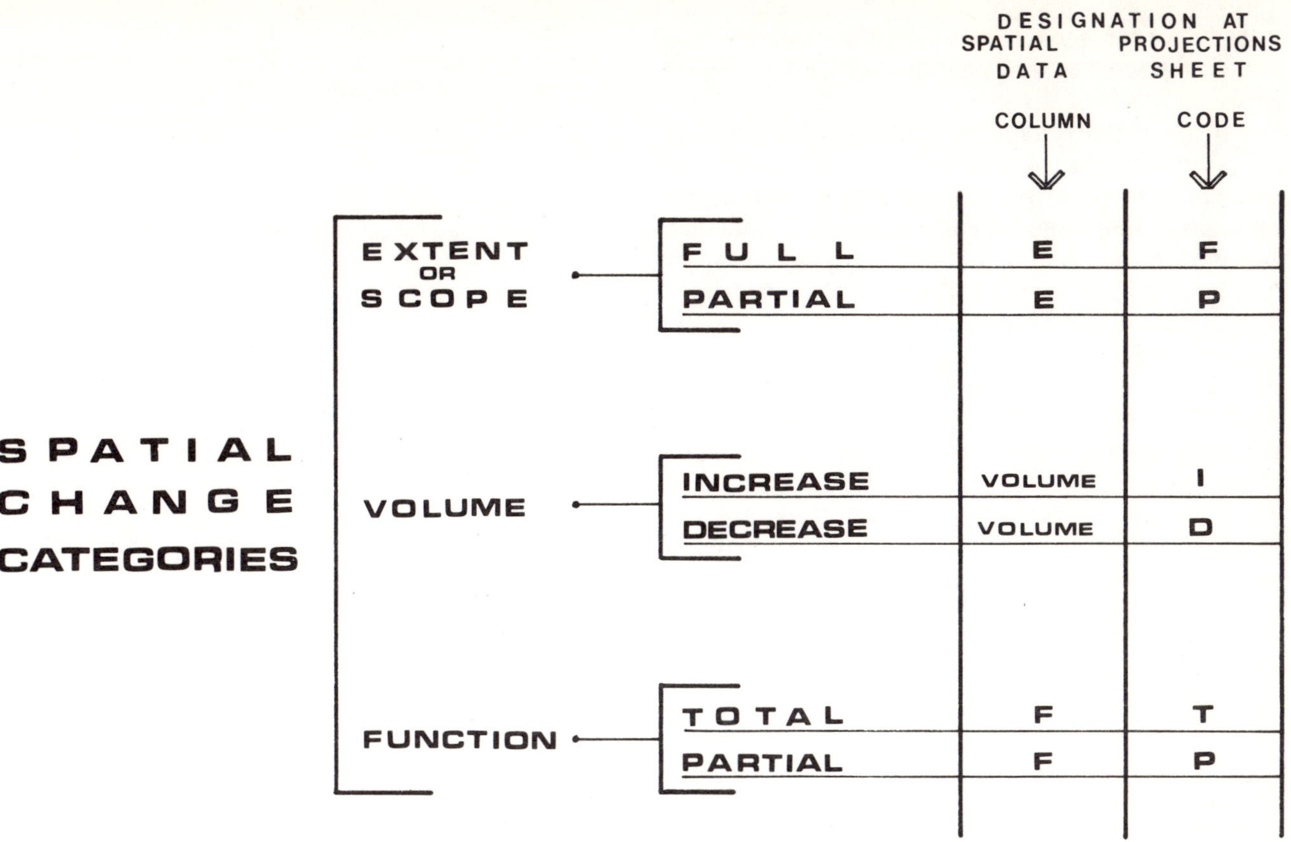

Figure 8.38

definite time span are established. In order to define the type of growth, we must determine whether it will occur within a specific site or whether the site parameters as currently developed will increase, in addition to growth within the building, in its services, etc.

The location and direction of any contemplated growth is also an essential component of its definition; that is, the determination of whether it will take place adjacent to a project, a project's site, or some distance removed.

Furthermore, functional performance can clarify not only possible restructurings of operational tasks and functions, but also foreseeable increases, decreases, allocation and redistribution within a program.

A Spatial Projections Data Sheet containing the above-mentioned information can be established as shown in Figure 8.37.

In this chart, the area increases or decreases should be recorded as positive or negative values. The functional performance of each division, component and element is noted in the "function" columns and identified as essential, complementary and services just as in Figure 8.14. It is very important to note any reclassifications or changes of use in the changes column.

Spatial change can be divided into several categories as illustrated in Figure 8.38. All of them are directly related to growth patterns, economic requirements, conditional uses.

Change alone, if it represents a complete redistribution of functional requirements, cannot be projected without being re-programmed. However, possible or foreseeable changes of a moderate type should be noted in the "changes" column of Figure 8.37.

If the extent or function of a specific change projection becomes total at a certain date, it is necessary to note the date or year by the corresponding "T" in column F at Figure 8.37, since this will determine at what point a reclassification of functional requirements will become necessary.

Growth and change projections are never exact figures. Usually they will represent a logical pattern of developments within projected confines of present conditions based on the data provided by a past performance which might or might not be indicative of what is to come. In spite of these possible inaccuracies, however, such projections are our only viable means of predicting future developments with the least amount of guesswork, complex mathematical procedures, or crystal balls.

No one can predict the future. The only thing we can be sure of encountering is change. That is probably the reason why we "plan" for it, and the basic motive behind our temporal analysis and its logical derivatives, growth, development, and change.

---

[17]As a rule of thumb the difference between a static element and a dynamic one is a percentage of operational time in a specific area. Dynamic elements are those which spend more than 25% of their operational time outside a specific location while still performing their functions. Transient elements are those not operating from any assigned location.

[18]Economic Feasibility Studies, although their scope could be considered a programming function in certain cases, are a specialized section which requires great financial expertise and should be executed by specialized consultants. A simplified example of an economic feasibility study can be found in Appendix D.

[19]Circulation factors which generally rank between 18% and 30% (multiplication factors or 1.2 thru 1.4) are generally added to the single occupant net area to actually determine the "standard net assignable area" figure. Refer to AIA Document D101 "The Architectural Area and Volume of Buildings."

[20]Based on Report No. 50 for the Federal Construction Council, Publication 1235, National Academy of Sciences, 1964, Task Group T-56.

[21]For simplifications and organization of these surveys refer to Subchapter 10.1.

9
Case Studies

The following case studies have been selected for their particular characteristics as illustrative examples of the procedures outlined in this text.

In the first case study, all steps and processes followed to arrive at a final program format will be covered, while the second case study consists basically of an introduction to the problem and a presentation of the final program format, since, as will become apparent, the format itself contains the necessary statements of procedure and methodologies employed in its development.

9.1 Case Study No. 1

Additions to and Remodelings of a Small Manufacturing Plant.

DEFINITION. A Small Manufacturing Plant of Outdoor Furniture and Playground Equipment.

PURPOSE. To provide adequate facilities for the continuation and development of this growing operation.

STEP NO. 1 — OBJECTIVES DEFINITION. After a preliminary analysis of the primary causes behind the project's statement of purpose, the following factors emerge as programmatic determinants:

1. Size and volume of the operation
2. Age and condition of existing building
3. Operational fragmentations were being contemplated
4. Upgrading of several production systems was being considered.

After this, a primary checklist of compliance factors reveals several existing code violations, several hazardous conditions within the complex and the need for project phasing in order to maintain the plant in full operation while the expansion is carried out. For a full layout of the results obtained in this first step, refer to the final program layout outlined in Step No. 4.

STEP NO. 2 — ANALYSIS OF EXISTING CONDITIONS. The detailed analysis of prevailing conditions covers a general study of the environmental conditions through the use of data collection forms as illustrated in Figures 9.1, 9.2 and 9.3, and a detailed description of the artificial conditions affecting the site. The structural and physical characteristics are covered by boring reports and a site survey.

The analysis of existing conditions regarding the spatial layout of divisional relationships, spatial components and their interactions is carried out by means of diagrammatic outlines and data collection forms as illustrated in Figures 9.4, 9.5, 9.6 and 9.7.

A regulatory survey executed at this point is only concerned with the compliance or non-compliance with code requirements in order to determine actual code violations affecting the project. (Figure 9.8 partially illustrates this process.)

The temporal analysis at this point is limited to a simple recording of the company's history.

STEP NO. 3 — ADDITIONS AND REMODELING PROGRAMMING. The spatial programming of this project is executed through a programming of the operational interaction on a divisional basis as illustrated in Figure 9.9.

Following this step, an actual programming of functional components and constitutive elements is determined (see Figures 9.9, 9.10 and 9.11).

At this point, a preliminary budget study based on the information previously established and a detailed regulatory survey are executed. Figures 9.12 and 9.13 partially illustrate the results of these studies.

Finally, based on the previous investigations regarding the operational development of the company, as well as the projections of its divisional reorganization, a summary of spatial projections is established (see Figure 9.14).

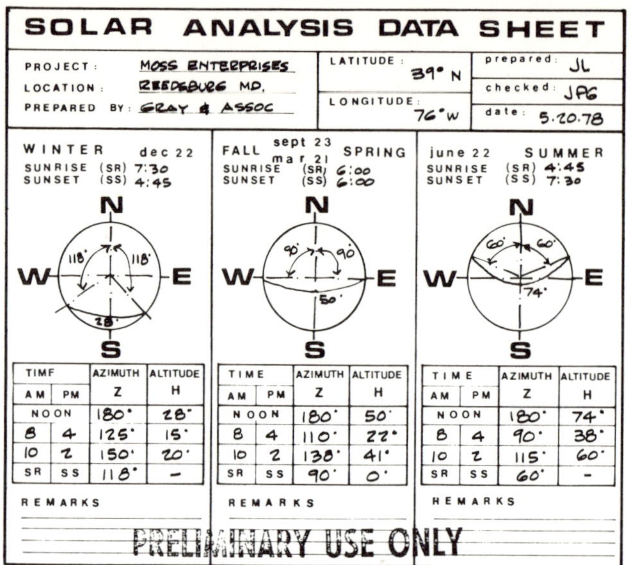

WINTER dec 22 — SUNRISE (SR) 7:30, SUNSET (SS) 4:45

TIME		AZIMUTH Z	ALTITUDE H
AM	PM		
NOON		180°	28°
8	4	125°	15°
10	2	150°	20°
SR	SS	118°	—

FALL sept 23 / mar 21 SPRING — SUNRISE (SR) 6:00, SUNSET (SS) 6:00

TIME		AZIMUTH Z	ALTITUDE H
AM	PM		
NOON		180°	50°
8	4	110°	22°
10	2	138°	41°
SR	SS	90°	0°

june 22 SUMMER — SUNRISE (SR) 4:45, SUNSET (SS) 7:30

TIME		AZIMUTH Z	ALTITUDE H
AM	PM		
NOON		180°	74°
8	4	90°	38°
10	2	115°	60°
SR	SS	60°	—

REMARKS

PRELIMINARY USE ONLY

Figure 9.1

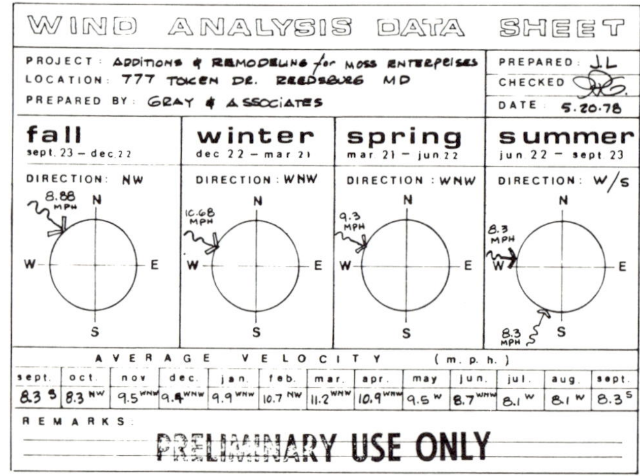

AVERAGE VELOCITY (m.p.h.)

sept.	oct.	nov.	dec.	jan.	feb.	mar.	apr.	may	jun.	jul.	aug.	sept.
8.3 S	8.3 NW	9.5 WNW	9.4 WNW	9.9 WNW	10.7 NW	11.2 WNW	10.9 WNW	9.5 W	8.7 WNW	8.1 W	8.1 W	8.3 S

REMARKS:

PRELIMINARY USE ONLY

Figure 9.2

TEMPERATURE DATA SHEET

	sept.	oct.	nov.	dec.	jan.	feb.	mar.	apr.	may	jun.	jul.	aug.
min. monthly total	58.4	46.3	36.3	27.8	24.2	26.3	33.3	42.8	52.2	61.6	66.5	65.4
max. monthly total	78.8	67.8	55.6	44.7	41.2	44.5	52.9	65	74	82.8	87	85.3
average monthly	68.6	57.1	46	36.3	32.7	35.4	43.1	53.9	63.1	72.2	76.8	75.4

REMARKS:

PRELIMINARY USE ONLY

Figure 9.3

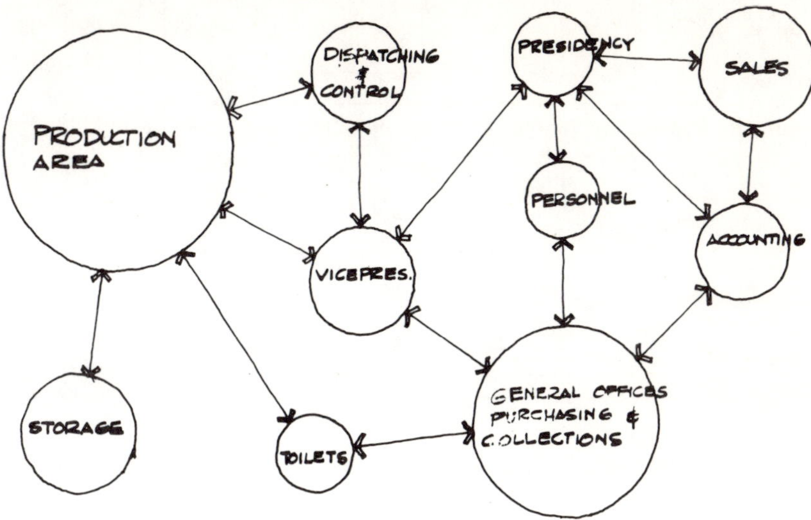

Figure 9.5

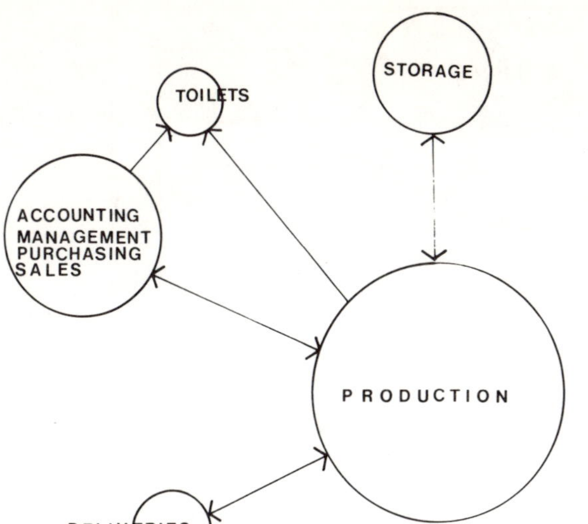

EXISTING DIVISIONAL RELATIONSHIPS DIAGRAM

Figure 9.4

SPATIAL COMPONENTS ANALYSIS DATA SHEET/ EXISTING FACILITIES																			JOB NO. 7804
PROJECT INDUSTRIAL BUILDING FOR MOSS ENTERPRISES — PREPARED J.L — DATE 5-25-78																			
LOCATION 777 TOKEN DR REEDSBURG. MD — CHECKED — PAGE 1 OF 1																			
ARCHITECT GRAY & ASSOCIATES INC — APPROVED																			
DIV.	COMPONENT DESCRIPTION	FUNCT.	CONSTITUTIVE ELEMENTS	FUNCT.	CLASS	ACTIVITY DESCRIPTION	LOCATION	ROOM Nº	Nº OF OCC. M	Nº OF OCC. F	SQ.FT./OCC	AREA sq. ft.	VOL. cu. ft.	EQUIP.	MECH. & ELEC.	ENVIROMT.	FT	OTHER (SPECIFY)	REMARKS
management, accounting, purchasing & sales	presidency	E	president	E	S	management	office	n.a.	1		−	150	1500				8		
	vicepresidency	E	vicepresident	E	S	management	office		1		−	150	1500				8		
	accounting	E	accountant	E	S	accounting	office			1	−	120	1200				8		
	general offices, purchasing & collections		secretaries	E	S	clerical			6	60		360	3600				8		
			purchaser	E	S	purchasing					−	64	642						
		E	employees	S	S	lounge					−	220	2200						
			files & storage	C	S	filing					−	75	750						
			officer	E	S	collections					−	75	750						
	personnel	E	director	E	T	personel				1	−	80	800						
	sales	E	salesman	E	T	sales	−		1		−	NONE	−				4		
			"	E	T	"	−		1		−	"	−				6		
			"	E	T	" .	−		1		−	"	−				6		
services	toilets & services	S	employees toilets	S	T	toilets	toilet rooms					200	1620				12		males 1wc 1ur 2 lavs / NOT TO females 2 wc 2 lavs / CODE
			men lockers	S	T	lockers	locker room					188	1880				12		
			storage	C	T	storage	storage room					100	1012				12		
production & deliveries	dispatching/ control	E	foreman	E	D	production & dispatching	office										12		−
	production area	E	cutters	E	S	materials cutting	factory												
			assemblers	E	S	assemblage													
			casters	E	S	concrete casting							7000	77000						
			finishers	E	S	finishing														
			metal workers	E	S	production														
			trucks	C	D	deliveries	parking lot												
storage	storage	C	miscellaneous	C	S	storage	factory												
			tools & flammables	E	T														
			materials	E	D														

Figure 9.6

Figure 9.7 — SPATIAL PROJECTIONS DATA SHEET 1B

PROJECT NO: 7804
DATE: 5.29.78
PAGE: 1 of 2
PREPARED: J.L.
CHECKED: (signature)
APPROVED: (signature)
PROJECT NAME: INDUSTRIAL BLDG. for MOSS ENTERPRISES
LOCATION: 777 TOKEN DR., REEDSBURG MD.
ARCHITECT: GRAY & ASSOC.

DIV	COMPONENT	F	CONSTITUTIVE ELEMENTS	F	LOCATION	MARK/ROOM	YEAR 1974 # OCC/OCC	YEAR 1974 s.f.	YEAR 1974 AREA	YEAR 1978 # OCC/OCC	YEAR 1978 s.f.	YEAR 1978 AREA
MANAGEMENT, ACCOUNTING	PRESIDENCY	E	PRESIDENT	E	OFFICE	NA	1	100	100	1	150	150
	VICE PRES.	E	VICE PRES.	E			−	80	80	−	150	150
	ACCOUNTING	E	ACCOUNTANT	E			0	0	0	1	120	120
			SECRETARIES	E			2	50	100	6	60	360
PURCHASING & SALES	GENERAL OFFICES		PURCHASER	E			0	0	0	1	64	64
	PURCHASING	E	EMPLOYEES GENERAL	S			0	0	0	10†	22	220
	Collections		FILES/STOR	C			10	NA	20	0	NA	75
			OFFICER	E			0	0	0	1	75	75
	PERSONNEL	E	DIRECTOR	E	→		0	0	0	1	80	80
			SALESMAN 1	E	−		−	60	60	1	0	0
	SALES	E	SALESMAN 2	E	−		0.5	0	0	−	0	0
			SALESMAN 3	E	−		0	0	0	0.5	0	0
SERV.	TOILETS & SERVICES	S	EMPLOYEES TOILETS	S	TOILETS		(1)	50	50 EA	(1)		120 EA
			MEN LOCKERS	S	LOCKER ROOM		4	30	120	20	20	400
			STORAGE	C	STORAGE		0	NA	100	0	NA	200
PRODUCTION & DELIVERIES	DISPATCHING CONTROL	E	FOREMAN	E	OFFICE		1	0	0	1	60	60
	PRODUCTION	E	CUTTERS	E	FACTORY		2	NA	(2)	5	NA	(3)
			ASSEMBLERS	E			2			8		
			CASTERS	E			−			4		
			FINISHERS	E			1			4		
			MTL. WORKERS	E			0			2		

TOTALS

REMARKS
(1) AS RQD. BY CODE.
(2) ALL INCLUDED WITHIN FACTORY AREA.
(3) ALL INCLUDED WITHIN FACTORY/STORAGE AREA.

Figure 9.7

Figure 9.8 — REGULATORY SURVEY – EXISTING CONDITIONS

PROJECT TITLE: ADDITIONS & REMODELING TO MOSS ENTERPRISES
PROJECT NO: 7804
DATE: 5.26.78
PAGE: 1 of 4
LOCATION: 777 TOKEN DR. REEDSBURG.
PREPARED: JL
CHECKED: (signature)
APPROVED:
ARCHITECT: GRAY & ASSOCIATES

C = COMPLIANCE
NC = NON COMPLIANCE
NA = NOT APPLICABLE.

*CODE REFERENCE

			Result
1.	Zoning Classification		C
2.	Permitted Uses	Primary	C
		Accessory	
		Conditional	
3.	Minimum Lot Dimensions	Area	C
		Width	C
		Depth	C
4.	Setback Requirements	Front	C
		Side(s)	C
		Rear	C
		Other	C
5.	Distance Between Buildings	On one lot	C
		On adjacent lots	C
6.	Height Requirements (note elevation datum point)	Max allowable height	C
		Max number of stories	C
7.	Floor Area Ratio (F.A.R.)		NA
8.	Open Space Ratio (O.S.R.) & Lot Coverage		C
9.	Signs		C

CODE REFERENCE:
D ____
T ____
N ____
C ____

Figure 9.8

PRELIMINARY LAYOUT for DIVISIONAL RELATIONSHIPS
4 DELIVERIES
PURCHASING INVENTORY RECEIVING 5
1 MANAGEMENT
TOILETS
6 PRODUCTION
PROMOTION
2 ACCOUNTING & COLLECTIONS
3 SALES
TOILETS
BATHROOMS & LOCKERS

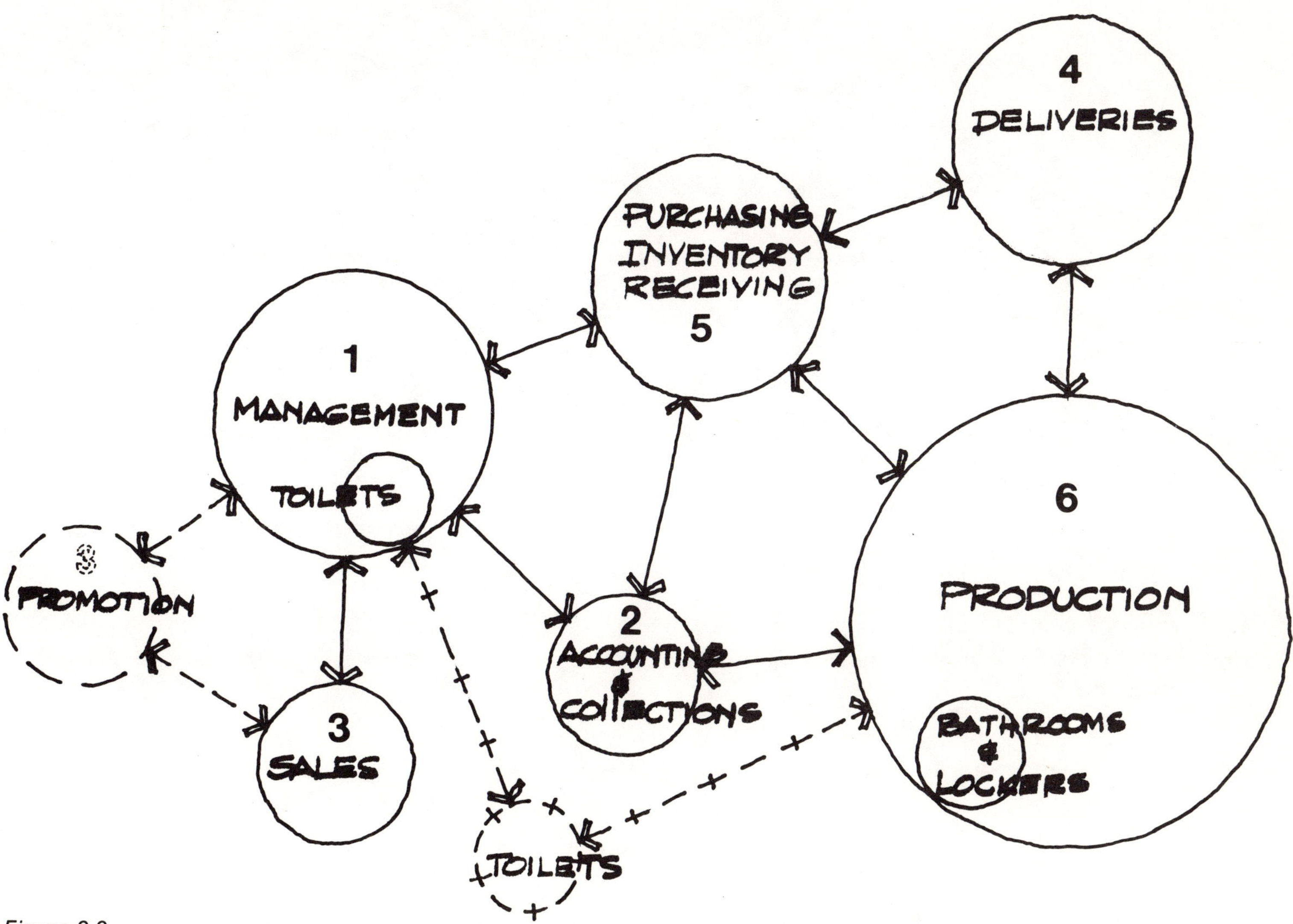

Figure 9.9

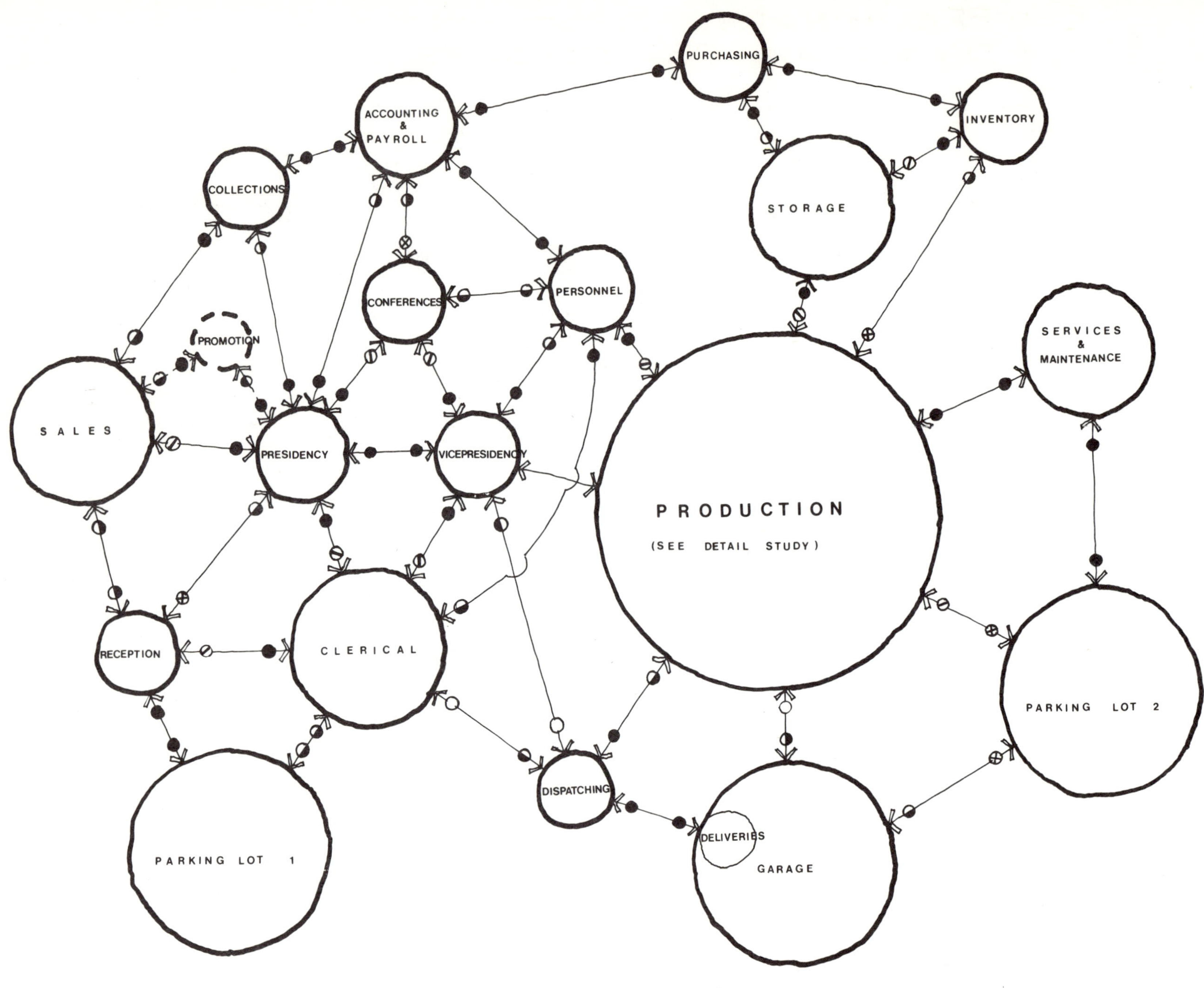

COMPONENTS SPATIAL RELATIONSHIPS DIAGRAM

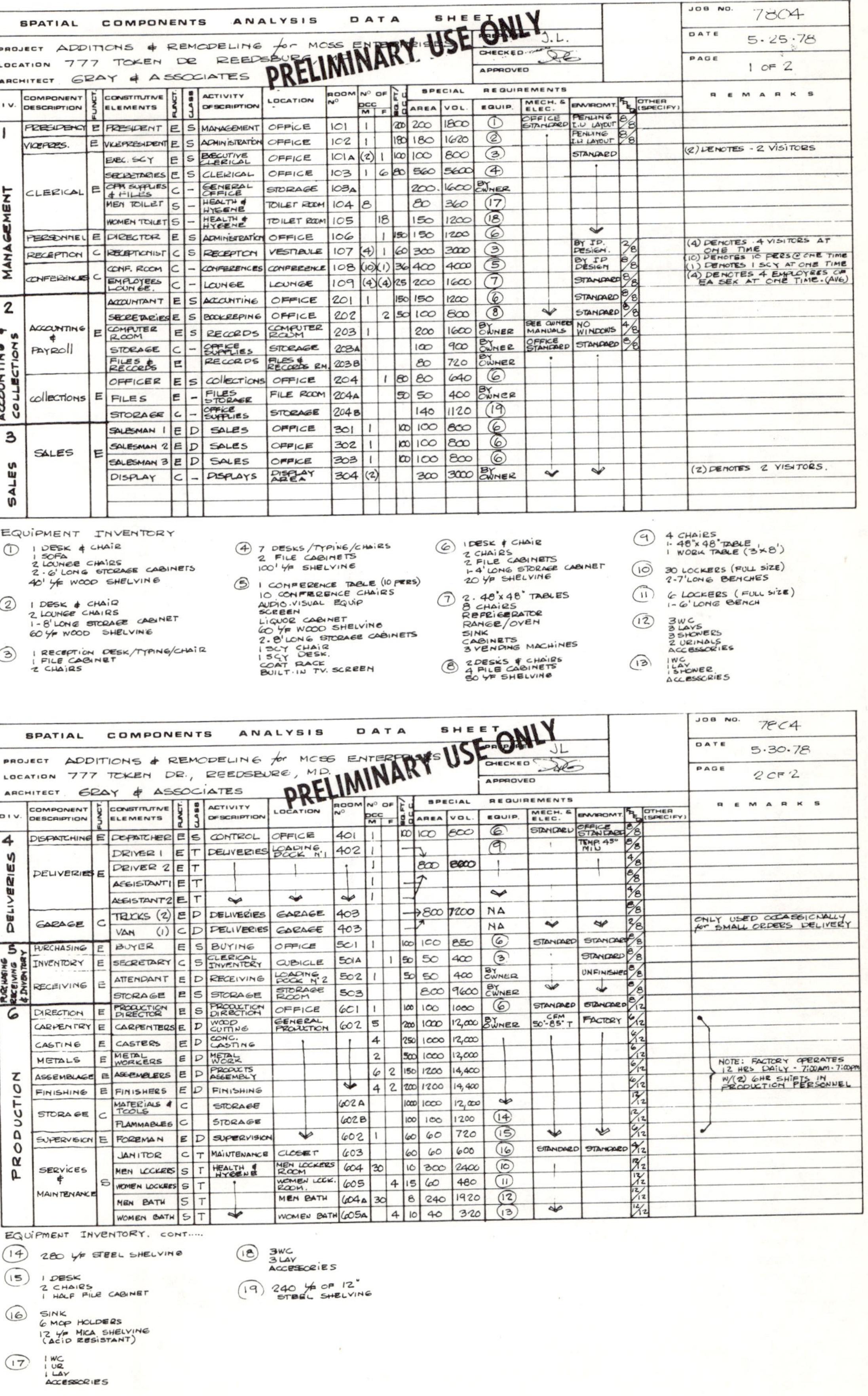

SPATIAL COMPONENTS ANALYSIS DATA SHEET — PRELIMINARY USE ONLY

JOB NO. 7804 DATE 5·25·78 PAGE 1 OF 2
PROJECT ADDITIONS & REMODELING for MOSS ENTERPRISES — CHECKED J.L.
LOCATION 777 TOKEN DR REEDSBURG
ARCHITECT GRAY & ASSOCIATES

DIV	COMPONENT DESCRIPTION	FNC	CONSTITUTIVE ELEMENTS	FNC	CL	ACTIVITY DESCRIPTION	LOCATION	ROOM N°	OCC M	OCC F	SQ.FT/OCC	AREA	VOL	EQUIP	MECH & ELEC	ENVIROMT	FRE	OTHER	REMARKS
1 MANAGEMENT	PRESIDENCY	E	PRESIDENT	E	S	MANAGEMENT	OFFICE	101	1		200	200	1800	(1)	OFFICE STANDARD	PENDING I.U LAYOUT	8/8		
	VICEPRES.	E	VICEPRESIDENT	E	S	ADMINISTRATION	OFFICE	102	1		180	180	1620	(2)		PENDING I.U LAYOUT	8/8		(2) DENOTES - 2 VISITORS
			EXEC. SCY	E	S	EXECUTIVE CLERICAL	OFFICE	101A	(2)	1	100	100	800	(3)		STANDARD			
	CLERICAL	E	SECRETARIES	E	S	CLERICAL	OFFICE	103	1	6	80	560	5600	(4)					
			OFR SUPPLIES & FILES	C	–	GENERAL OFFICE	STORAGE	103A				200	1600	BY OWNER					
			MEN TOILET	S	–	HEALTH & HYGENE	TOILET ROOM	104	8			80	360	(17)					
			WOMEN TOILET	S	–	HEALTH & HYGENE	TOILET ROOM	105		18		150	1200	(18)					
	PERSONNEL	E	DIRECTOR	E	S	ADMINISTRATION	OFFICE	106		1	150	150	1200	(6)					
	RECEPTION	C	RECEPTIONIST	C	S	RECEPTION	VESTIBULE	107	(4)	1	60	300	3000	(3)		BY I.D. DESIGN	2/8		(4) DENOTES 4 VISITORS AT ONE TIME
	CONFERENCES	C	CONF. ROOM	C	–	CONFERENCES	CONFERENCE	108	(10)	(1)	36	400	4000	(5)		BY I.D. DESIGN	8/8		(10) DENOTES 10 PERS @ ONE TIME / (1) DENOTES 1 SCY AT ONE TIME
			EMPLOYEES LOUNGE	C	–	LOUNGE	LOUNGE	109	(4)	(4)	25	200	1600	(7)		STANDARD	8/8		(4) DENOTES 4 EMPLOYEES OF EA SEX AT ONE TIME (AVG)
2 ACCOUNTING & COLLECTIONS	ACCOUNTING & PAYROLL	E	ACCOUNTANT	E	S	ACCOUNTING	OFFICE	201	1		150	150	1200	(6)		STANDARD	8/8		
			SECRETARIES	E	S	BOOKKEEPING	OFFICE	202		2	50	100	800	(8)		STANDARD	8/8		
			COMPUTER ROOM	E	S	RECORDS	COMPUTER ROOM	203	1			200	1600	BY OWNER	SEE OWNER MANUALS	NO WINDOWS	4/8		
			STORAGE	C	–	OFFICE SUPPLIES	STORAGE	203A				100	900	BY OWNER	OFFICE STANDARD	STANDARD	8/8		
			FILES & RECORDS	E		RECORDS	FILES & RECORDS RM.	203B				80	720	BY OWNER					
	COLLECTIONS	E	OFFICER	E	S	COLLECTIONS	OFFICE	204		1	80	80	640	(6)					
			FILES	E	–	FILES STORAGE	FILE ROOM	204A			50	50	400	BY OWNER					
			STORAGE	C	–	OFFICE SUPPLIES	STORAGE	204B				140	1120	(19)					
3 SALES	SALES	E	SALESMAN 1	E	D	SALES	OFFICE	301	1		100	100	800	(6)					
			SALESMAN 2	E	D	SALES	OFFICE	302	1		100	100	800	(6)					
			SALESMAN 3	E	D	SALES	OFFICE	303	1		100	100	800	(6)					
			DISPLAY	C	–	DISPLAYS	DISPLAY AREA	304	(2)			300	3000	BY OWNER					(2) DENOTES 2 VISITORS.

EQUIPMENT INVENTORY

(1) 1 DESK & CHAIR / 1 SOFA / 2 LOUNGE CHAIRS / 2·6' LONG STORAGE CABINETS / 40' Y/F WOOD SHELVING

(2) 1 DESK & CHAIR / 2 LOUNGE CHAIRS / 1-8' LONG STORAGE CABINET / 60 Y/F WOOD SHELVING

(3) 1 RECEPTION DESK/TYPING/CHAIR / 1 FILE CABINET / 2 CHAIRS

(4) 7 DESKS/TYPING/CHAIRS / 2 FILE CABINETS / 100' Y/F SHELVING

(5) 1 CONFERENCE TABLE (10 PERS) / 10 CONFERENCE CHAIRS / AUDIO·VISUAL EQUIP / SCREEN / LIQUOR CABINET / 60 Y/F WOOD SHELVING / 2·8' LONG STORAGE CABINETS / 1 SCY CHAIR / 1 SCY DESK / COAT RACK / BUILT-IN TV SCREEN

(6) 1 DESK & CHAIR / 2 CHAIRS / 2 FILE CABINETS / 1-4' LONG STORAGE CABINET / 20 Y/F SHELVING

(7) 2·48"x48" TABLES / 8 CHAIRS / REFRIGERATOR / RANGE/OVEN / SINK / CABINETS / 3 VENDING MACHINES

(8) 2 DESKS & CHAIRS / 4 FILE CABINETS / 50 Y/F SHELVING

(9) 4 CHAIRS / 1·48"x48" TABLE / 1 WORK TABLE (3'x8')

(10) 30 LOCKERS (FULL SIZE) / 2-7' LONG BENCHES

(11) 6 LOCKERS (FULL SIZE) / 1-6' LONG BENCH

(12) 3 WC / 3 LAV / 3 SHOWERS / 2 URINALS / ACCESSORIES

(13) 1 WC / 1 LAV / 1 SHOWER / ACCESSORIES

SPATIAL COMPONENTS ANALYSIS DATA SHEET — PRELIMINARY USE ONLY

JOB NO. 7804 DATE 5·30·78 PAGE 2 OF 2
PROJECT ADDITIONS & REMODELING for MOSS ENTERPRISES — CHECKED JL
LOCATION 777 TOKEN DR., REEDSBURG, MD.
ARCHITECT GRAY & ASSOCIATES

DIV	COMPONENT DESCRIPTION	FNC	CONSTITUTIVE ELEMENTS	FNC	CL	ACTIVITY DESCRIPTION	LOCATION	ROOM N°	OCC M	OCC F	SQ.FT/OCC	AREA	VOL	EQUIP	MECH & ELEC	ENVIROMT	FRE	OTHER	REMARKS
4 DELIVERIES	DISPATCHING	E	DISPATCHER	E	S	CONTROL	OFFICE	401	1		100	100	800	(6)	STANDARD	OFFICE STANDARD	8/8		
	DELIVERIES	E	DRIVER 1	E	T	DELIVERIES	LOADING DOCK N°1	402	1					(9)		TEMP 45° MIN	8/8		
			DRIVER 2	E	T				1			800	8000				4/8		
			ASSISTANT 1	E	T				1								8/8		
			ASSISTANT 2	E	T				1								4/8		
	GARAGE	C	TRUCKS (2)	E	D	DELIVERIES	GARAGE	403				800	7200	NA			8/8		ONLY USED OCCASIONALLY for SMALL ORDERS DELIVERY
			VAN (1)	C	D	DELIVERIES	GARAGE	403						NA			2/8		
5 PURCHASING RECEIVING & INVENTORY	PURCHASING	E	BUYER	E	S	BUYING	OFFICE	5C1	1		100	100	850	(6)	STANDARD	STANDARD	8/8		
	INVENTORY	E	SECRETARY	C	S	CLERICAL INVENTORY	CUBICLE	501A		1	50	50	400	(3)		STANDARD	8/8		
	RECEIVING	E	ATTENDANT	E	D	RECEIVING	LOADING DOCK N°2	502	1		50	50	400	BY OWNER		UNFINISHED	8/8		
			STORAGE	E	S	STORAGE	STORAGE ROOM	503				800	9600	BY OWNER			8/8		
6 PRODUCTION	DIRECTION	E	PRODUCTION DIRECTOR	E	S	PRODUCTION DIRECTION	OFFICE	601	1		100	100	1000	(6)	STANDARD	STANDARD	6/12		
	CARPENTRY	E	CARPENTERS	E	D	WOOD CUTTING	GENERAL PRODUCTION	602	5		200	1000	12,000	BY OWNER	CFM 50°-85° T	FACTORY	6/12		NOTE: FACTORY OPERATES 12 HRS DAILY - 7:00AM - 7:00PM W/(2) 6HR SHIFTS IN PRODUCTION PERSONNEL
	CASTING	E	CASTERS	E	D	CONC. CASTING			4		250	1000	12,000				6/12		
	METALS	E	METAL WORKERS	E	D	METAL WORK			2		500	1000	12,000				6/12		
	ASSEMBLAGE	E	ASSEMBLERS	E	D	PRODUCTS ASSEMBLY			6	2	150	1200	14,400				6/12		
	FINISHING	E	FINISHERS	E	D	FINISHING			4	2	200	1200	14,400				6/12		
	STORAGE	C	MATERIALS & TOOLS	C		STORAGE		602A				1000	12,000				12/12		
			FLAMMABLES	C		STORAGE		602B				100	1200	(14)			6/12		
	SUPERVISION	E	FOREMAN	E	D	SUPERVISION		602	1		60	60	720	(15)			6/12		
	SERVICES & MAINTENANCE	S	JANITOR	C	T	MAINTENANCE	CLOSET	603			60	60	600	(16)	STANDARD	STANDARD	6/12		
			MEN LOCKERS	S	T	HEALTH & HYGENE	MEN LOCKERS ROOM	604	30		10	300	2400	(10)			12/12		
			WOMEN LOCKERS	S	T		WOMEN LCK. ROOM	605		4	15	60	480	(11)			12/12		
			MEN BATH	S	T		MEN BATH	604A	30		8	240	1920	(12)			12/12		
			WOMEN BATH	S	T		WOMEN BATH	605A		4	10	40	320	(13)			12/12		

EQUIPMENT INVENTORY. CONT.....

(14) 280 Y/F STEEL SHELVING

(15) 1 DESK / 2 CHAIRS / 1 HALF FILE CABINET

(16) SINK / 6 MOP HOLDERS / 12 Y/F MICA SHELVING (ACID RESISTANT)

(17) 1 WC / 1 UR / 1 LAV / ACCESSORIES

(18) 3 WC / 3 LAV / ACCESSORIES

(19) 240 Y/F OR 12" STEEL SHELVING

Figure 9.11

BUDGET ESTIMATE FORM		PROJECT NO. 7804
PROJECT INDUSTRIAL BLDG. for MOSS ENTERPRISES		DATE 5-30-78
LOCATION REEDSBURG MD.		PREPARED BY J.L & T.W
GROSS AREA 17620 · NET AREA 14530 · BLDG TYPE IND.		EFFICIENCY 80%
ITEM		COST
A	BUILDING COST (GROSS AREA x SQ.FT. UNIT COST)	300,000.00
B	SITE DEVELOPMENT	35,000.00
C	FIXED EQUIPMENT	115,000.00
D	HEATING VENTILATING & AIR CONDITIONING	49,000.00
E	PLUMBING & FIRE PROTECTION	25,000.00
F	ELECTRICAL	41,000.00
G	TOTAL BUILDING COSTS A+B+C+D+E+F=G SUBTOTAL 1	565,000.00
H	SITE ACQUISITION	NA (1)
I	UNUSUAL SITE CONDITIONS	-0-
J	OFF SITE WORK	-0-
K	GROWTH & CHANGE PROVISIONS	35,000.00
L	LANDSCAPING	5,000.00
M	FURNISHINGS	32,000.00
N	ALLOWANCES / CONTINGENCIES	60,000.00
O	MISCELLANEOUS	5,000.00
P	ADDITIONAL BUDGET ITEMS SUBTOTAL 2	137,000.00
Q	PROFESSIONAL FEES	34,000.00
R	SURVEYS AND INSURANCE	2,500.00
S	LEGAL, ACCOUNTING AND ADMINISTRATIVE COSTS	10,000.00
T	LEASING, ADVERTISING AND PROMOTION	-0-
U	FINANCING COSTS AND TAXES	17,000.00
V	OWNER'S BUDGET ITEMS SUBTOTAL 3	63,500.00
W	TOTAL BUDGET REQUIRED G+P+V=W	$765,500.00 USE $770,000.00

NOTES:
(1) COST OF PARCELS A & B NOT AVAILABLE AT PRESENT TIME. OWNER INDICATES TRANSACTION WOULD TAKE PLACE INDEPENDENTLY OF THIS ANALYSIS SHOULD ACQUISITION BECOME NECCESSARY.

PRELIMINARY USE ONLY

Figure 9.12

4. Type of Construction
- Floors
  - Type: NON-COMBUSTIBLE (NC) - 2HR
  - Max Height: 2 STORIES
  - Area/Floor: 20,000
- Basement(s) Classification
  - Type: NA
  - Max Height: NA
  - Area/Floor: NA

5. Exit Requirements
- No. of Occupants/Floor: 266
- No. of Exits/Floor: 4
  - Unit/Exit Width: 40"/PER - UNSPRINKLERED, 50"/PER - SPRINKLERED
  - Corridors: MIN. 4'0"
- Widths of Exits
  - Stairs: 4'0"
  - Doors: 3'4"
  - Ramps Width: 4'0"
  - Ramps Slope: 1:12 W/5' LANDINGS @ EA 30'
- Distance to exits: 100' UNSPRINKLERED 150' SPRINKLERED
- Dead Ends: MAX. 20'
- Barrier Free Requirements: SEE SECTS 52.04 & 55.07
- Door Swings: EXITS - DIRECTION OF TRAVEL
- Special Requirements: —

6. Fire Protection
- Structural Frame
  - Columns: NC-2
  - Floors: NC-2
  - Ceilings: -
  - Walls & Partitions: NC-0
    - Exterior Walls: NC-2
    - Fire Walls: 4HR
    - Party Walls: NA
    - Exit Enclosures: 2HR
    - Rated Enclosures: 2HR
    - Shafts: 2HR
    - Parapets: 2HR
    - Partitions: 0
      - Load Bearing: 1HR
      - Non Bearing: 0
      - Corridors: 1HR
  - Roof Structure: NC-1
  - Roof Covering: CLASS A
  - Vertical Openings: 2HR
  - Stairways & Smoke Proof Enclosures: 2HR
  - Doors, Windows, Glass: NA
  - Skylights: NA
  - Attic Separations: 2HR
  - Cantilevers: NA
  - Fire Extinguishers: 1/2500 SQ.FT
  - Standpipes: NONE RQD.

PRELIMINARY USE ONLY

3

Figure 9.13

SPATIAL PROJECTIONS PRELIMINARY USE ONLY

PROJECT NAME ADDITIONS & REMODELING for MOSS ENTERPRISES
LOCATION 777 TOKEN DR. REEDSBURG MD.
ARCHITECT GRAY & ASSOCIATES

PREPARED J.L.
CHECKED [signature]
APPROVED

PROJECT No 7804
DATE 6-4-78
PAGE 1 OF 3

DIV.	COMPONENT	F	CONSTITUTIVE ELEMENTS	F	LOCATION	MARK/ROOM*	1979 # occ	1979 s.f. occ.	1979 AREA	1985 # occ	1985 s f occ	1985 AREA	1990 # occ	1990 s f occ	1990 AREA	1995 # occ	1995 s f occ	1995 AREA	2000 # occ	2000 s f occ	2000 AREA	GROWTH TYPE	GROWTH PLACEMENT	GROWTH L/D	CHANGES E	CHANGES VOLUME	CHANGES F
1 MANAGEMENT	PRESIDENCY	E	PRESIDENT	E	OFFICE	101	1	200	200												→	—	BUILDING	—	—	—	—
	VICE PRESIDENCY	E	VICE PRESIDENT	E	OFFICE	102	1	180	180												→	—		—	—	—	—
	CLERICAL	E	EX. SCY	E	OFFICE	101A	3	30	100	6	30	200									→	—		ADJ/H	P	I	P
			SECRETARIES	E	OFFICE	103	7	80	560	10	80	800			→	12	100	1200			→	A		ADJ/H	—	I	—
			OFF. SUPPLIES & FILES	C	STORAGE	103A	—	—	200	—	—	250	—	—	300			→	—	—	350	A		ADJ/H	—	I	—
			MEN TOILET	S	TOILET RM	104	(8)	—	80	(10)	—	80									→	—		—	—	—	—
			WOMEN TOILET	S	TOILET RM	105	(8)	—	150	(22)	—	150	(24)	—	150	(30)	—	200			→	—		ADJ/H	—	I	—
	PERSONNEL	E	DIRECTOR	E	OFFICE	106	1	150	150												→	—		—	—	—	—
	RECEPTION	C	RECEPTIONIST	C	VESTIBULE	107	5	60	300												→	—		—	—	—	—
	CONFERENCES	C	CONF. ROOM	C	CONFERENCE	108	(11)	36	400												→	—		—	—	—	—
			EMPLOYEES LOUNGE	C	LOUNGE	109	(8)	25	200			→	(12)	25	300						→	—		ADJ/H	—	I	—
2 ACCOUNTING & COLLECTIONS	ACCOUNTING & PAYROLL	E	ACCOUNTANT	E	OFFICE	201	1	150	150												→	—		—	—	—	—
			SECRETARIES	E	OFFICE	202	2	50	100									→	3	50	150	—		ADJ/H	—	I	—
			COMPUTER	E	COMPUTER ROOM	203	1	—	200												→	—		—	—	—	—
			STORAGE	C	STORAGE RM	203A	—	—	100												→	—		—	—	—	—
			FILES & RECORDS	E	RECORDS ROOM	203B	—	—	80			→	—	—	100			→	—	—	120	A		ADJ/H	—	I	—
	COLLECTIONS	E	OFFICER	E	OFFICE	204	1	80	80												→	—		—	—	—	P
			SECRETARY	C	CUBICLE	204C	NA	→		1	50	50									→	—		ADJ/H	L	—	P
			FILES	E	FILE ROOM	204A	—	—	50						→	—	—	80			→	—		—	—	I	—
			STORAGE	C	STORAGE RM.	204B	—	—	140						→	—	—	110			→	—		—	—	D	—
																						—	↓	—	—	—	—
TOTALS							23		3420	40		3860	40		4030	42		4480	43		4600						

REMARKS
1. CIRCLED NUMBERS CANNOT BE ADDED IN THE "TOTALS" CALCULATIONS SINCE THEY ARE INCLUDED IN DEPARTAMENTAL COUNTS
2. TOILET ROOMS REQUIREMENTS MIGHT VARY DEPENDING ON FUTURE CODE REQUIREMENTS

Figure 9.14

STEP NO. 4 — PROGRAMMING REQUIREMENTS. Program requirements for Moss Enterprises, Inc., Project No. 7804, Date: August, 1978. *Purpose:* to provide adequate facilities for the growing operation of this outdoor furniture and playground equipment manufacturing plant.

1. Objectives definition.

  1.1 *Natural:* existing site/see accompanying survey (not included in this text) with the possibility of Parcels A and B to be included within the project limits. Greater protection from weather than presently available is desired; seasonal variations in activities and workload, increased volume during spring and summer months; energy conservation should be considered a determining factor in design approach; functional placement of elements as they now exist is not satisfactory; accesses and site circulation are presently acceptable, although improvements in parking facilities should be considered; temperature, light and sound should be considered standard for the building type; sewer, water and power are presently available at the site; polluting waste of the operation must be controlled.

  1.2 *Human:* activities performed within the project involve the manufacturing, sales, distribution, maintenance and repair of outdoor furniture and playground equipment, as well as related functions of management, bookkeeping and purchasing which are necessary to the operation. Human elements, equipment and products interact within the complex; no change in use for the new facilities is contemplated at this time; still, the need for flexibility is apparent since growth and change are indicated by future developments such as increased sales and production activities, and divisional relocations are foreseeable in the near future, namely in the case of the sales division. Existing personnel and manpower requirements appear satisfac-

tory; measurements and scale are to be compatible with the areas involved (see spatial analysis for further definition); the ratio of male/female employees varies from high in the production departments to low in the office and clerical areas; age group is between 20-45 years; great hazards are foreseeable mainly because of the use of paints, stains, etc., as well as heavy machinery, in the production department. Health and hygiene adequacy as well as security are imperative; barrier-free requirements necessary throughout; privacy is desired in the offices portion; building character and architectural style should be compatible with functions performed; owner favors modern and progressive image for building architecture, although creativity and innovative ideas or approaches are also encouraged.

  1.3 *Regulatory:* applicable codes — Building, Department of Health, Fire, Zoning (the facilities are presently operating under a variance grant which will not affect the development of the new building but could be reversed if new concessions are sought), Traffic and Highway.

  1.4 *Economic:* budgeting and economic studies are defined within the economic analysis segment. Building character and quality must not exceed limitations imposed by these requirements under any circumstances; aesthetics and image will inevitably become secondary issues as opposed to cost control and project affordability; lowest possible operating costs are desired; life cycle costs study not required.

  1.5 *Temporal:* scheduling and phasing are essential program elements, since the operation must continue as the new facilities are built. It is advisable to consider the seasonal character of this operation and to program construction during the winter months so that most major operational activities can be moved to the new building before or after the overload period of spring and summer occurs; change is projected through the intended transfer of the sales division to an off-site location in order to provide room for production expansion; although the company's growth within the past two years has been substantial, this growth ratio will gradually decrease as the new facilities become productive; yet, the need for room for future expansions on a departmental basis must be taken into account.

2. Site Analysis

  2.1 *Natural characteristics.*

    2.11 Structure: see boring reports (not included in this text).

    2.12 Physical: see survey (not included in this text).

    2.13 Environmental: climate can be considered near humid subtropical with relative humidity often reaching 100% and an average monthly temperature record of 87 degrees in July to 24.2 degrees in January; refer to the temperature, precipitation and snow-frost data included in this report for further explanatory material; for existing landscape and related data refer to the site survey. (See Figures 9.15, 9.16, 9.17 and 9.18).

  2.2 *Artificial conditions:* not applicable; covered within other segments of this document.

3. Spatial Analysis

  3.1 *Divisional fragmentation and relationships diagram:* the following diagram constitutes a graphic representation of the divisional relationships as analyzed and programmed for this project; the values indicated actually represent the required functional characteristics the operational relationships are

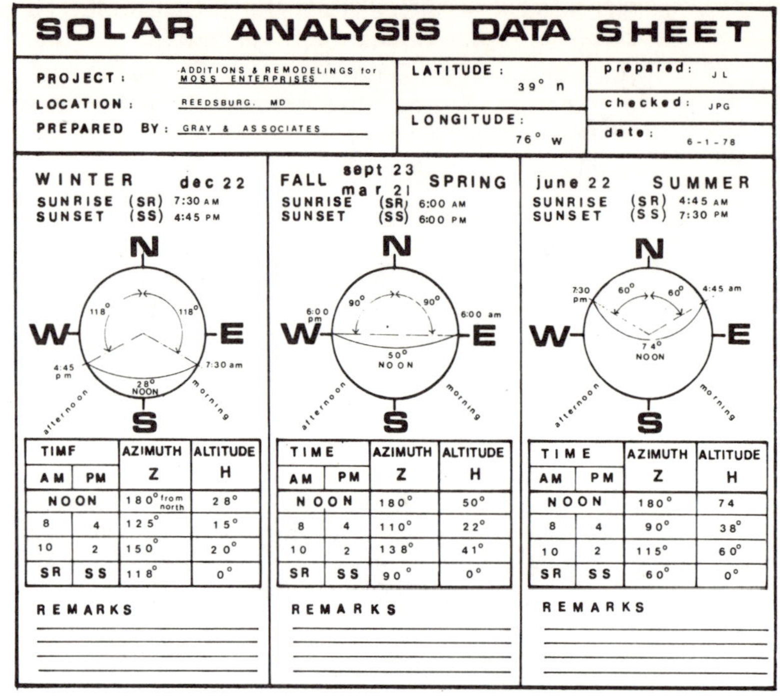

WINTER — dec 22

SUNRISE (SR) 7:30 AM
SUNSET (SS) 4:45 PM

TIME		AZIMUTH Z	ALTITUDE H
AM	PM		
NOON		180° from north	28°
8	4	125°	15°
10	2	150°	20°
SR	SS	118°	0°

REMARKS

FALL sept 23 / mar 21 — SPRING

SUNRISE (SR) 6:00 AM
SUNSET (SS) 6:00 PM

TIME		AZIMUTH Z	ALTITUDE H
AM	PM		
NOON		180°	50°
8	4	110°	22°
10	2	138°	41°
SR	SS	90°	0°

REMARKS

june 22 — SUMMER

SUNRISE (SR) 4:45 AM
SUNSET (SS) 7:30 PM

TIME		AZIMUTH Z	ALTITUDE H
AM	PM		
NOON		180°	74
8	4	90°	38°
10	2	115°	60°
SR	SS	60°	0°

REMARKS

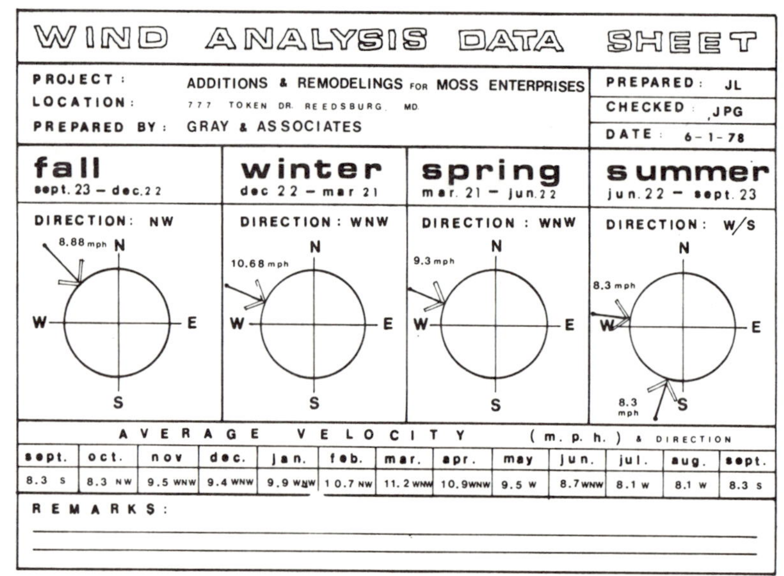

fall sept. 23 – dec. 22 — DIRECTION: NW — 8.88 mph

winter dec 22 – mar 21 — DIRECTION: WNW — 10.68 mph

spring mar. 21 – jun. 22 — DIRECTION: WNW — 9.3 mph

summer jun. 22 – sept. 23 — DIRECTION: W/S — 8.3 mph / 8.3 mph

AVERAGE VELOCITY (m.p.h.) & DIRECTION

sept.	oct.	nov	dec.	jan.	feb.	mar.	apr.	may	jun.	jul.	aug.	sept.
8.3 S	8.3 NW	9.5 WNW	9.4 WNW	9.9 WNW	10.7 NW	11.2 WNW	10.9 WNW	9.5 W	8.7 WNW	8.1 W	8.1 W	8.3 S

REMARKS:

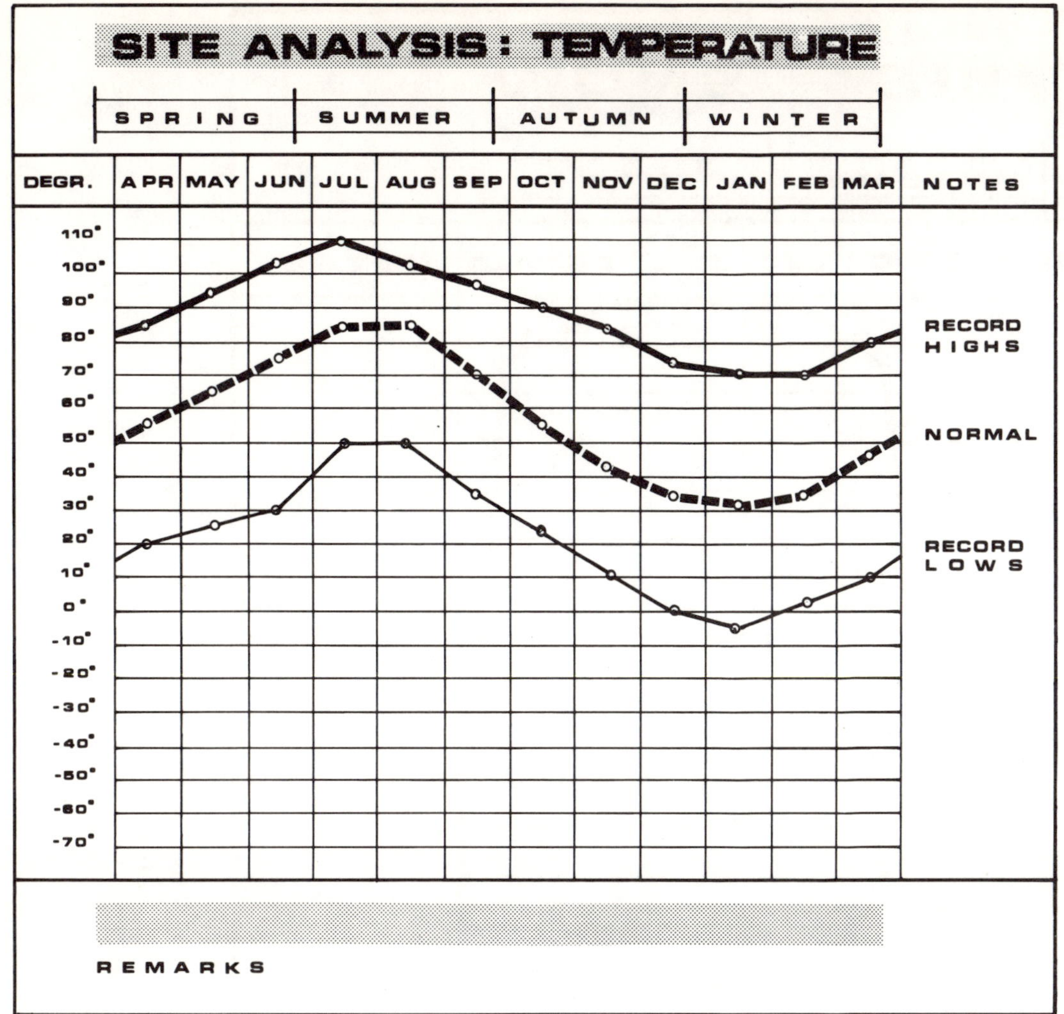

Figure 9.16

expected to achieve through their spatial layout. (See Figure 9.19)

3.2 *Spatial components characteristics and relationships:* the items illustrated in Figures 9.20 through 9.24 are a detailed breakdown of the spatial characteristics of the project's operational components as well as a representation of their interrelationships and materials flow diagram.

4. Economic Analysis

The outlined budget estimate in Figure 9.25 has been based on square and cubic feet costs approximations of recently completed construction projects of a type and quality of construction similar to this one; adjustments based on possible cost escalations due to inflation, etc., have been projected to reflect increases up to December, 1978; it is estimated at this point that construction would be scheduled to begin in November, 1978.

5. Regulatory Survey

The survey shown in Figure 9.26 outlines the essential regulatory conditions and items which will have to be met should the project proceed as presently programmed; should any change in occupancy or intended use take place, the contents of this survey must be revised accordingly in their entirety.[22]

6. Growth and Change Projections

The growth and change projections recorded in this document (see Figures 9.27, 9.28 and 9.29) represent the logical operational developments of the analyzed functions within an organized sequence of expected events; changes and alterations due to unpredictable circumstances will, logically, cause adjustments and revisions to the figures and information contained in this table. Therefore, these projections represent only generalized approximations, not a rigorous plan of operational development.

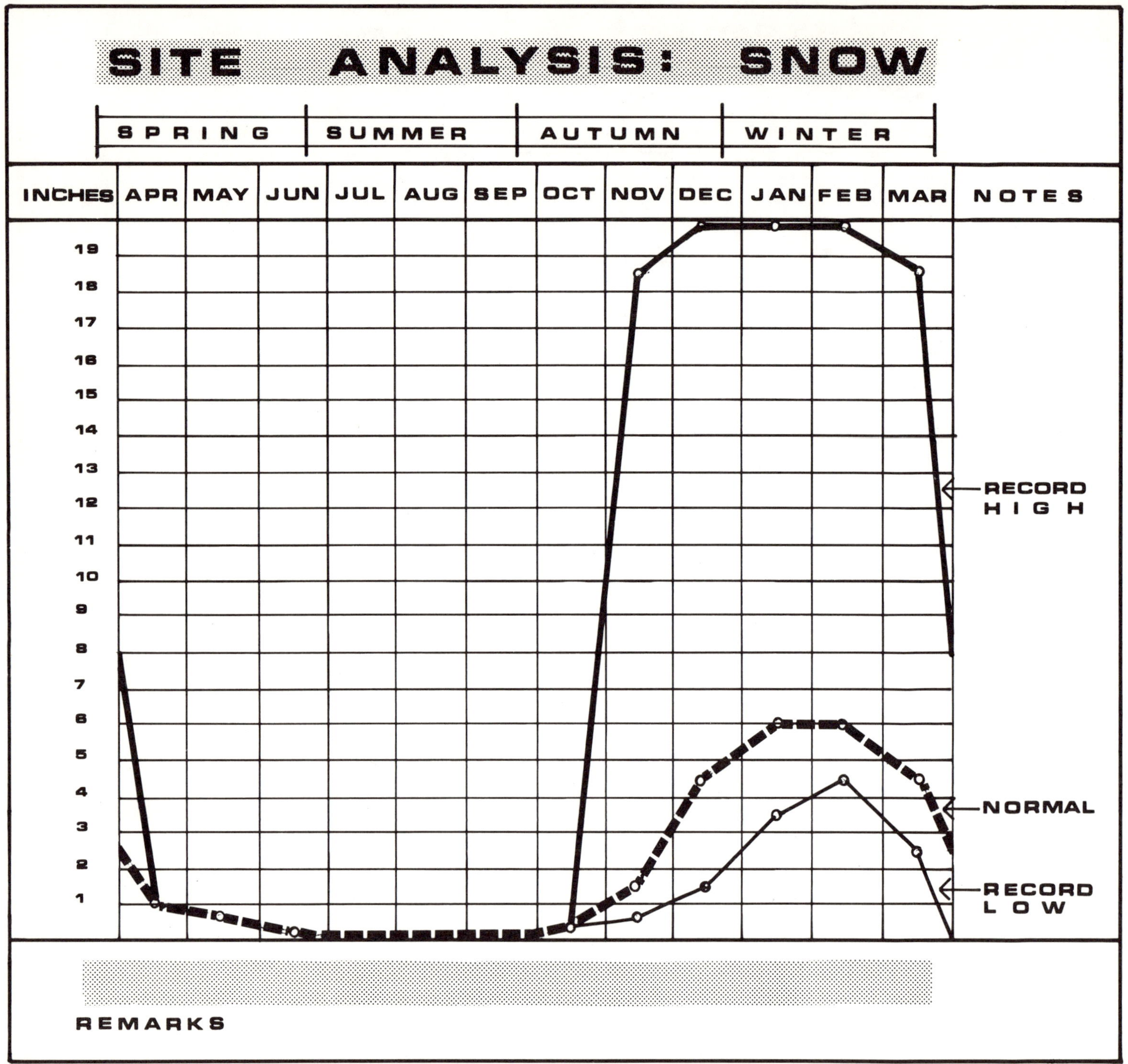

Figure 9.17

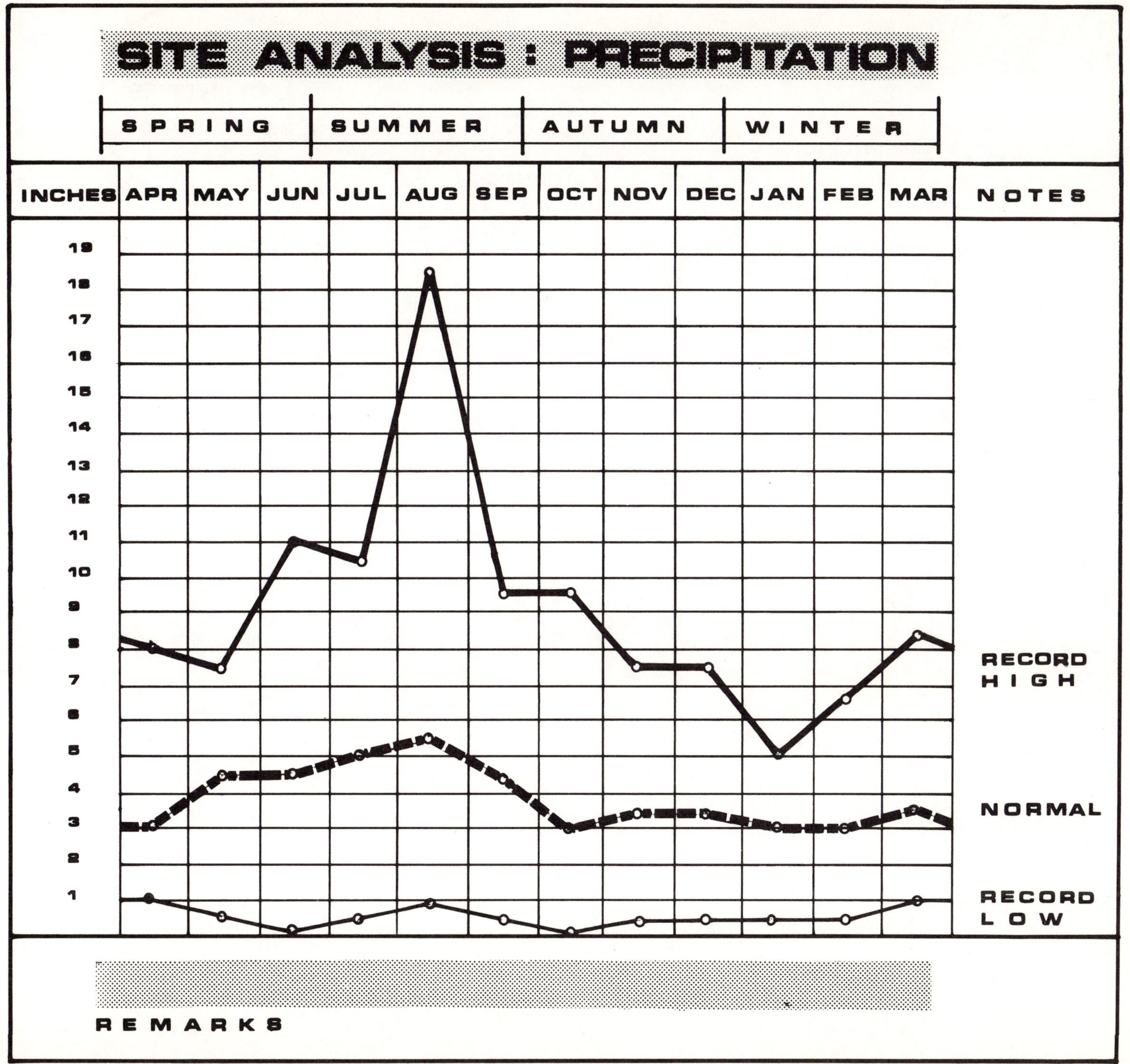

Figure 9.18

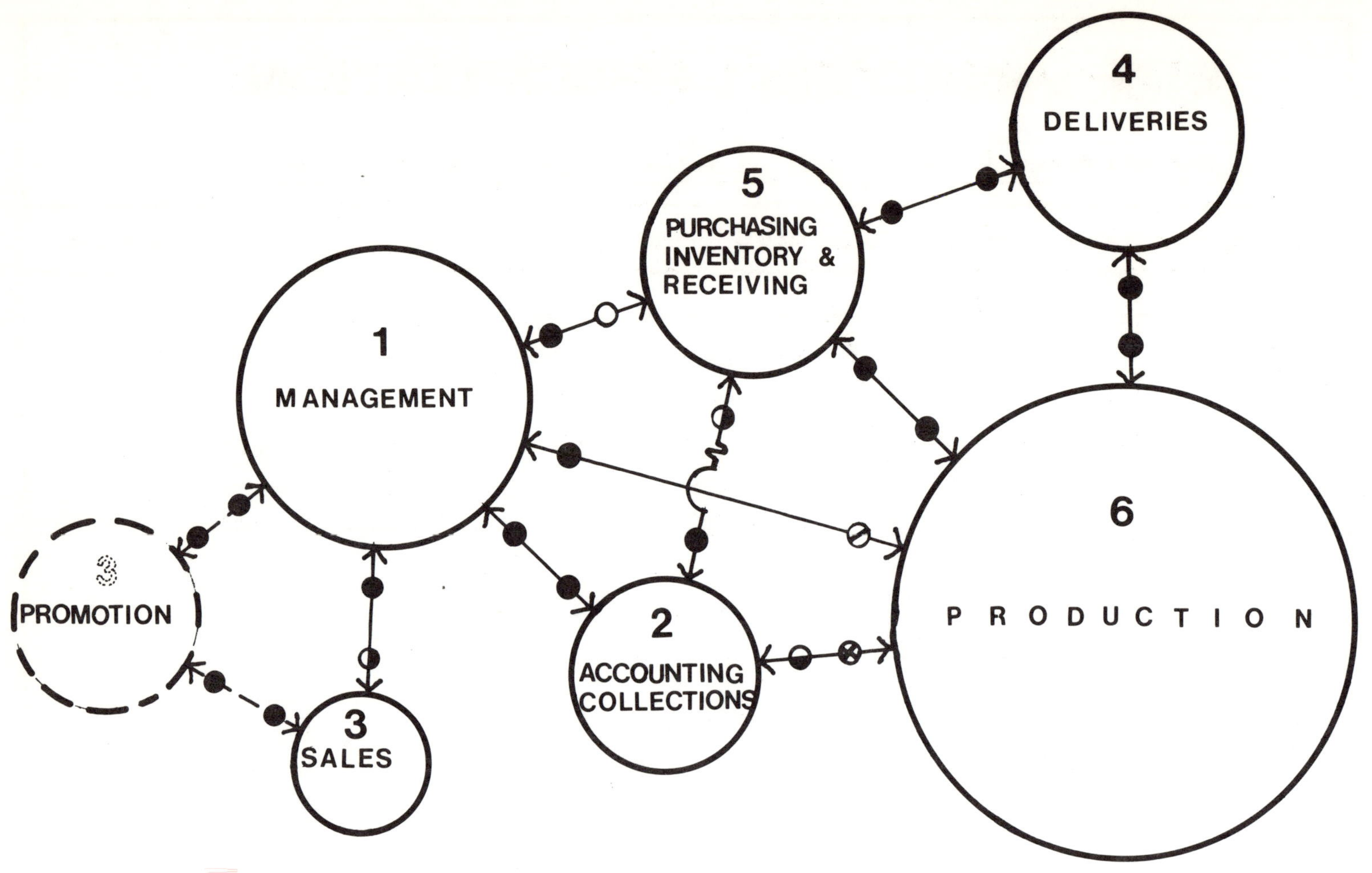

DIVISIONAL RELATIONSHIPS DIAGRAM

Figure 9.19

SPATIAL COMPONENTS ANALYSIS DATA SHEET

JOB NO. 7804

PROJECT ADDITIONS & REMODELINGS for MOSS ENTERPRISES
LOCATION 777 TOKEN DR REEDSBURG MD.
ARCHITECT GRAY & ASSOCIATES

PREPARED J.L. **CHECKED** J.P.G. **APPROVED**
DATE 7-2-78 **PAGE** 1 OF 2

DIV.	COMPONENT DESCRIPTION	FUNCT.	CONSTITUTIVE ELEMENTS	FUNCT.	CLASS	ACTIVITY DESCRIPTION	LOCATION	ROOM N°	N° OF OCC. M	N° OF OCC. F	SQ.FT/OCC	AREA	VOL.	EQUIP.	MECH. & ELEC.	ENVIROMT.	FREQ	OTHER (SPECIFY)	REMARKS
1	PRESIDENCY	E	PRESIDENT	E	S	MANAGEMENT	OFFICE	101	1		200	200	1800	(1)	OFFICE STANDARD	PENDING I.D. LAYOUTS	8/8	NONE	
	VICEPRESIDENCY	E	VICEPRESIDENT	E	S	ADMINISTRATION	OFFICE	102	1		180	180	1620	(2)		PENDING I.D. LAYOUTS			
	CLERICAL	E	EXEC. S'CY.	E	S	EXECUTIVE CLERICAL	OFFICE	101A	(2)	1	100	100	800	(3)		STANDARD			(2) DENOTES 2 VISITORS
			SECRETARIES	E	S	CLERICAL	OFFICE	103	1	6	80	560	5600	(4)					
			OFF. SUPPLIES	C	T	GEN. OFFICE	STORAGE	103A				200	1600	BY OWNER					
			MEN TOILET	S	T	HYGENE	TOILET ROOM	104	8			80	560	(17)					
			WOMEN TOILET	S	T	HYGENE	TOILET ROOM	105		18		150	1050	(18)				↓	
			EMPLOYEES LOUNGE	S	T	LOUNGING	LOUNGE	109	(4)	(4)	25	200	1600	(7)				OPEN TO EXT. TERRACE	(4) DENOTES 4 VISITORS AT ONE TIME
	PERSONNEL	E	DIRECTOR	E	S	ADMINISTRATION	OFFICE	106		1	150	150	1200	(6)		↓		NONE	
	RECEPTION	C	RECEPTIONIST	C	S	RECEPTION	VESTIBULE	107	(4)	1	60	300	3000	(3)		BY I.D. LAYOUT			(4) SAME AS ABOVE
	CONFERENCES	C	CONFERENCE RM.	C	T	CONFERENCES	CONFERENCE RM.	108	(10)	(1)	36	400	3200	(5)		BY I.D. LAYOUT	2/8		(10) & (1) SAME AS ABOVE
2	ACCOUNTING & PAYROLL	E	ACCOUNTANT	E	S	ACCOUNTING	OFFICE	201	1		150	150	1200	(6)		STANDARD	8/8		
			SECRETARIES	E	S	BOOKKEEPING	OFFICE	202		2	50	100	800	(8)	↓	STANDARD	8/8		
			COMPUTER/OPERATOR	E	S	RECORDS	COMPUTER ROOM	203	1		200	200	1600	BY OWNER	SEE OWNER'S MANUAL	NO WINDOWS	4/8		
			STORAGE	C	T	OFFICE SUPPLIES	STORAGE	203A				100	900		OFFICE STANDARD	STANDARD	8/8		
			FILES RECORDS	E	T	RECORDS STORAGE	FILES RECORDS ROOM	203B				80	720	↓					
	COLLECTIONS	E	OFFICER	E	S	COLLECTIONS	OFFICE	204		1	80	80	640	(6)					
			FILES	E	D	FILE STORAGE	FILE ROOM	204A			50	50	400	BY OWNER					
			STORAGE	C	T	OFFICE SUPPLIES	STORAGE	204B				140	1120	(19)					
3	SALES	E	SALESMAN 1	E	D	SALES	OFFICE	301	1		100	100	800	(6)					
			SALESMAN 2	E	D	↓		302	1		100	100	800	(6)					
			SALESMAN 3	E	D	↓	↓	303	1		100	100	800	(6)					
			DISPLAY	C	T	DISPLAYS	DISPLAY AREA	304	(2)			300	3000	BY OWNER	↓	↓	↓	↓	

EQUIPMENT INVENTORY

(1)
1 DESK & CHAIR
1 SOFA
2 LOUNGE CHAIRS
2·6' LONG STORAGE CABINETS
40 L.F. WOOD SHELVING

(2)
1 DESK & CHAIR
2 LOUNGE CHAIRS
1·8' LONG STORAGE CABINET
60 L.F. WOOD SHELVING

(3)
1 RECEPTION DESK / TYPING / CHAIR
1 FILE CABINET
2 CHAIRS

(4)
7 DESKS TYPING CHAIRS
2 FILE CABINETS
100 L.F. SHELVING

(5)
1 CONFERENCE TABLE (10) PERS.
10 CONF. CHAIRS
AUDIOVISUAL EQUIP (BY OWNER)
PROJECTION SCREEN
LIQUOR CABINET
60 L.F. WOOD SHELVING
2·8' LONG STORAGE CABINETS
1 SC'Y. DESK & CHAIR
COAT RACK
BUILT·IN T.V. SCREEN

(6)
1 DESK & CHAIR
2 CHAIRS
2 FILE CABINETS
1·4' LONG STORAGE CABINET
20 LF SHELVING

(7)
2 4'x4' TABLES
8 CHAIRS
REFRIGERATOR
RANGE/OVEN
SINK
CABINETS
3 VENDING MACHINES

(8)
2 DESKS & CHAIRS
4 FILE CABINETS
50 LF SHELVING

(9)
4 CHAIRS
1·4'x4' TABLE
1·3'x8' WORK TABLE

(10)
30 LOCKERS
2·7' LONG BENCHES

(11)
6 LOCKERS
1·6' LONG BENCH

(12)
3 WC
3 LAVS
3 SHOWERS
2 URINALS
ACCESSORIES

(13)
1 WC
1 LAV
1 SHOWER
ACCESSORIES

(14) 280 L.F. SHELVING

Figure 9.20

SPATIAL COMPONENTS ANALYSIS DATA SHEET

JOB NO. 7804

PROJECT	ADDITIONS & REMODELINGS for MOSS ENTERPRISES	PREPARED	JL	DATE	7-2-78
LOCATION	777 TOKEN DR REEDSBURG, MD	CHECKED	JPG		
ARCHITECT	GRAY & ASSOCIATES	APPROVED		PAGE	2 OF 2

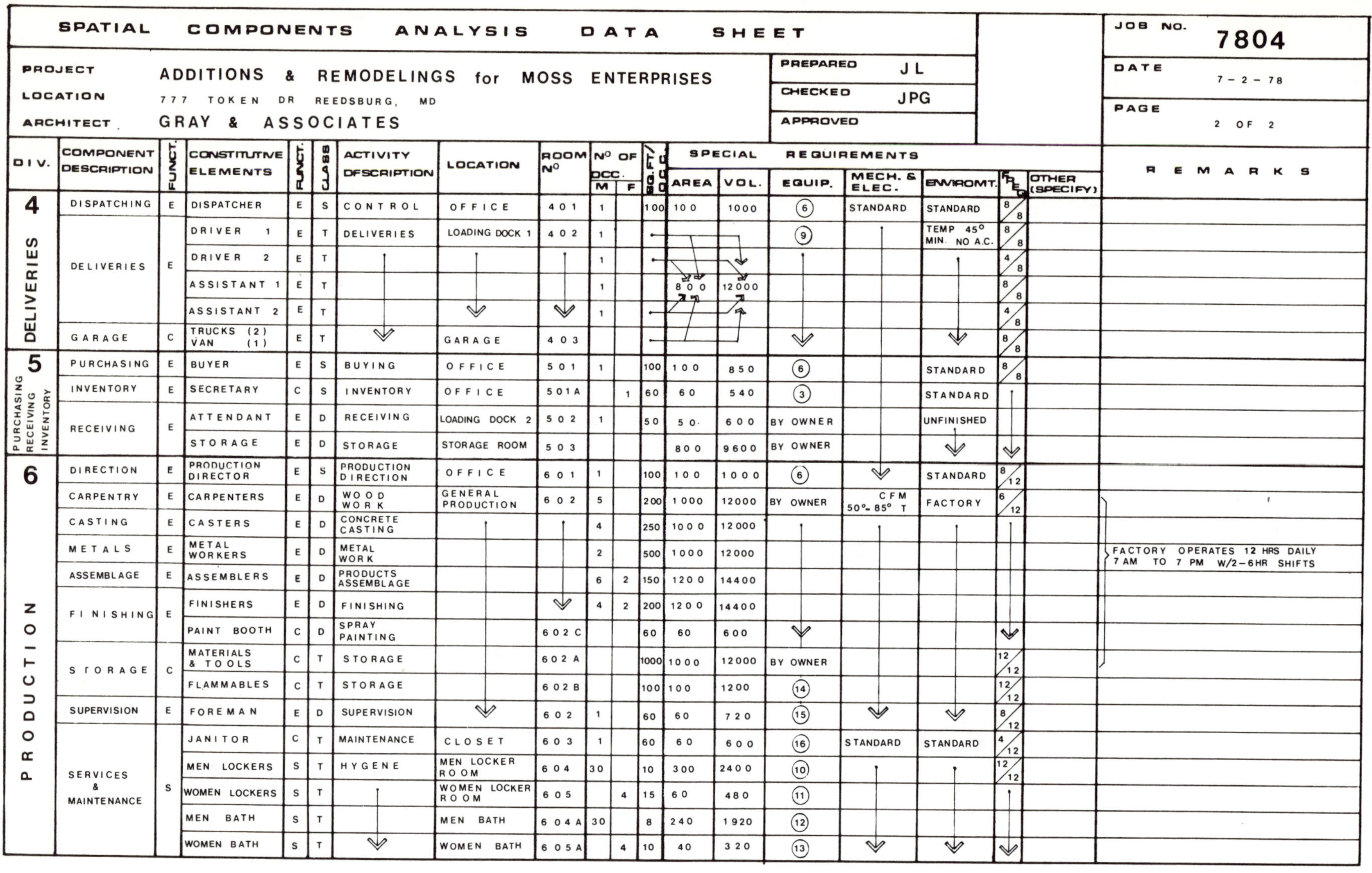

DIV.	COMPONENT DESCRIPTION	FUNCT.	CONSTITUTIVE ELEMENTS	FUNCT.	CLASS	ACTIVITY DESCRIPTION	LOCATION	ROOM Nº	Nº OF OCC. M	Nº OF OCC. F	SQ.FT/OCC	AREA	VOL.	EQUIP.	MECH. & ELEC.	ENVIROMT.	FREQ	OTHER (SPECIFY)	REMARKS
4	DISPATCHING	E	DISPATCHER	E	S	CONTROL	OFFICE	401	1		100	100	1000	(6)	STANDARD	STANDARD	8/8		
	DELIVERIES	E	DRIVER 1	E	T	DELIVERIES	LOADING DOCK 1	402	1					(9)		TEMP 45° MIN. NO A.C.	8/8		
			DRIVER 2	E	T				1								4/8		
			ASSISTANT 1	E	T				1			800	12000				8/8		
			ASSISTANT 2	E	T				1								4/8		
	GARAGE	C	TRUCKS (2) VAN (1)	E	T		GARAGE	403									8/8		
5	PURCHASING	E	BUYER	E	S	BUYING	OFFICE	501	1		100	100	850	(6)		STANDARD	8/8		
	INVENTORY	E	SECRETARY	C	S	INVENTORY	OFFICE	501A		1	60	60	540	(3)		STANDARD			
	RECEIVING	E	ATTENDANT	E	D	RECEIVING	LOADING DOCK 2	502	1		50	50.	600	BY OWNER		UNFINISHED			
			STORAGE	E	D	STORAGE	STORAGE ROOM	503				800	9600	BY OWNER					
6	DIRECTION	E	PRODUCTION DIRECTOR	E	S	PRODUCTION DIRECTION	OFFICE	601	1		100	100	1000	(6)		STANDARD	8/12		
	CARPENTRY	E	CARPENTERS	E	D	WOOD WORK	GENERAL PRODUCTION	602	5		200	1000	12000	BY OWNER	CFM 50°-85° T	FACTORY	6/12		
	CASTING	E	CASTERS	E	D	CONCRETE CASTING			4		250	1000	12000						
	METALS	E	METAL WORKERS	E	D	METAL WORK			2		500	1000	12000						FACTORY OPERATES 12 HRS DAILY 7 AM TO 7 PM W/2-6HR SHIFTS
	ASSEMBLAGE	E	ASSEMBLERS	E	D	PRODUCTS ASSEMBLAGE			6	2	150	1200	14400						
	FINISHING	E	FINISHERS	E	D	FINISHING			4	2	200	1200	14400						
			PAINT BOOTH	C	D	SPRAY PAINTING		602C			60	60	600						
	STORAGE	C	MATERIALS & TOOLS	C	T	STORAGE		602A			1000	1000	12000	BY OWNER			12/12		
			FLAMMABLES	C	T	STORAGE		602B			100	100	1200	(14)			12/12		
	SUPERVISION	E	FOREMAN	E	D	SUPERVISION		602	1		60	60	720	(15)			8/12		
	SERVICES & MAINTENANCE	S	JANITOR	C	T	MAINTENANCE	CLOSET	603	1		60	60	600	(16)	STANDARD	STANDARD	4/12		
			MEN LOCKERS	S	T	HYGENE	MEN LOCKER ROOM	604	30		10	300	2400	(10)			12/12		
			WOMEN LOCKERS	S	T		WOMEN LOCKER ROOM	605		4	15	60	480	(11)					
			MEN BATH	S	T		MEN BATH	604A	30		8	240	1920	(12)					
			WOMEN BATH	S	T		WOMEN BATH	605A		4	10	40	320	(13)					

EQUIPMENT INVENTORY (cont....)

(15) 1 DESK / 2 CHAIRS / 1 HALF FILE CABINET

(16) SINK / 6 MOP HOLDERS / 12 L.F. MICA SHELVING (ACID RESISTANT)

(17) 1 WC / 1 URINAL / 1 LAV / ACCESSORIES

(18) WC / LAVS / ACCESSORIES

(19) 240 L.F. OF 12" STEEL SHELVING

Figure 9.21

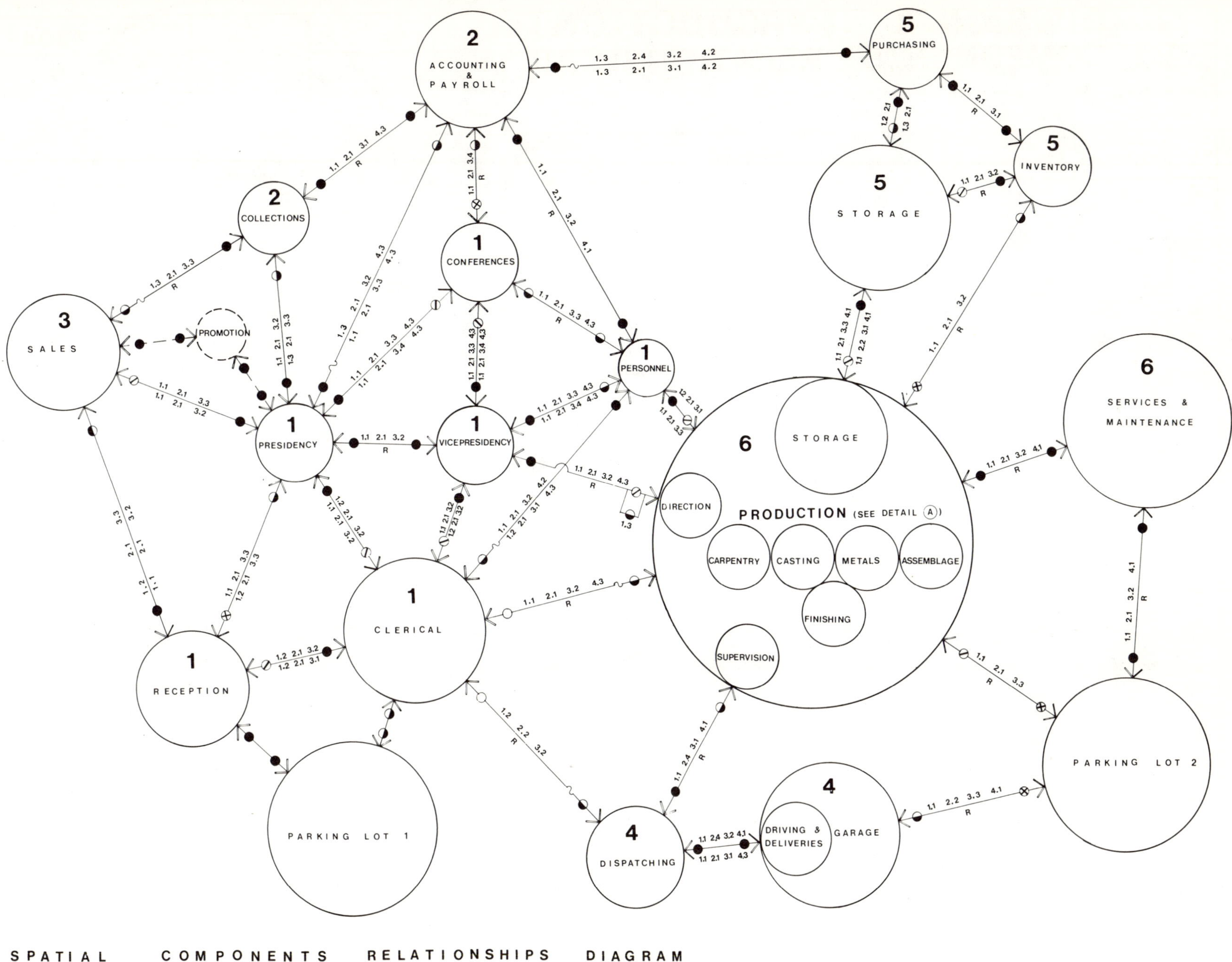

SPATIAL   COMPONENTS   RELATIONSHIPS   DIAGRAM

Figure 9.22

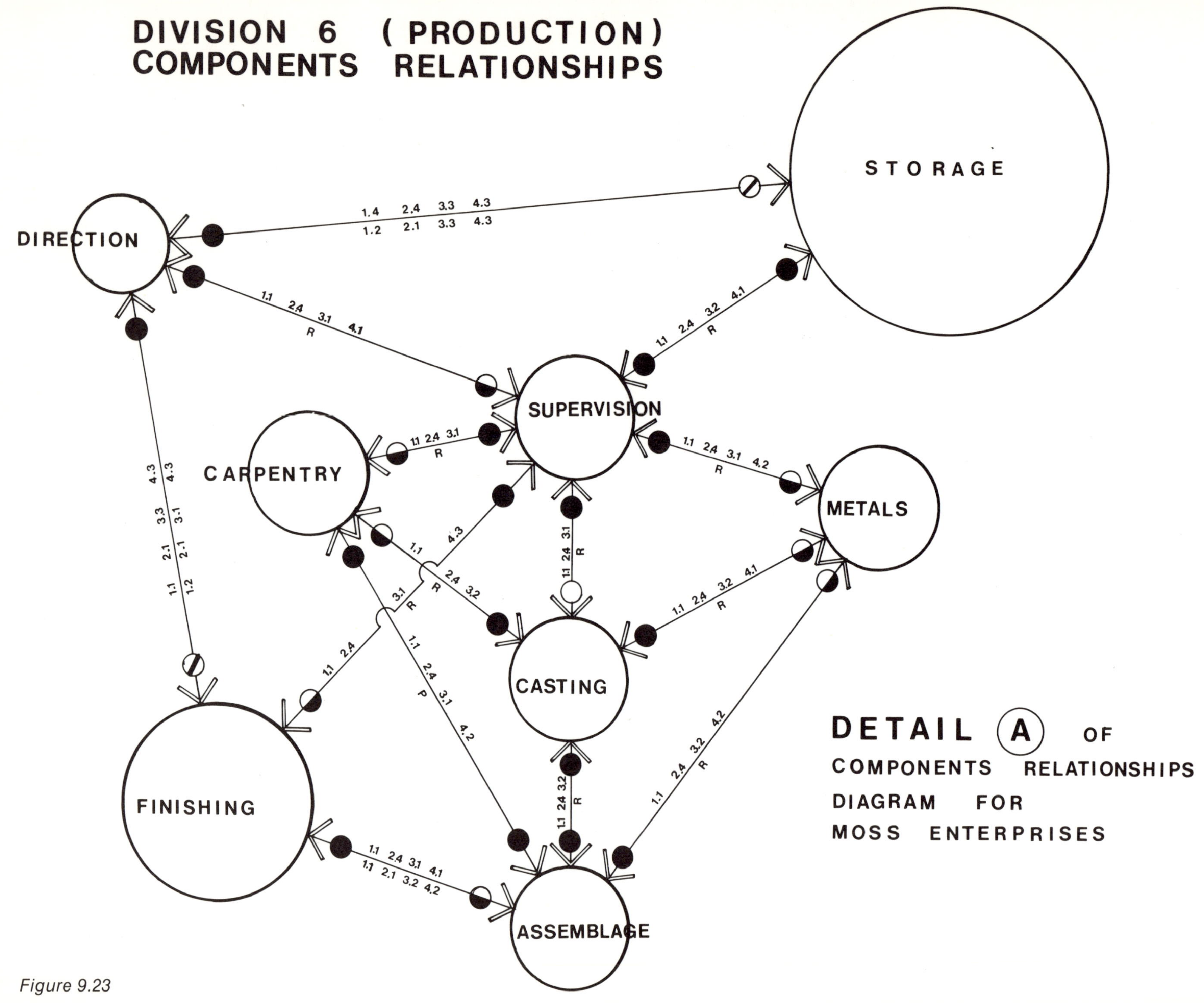

DIVISION 6 (PRODUCTION)
COMPONENTS RELATIONSHIPS
STORAGE
DIRECTION
SUPERVISION
CARPENTRY
METALS
CASTING
FINISHING
ASSEMBLAGE
DETAIL A OF
COMPONENTS RELATIONSHIPS
DIAGRAM FOR
MOSS ENTERPRISES

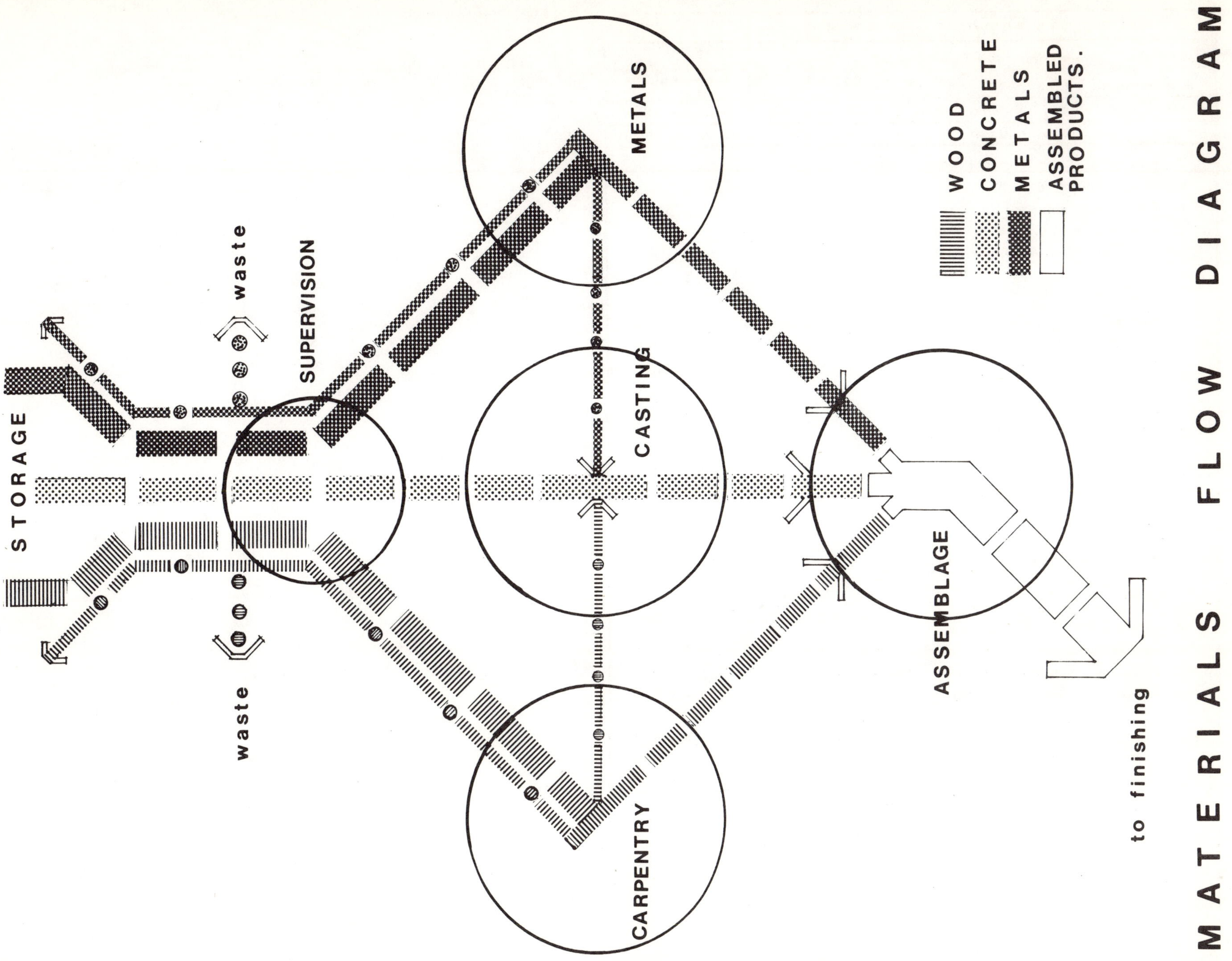

MATERIALS FLOW DIAGRAM
WOOD
CONCRETE
METALS
ASSEMBLED PRODUCTS.
STORAGE
SUPERVISION
waste
waste
METALS
CASTING
CARPENTRY
ASSEMBLAGE
to finishing

Figure 9.24

BUDGET ESTIMATE FORM	PROJECT NO. 7 8 0 4
PROJECT **MOSS ENTERPRISES**	DATE 7 - 2 - 78
LOCATION REEDSBURG M D	PREPARED BY **T W**
GROSS AREA 18,000 sf / NET AREA 14,500 sf / BLDG TYPE **IND.**	EFFICIENCY 80 %

ITEM	COST
A BUILDING COST (GROSS AREA X SQ.FT. UNIT COST)	300,000.00
B SITE DEVELOPMENT	40,000.00
C FIXED EQUIPMENT	120,000.00
D HEATING VENTILATING & AIR CONDITIONING	50,000.00
E PLUMBING & FIRE PROTECTION	25,000.00
F ELECTRICAL	45,000.00
G TOTAL BUILDING COSTS SUBTOTAL 1 A + B + C + D + E + F = G	$ 580,000.00
H SITE ACQUISITION	NOT APPLICABLE [1]
I UNUSUAL SITE CONDITIONS	0.00
J OFF SITE WORK	0.00
K GROWTH & CHANGE PROVISIONS	40,000.00
L LANDSCAPING	5,000.00
M FURNISHINGS	40,000.00
N ALLOWANCES / CONTINGENCIES	60,000.00
O MISCELLANEOUS	10,000.00
P ADDITIONAL BUDGET ITEMS SUBTOTAL 2	$ 155,000.00
Q PROFESSIONAL FEES	40,000.00
R SURVEYS AND INSURANCE	3,000.00
S LEGAL, ACCOUNTING AND ADMINISTRATIVE COSTS	8,000.00
T LEASING, ADVERTISING AND PROMOTION	0.00
U FINANCING COSTS AND TAXES	18,000.00
V OWNER'S BUDGET ITEMS SUBTOTAL 3	$ 69,000.00
W TOTAL BUDGET REQUIRED G + P + V = W	$ 804,000.00

NOTES:

(1) PENDING PRELIMINARY LAYOUTS

Figure 9.25

144

4. Type of Construction

Floors	
Type	NON COMBUSTIBLE (NC) 2 hr
Max Height	2 STORIES
Area/Floor	20,000 sf

Basement(s) Classification	
Type	N A
Max Height	N A
Area/Floor	N A

5. Exit Requirements

No. of Occupants/Floor	270
No. of Exits/Floor	4
Unit/Exit Width	40" 100 PERS UNSPRINKL'D / 50" 100 PERS SPRINKLERED
Corridors	4'-0" MIN.

Widths of Exits

Stairs	4'-0"
Doors	3'-4"
Ramps Width	4'-0"
Ramps Slope	1:12 W/5' LANDINGS AT EACH 30'
Distance to exits	100' UNSPRINKLERED 150' SPRINKLERED
Dead Ends	MAX. 20'
Barrier Free Requirements	SEE SECTS. 52.04 & 55.07
Door Swings	EXITS DIRECTION OF TRAVEL
Special Requirements	—

6. Fire Protection

Structural Frame

Columns	NC–2
Floors	NC–2
Ceilings	—
Walls & Partitions	NC–0
Exterior Walls	NC–2
Fire Walls	4 hr
Party Walls	NA
Exit Enclosures	2 hr
Rated Enclosures	2 hr
Shafts	2 hr
Parapets	2 hr
Partitions	0
Load Bearing	1 hr
Non Bearing	0
Corridors	1 hr
Roof Structure	NC–1
Roof Covering	CLASS A
Vertical Openings	2 hr
Stairways & Smoke Proof Enclosures	2 hr
Doors, Windows, Glass	NA
Skylights	NA
Attic Separations	2 hr
Cantilevers	NA
Fire Extinguishers	1/2500 sq. ft.
Standpipes	NONE REQUIRED

3

Figure 9.26

PROJECT Nº 7804

PROJECT NAME	ADDITIONS & REMODELING for MOSS ENTERPRISES	PREPARED JL	DATE 6-15-78
LOCATION	777 TOKEN RD REEDSBURG MD	CHECKED JPG	
ARCHITECT	GRAY & ASSOCIATES	APPROVED	PAGE 1 of 3

DIV.	COMPONENT	F	CONSTITUTIVE ELEMENTS	F	LOCATION	MARK/ ROOM	1979 #occ	1979 s.f. occ	1979 AREA	1985 #occ	1985 sf occ	1985 AREA	1990 #occ	1990 sf occ	1990 AREA	1995 #occ	1995 sf occ	1995 AREA	2000 #occ	2000 sf occ	2000 AREA	GROWTH TYPE	PLACEMENT	L/D	CHANGES E	VOLUME	yr F:
1	presidency	e	president	e	office	101	1	200	200												→	−	building	−	−	−	−
	vicepresidency	e	vicepresident	e	office	102	1	180	180												→	−		−	−	−	−
	clerical	e	executive secretary	e	office	101A	3	30	100	6	30	200									→	−		adj / h	P	I	P 1985
			secretaries	e	office	103	7	80	560	10	80	800			→	12	100	1200			→	A		adj / h	−	I	−
			office supplies	c	storage	103A	−	−	200	−	−	250	−	−	300			→	−	−	350	A		adj / h	−	I	−
			men toilet	s	toilet room	104	(8)	−	80	(10)	−	80									→	−		−	−	−	−
			women toilet	s	toilet room	105	(18)	−	150	(22)	−	150	(24)	−	150	(30)	−	200			→	−		adj / h	−	I	−
			employees lounge	c	lounge	109	(8)	25	200			→	(12)	25	300						→	−		adj / h	−	I	−
	personnel	e	director	e	office	106	1	150	150												→	−		−	−	−	−
	reception	c	receptionist	c	vestibule	107	5	60	300												→	−		−	−	−	−
	conferences	c	conference room	c	conference	108	(11)	36	400												→	−		−	−	−	−
2	accounting & payroll	e	accountant	e	office	201	1	150	150												→	−		−	−	−	−
			secretaries	e	office	202	2	50	100			→				3	50	150				−		adj / h	−	I	−
			computer	e	computer room	203	1	−	200												→	−		−	−	−	−
			storage	c	storage room	203A	−	−	100												→	−		−	−	−	−
			files & records	e	records room	203B	−	−	80			→	−	−	100			→	−	−	120	A		adj / h	−	I	−
	collection	e	officer	e	office	204	1	80	80												→	−		−	−	−	P 1985
			secretary	c	cubicle	204C	na			1	50	50									→	−		adj / h	−	−	P 1985
			files	e	files room	204A	−	−	50			→				−	−	80			→	−		adj / h	−	I	−
			storage	c	storage room	204B	−	−	140			→				−	−	110			→	−	↓	−	−	D	−
																						−		−	−	−	−
TOTALS							23		3420	40		3860	40		4030	42		4480	43		4600						

REMARKS

1 circled numbers are included in departamental counts.

2 toilets rqmts. may vary depending on future regulations.

Figure 9.27

SPATIAL PROJECTIONS DATA SHEET

PROJECT N° **7804**

PROJECT NAME	ADDITIONS & REMODELING for MOSS ENTERPRISES	PREPARED	JL	DATE	6 – 15 – 78
LOCATION	777 TOKEN RD REEDSBURG, MD	CHECKED	JPG		
ARCHITECT	GRAY & ASSOCIATES	APPROVED		PAGE	2 of 3

DIV.	COMPONENT	F	CONSTITUTIVE ELEMENTS	F	LOCATION	MARK/ROOM#	YEAR 1979 # occ	s.f. occ	AREA	YEAR 1985 # occ	sf occ	AREA	YEAR 1990 # occ	sf occ	AREA	YEAR 1995 # occ	sf occ	AREA	YEAR 2000 # occ	sf occ	AREA	GROWTH TYPE	PLACEMENT	L / D	CHANGES E	VOLUME	F yr:
3 SALES	sales	e	director	e	office	305	na	→	→	1	120	120	→					→			→	—	bldg. or region	adj / h/v	F	I	T 1985
			salesmen	e	offices	301–3	3	100	300	5(3)	na	300	→					→	10(3)	na	300	—	region or natl.	remote	P	I	T 1985
			secretaries	c	office	306	na	→	→	1	60	60	→					→	2	60	120	—	bldg. or region	adj / h/v	—	—	—
			display	c	display area	304	—	—	300	→			→					→	—	—	600	—	↓	↓	P	I	P 1985
	promotion (4)	c	director	e	office	307	na	→	→	1	80	80	→						→			—	↓		F	I	T 1985
			secretary	c	cubicle	307A	na	→	→	1	50	50	→						→			—	↓	↓	F	I	T 1985
4 DELIVERIES	dispatching	e	dispatcher	e	office	401	1	100	100	→									→			—	building	adj / h	—	—	—
	driving and deliveries	e	drivers	e	loading dock	402	1.5	—	↑	2	—	↑	3	—	↑							—			—	I	—
			assistants	c	loading dock	402	1.5	—	800	2	—	800	3	—	1000	→					→	—			—	I	—
	garage	e	trucks	e	garage	403	2	—		—	—		3	—								—			—	I	—
			vans	c	garage	403	1	—	↓	—	—	↓	—	—	↓							—		↓	—	I	—
5 PURCHASING RECEIVING INVENTORY	purchasing	e	buyer	e	office	501	1	100	100	→									→			—			P	—	P 1985
			secretary	c	cubicle	501B	na	—	—	1	60	60	→						→			—			—	—	—
	inventory	e	secretary	e	office	501A	1	50	50	1	80	80	→						→			—			P	—	P 1985
	receiving	e	attendant	e	loading dock	502	1	50	50	→									→			—			—	—	—
			storage	e	storage room	503	—	—	800	→			—	—	1200	→			—	—	1800	G	↓	↓	—	I	—
TOTALS							**13**		**2500**	**15**		**2900**	**17**		**3500**	**17**		**3500**	**23**		**4460**						

REMARKS

3 additional salesmen will not be located in house.

4 promotion department location might be on building or at another location

Figure 9.28

					PROJECT N°	**7804**
PROJECT NAME	ADDITIONS & REMODELING for MOSS ENTERPRISES	PREPARED	JL		DATE	6 – 15 – 78
LOCATION	777 TOKEN RD. REEDSBURG, MD	CHECKED	JPG			
ARCHITECT	GRAY & ASSOCIATES	APPROVED			PAGE	3 of 3

DIV.	COMPONENT	F	CONSTITUTIVE ELEMENTS	F	LOCATION	MARK/ROOM#	1979 #OCC	1979 s.f.OCC	1979 AREA	1985 #OCC	1985 sfOCC	1985 AREA	1990 #OCC	1990 sfOCC	1990 AREA	1995 #OCC	1995 sfOCC	1995 AREA	2000 #OCC	2000 sfOCC	2000 AREA	GROWTH TYPE	PLACEMENT	L / D	CHANGES E	VOLUME	F
6	direction	e	production director	e	office	601	1	100	100												→	−	building	adj / h	none	−	none
	carpentry	e	carpenters	e	production area	602	5	200	1000	7	200	1400			→	10	200	2000			→	A				I	I
	casting	e	casters	e			4	250	1000	6	250	1500			→	8	200	1600			→	A				I	
	metals	e	metal workers	e			2	500	1000	4	500	2000									→	A				I	
	assemblage	e	assemblers	e			8	150	1200	10	150	1500			→	12	150	1800			→	A				I	
	finishing	e	finishers	e			6	200	1200	7	200	1400			→	8	200	1600			→	A				I	
	storage	c	materials & tools	c		602A	−	−	1000			→				−	−	2000			→	—				I	
			flammables	c		602B	−	−	100			→				−	−	200			→	—				I	
	supervision	e	foreman	e	↓	602	1	60	60												→	—				—	
	services & maintenance	s	janitor	c	closet	603	1	−	60												→	—				—	
			men lockers	s	locker room	604	30	10	300			→				40	10	400			→	A				I	
			women lockers	s	locker room	605	4	15	60			→				10	15	150			→	A				I	
			men bath	s	men bath	604A	30	8	240			→				40	8	320			→	A				I	
			women bath	s	women bath	605A	4	10	40			→				10	10	100			→	A	↓	↓	↓	I	↓
TOTALS							28		7360	37		9760	37		9760	45		12,390	45		12390						

Figure 9.29

9.2 Case Study No. 2

Spatial Analysis Survey of Westchester County Government Offices.

The following project description has been based on the actual study contracted by Westchester County government officials and Board of Supervisors for the space needs analysis and determination of several governmental offices housed within the county courthouse building.

PROJECT DESCRIPTION. The need for additional space for the county official department and some of its divisions, as well as the county treasurer department and the county clerk and accounting divisions, has been identified in several preliminary interdepartmental questionnaires and space needs studies executed by the administrative sector of the county official's department.

Since the original building was designed to accomodate growth, the most feasible way of providing additional space is by expanding the present county courthouse. Parking around the existing building is at a critical point, and any spatial expansion will generate more traffic. Additional parking spaces could be provided by constructing a new attendant parking ramp. A parking study and conclusions, however, are not to be considered part of this project.

The work will include the following phases:

Phase I. Conduct a facility evaluation of the existing building and determine the spatial needs of the following departments or divisions:
—County Clerk Department
  Accounting Division
—County Official Department
  Administrative Sector
    Personnel Division
    Purchasing Division
    Engineering Division
      Public Works Subdivision
      Surveyor Subdivision
      Zoning and Planning Subdivision
    Data Processing Division
—County Treasurer Department

Phase II. Determine alternative schematic designs for optimum operational and functional purposes.

Phase III. Prepare contract documents and provide construction administration services.

In presenting this particular case study, we will refer only to Phase I of the project.

To further determine the extent and procedure of this phase, the portion of the professional design services contract which covered Phase I was established as follows:

SPACE NEEDS STUDY. The professional designer should study the spatial needs requirement of the county officials listed under the project description portion of this document, and project the future spatial needs for these departments or divisions through the year 1990.

The professional designer shall interview administrative staff and identify present problems and objectives, formulate a questionnaire regarding functional requirements, determine growth and expansion requirements, review and verify all data with administrative officials and compile all data into a booklet form. The contents of this report shall include the following:

A. Spatial survey reports of each department
B. Total project area and personnel summary, including net assigned area and gross area of the project
C. Identification of existing problem areas and outline of future functional requirements
D. Conclusions and recommendations

As we can see, the complete procedure and spatial analysis scope in this case were not only pre-established but also made a matter of contractual agreement. Therefore, the procedures employed are aimed specifically at the detailed provision of each item outlined in the contract.

The following pages contain an abridged version of the final document executed under this particular contract. Although the steps followed in this case do not differ essentially from those of Case Study No. 1, the actual programming of elements is outlined in the spatial projections data sheet since, as established in the project definition, the spatial requirements determinations would derive from the projections study. The areas indicated as programmed square footages for specific functions in the spatial relationships diagrams are therefore based on the 1980 space projection established for each department.

In order to simplify and abbreviate the extent of this study, only two particular department's summaries have been included in this text's outline of the final spatial needs study report. In addition, a lengthy structural report corroborating the feasibility of the proposed expansion has been omitted.

WESTCHESTER COUNTY GOVERNMENT SPATIAL NEEDS REPORT

TABLE OF CONTENTS

INTRODUCTION

In January of 1978, the professional design firm of Cole, Cass & Rawlings, Inc. was retained by the Westchester County Board of Supervisors to conduct a Space Needs Study of the county courthouse building as a result of identified needs for additional space in various county government departments.

This need has been outlined in detail in the project description of the county courthouse building.

The primary identification of a need for expansion regarding certain administrative functions of the county government was made through a 1976 Space Needs Questionnaire executed by the administrative sector of the Westchester County Official Department and designed to update a previous Space Needs Survey formulated in 1974.

In this second instance, however, no detailed study was conducted regarding the spatial analysis of each Westchester County Government Department or Division to determine and support a viable course of action to remedy identified spatial needs deficiencies or inadequacies.

The resulting detailed Spatial Analysis is the subject of this report.

SCOPE

This document presents the spatial analysis and determination of the spatial needs for the various Westchester County Government Departments and Divisions presently housed in the county courthouse building by means of an effective spatial analysis study.

The departments and divisions designated for review in this particular study are:
—County Clerk Department
  Accounting Division
—County Official Department
  Administrative Sector
    Personnel Division
    Purchasing Division
    Engineering Division
      Public Works Subdivision
      Zoning and Planning Subdivision
    Data Processing Division
—County Treasurer Department

ORGANIZATIONAL DIAGRAM

A Graphic representation which illustrates the organization and hierarchies of the county government departments and divisions can be outlined as shown in Figure 9.30.

SPATIAL ANALYSIS SUMMARY

In order to summarize the results obtained in this report and give a synthesis of results at the outset, we provide the table (Figure 9.31) containing the full range of total areas and occupancy projections and general tabulations of net to departmental gross[23] or building gross ratios and conversions.

PART I — SPATIAL ANALYSIS

Introduction. A spatial analysis study of the administrative functions included in this section presents numerous sets of variables and special conditions which cannot be understood properly unless all the corresponding interdepartmental relationships are taken into account throughout every step of the analysis procedure.

Furthermore, because of the close relationships of several surveyed departments included under this heading with other departments or divisions of the county government, any attempt to fully describe every possible factor which might affect spatial developments or projections would probably result in a report of such magnitude and complexity that its primary purpose would be imperiled by its own comprehensiveness.

Therefore, in outlining the primary tasks of this study, the main concerns have been carefully considered and determined so as to provide a direct and concise layout of spatial needs in relation to the operational performance they represent, without extensive and intricate procedures or speculative approaches but with a clear determination of those essential factors which form the basis for spatial studies.

The clear definition of any problem contains within itself the primary postulate for its solution, since a problem definition not only establishes wants and needs, but also contains a description of function and purpose and a statement of intent.

In this light, this report contains not only the procedures for identification of problems outlined by its departmental surveys section, but also the means of identifying the germ of their solution.

Definition and Purpose. This Spatial Needs Report represents the process of spatial and functional analysis of the specific Westchester County Government operations as presently executed within the departments defined in the statement of the project's scope. The purpose of this study is the final determination of identified spatial and environmental needs for these government operations as related to the time and place in which the functions which structure those specific operations must be performed.

This procedure, however, DOES NOT CONSTITUTE AN OPERATIONAL PROGRAM, but rather a detailed analysis related only to the spatial functionalism of the surveyed operations. Therefore, the results of this study DO NOT CONSTITUTE A PERFORMANCE EVALUATION OF THE OPERATIONS THEMSELVES, but a performance statement related to the spatial parameters and environments in which their integral functions are performed.

Therefore, the major areas of concern in

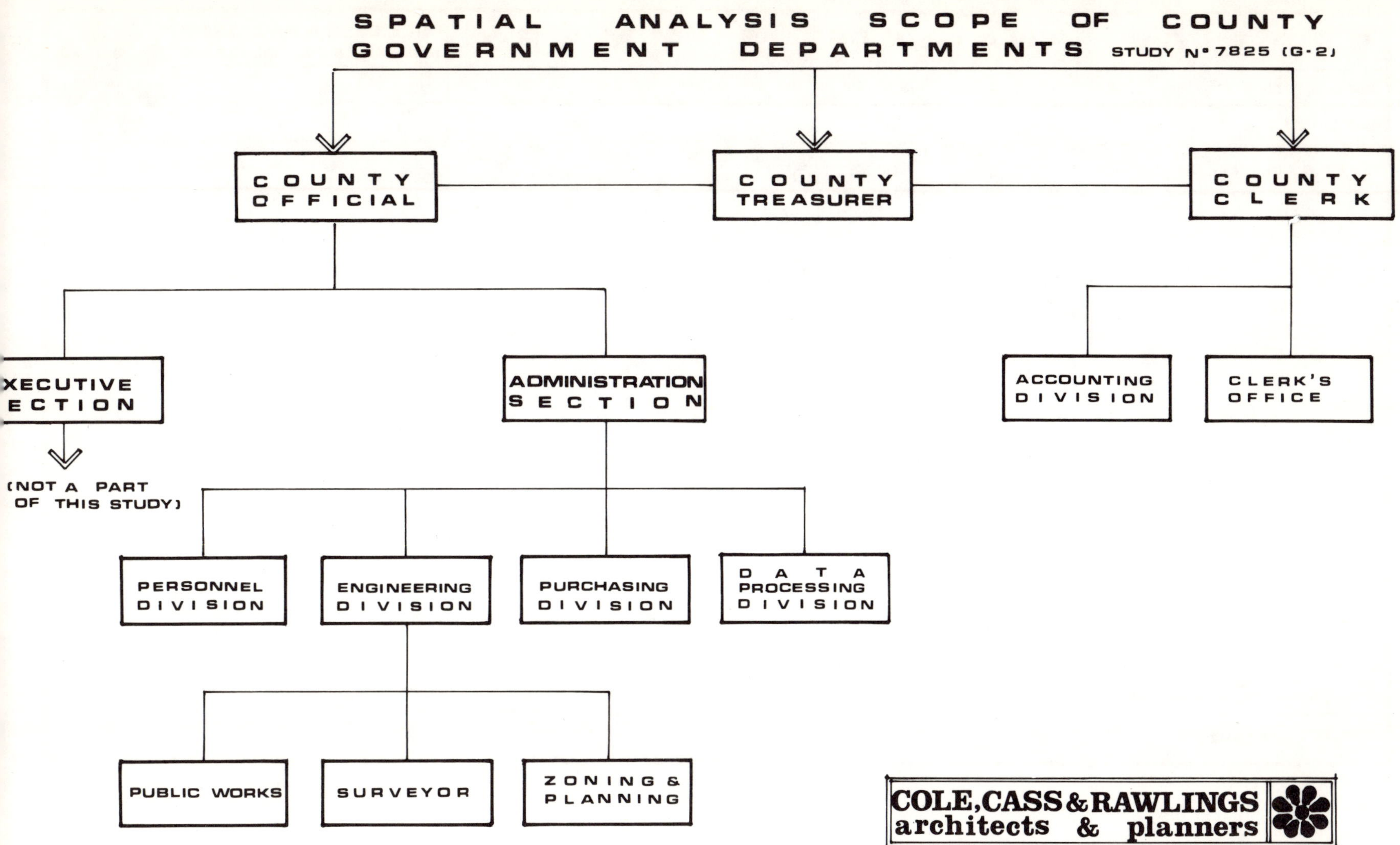

ORGANIZATIONAL DIAGRAM

ure 9.30

151

S P A T I A L   A N A L Y S I S   S U M M A R Y

COLE, CASS & RAWLINGS
architects & planners

PROJECT: SPATIAL ANALYSIS SURVEY OF WESTCHESTER COUNTY GOVERNMENT OFFICES

LOCATION: 11073 JEFFERSON DRIVE, GREEN OAKS, MO.

PROJECT № 7825 [G2]

DATE

PREPARED BY JL

REVIEWED BJC

DEPARTMENT / DIVISION	1980 [1]				1985 [1]				1990 [1]			
	OCC	NET [2]	D.GR. [3]	E.R. [4]	OCC	NET [2]	D.GR. [3]	E.R. [4]	OCC	NET [2]	D.GR. [3]	E.R. [4]
COUNTY CLERK DEPARTMENT	12.5	1900	2500	1.31	14	1700	2200	1.3	15	1800	2400	1.33
ACCOUNTING DIVISION	14	2300	3000	1.3	16	2500	3300	1.32	16	2500	3300	1.32
COUNTY OFFICIAL DEPARTMENT	30	4200	5500	1.31	30	4200	5500	1.31	34	5000	6500	1.3
ADMINISTRATIVE SECTOR	5.5	1000	1300	1.3	6	1000	1300	1.3	6	1000	1300	1.3
PERSONNEL DIVISION	25	1700	2200	1.3	28.5	1900	2500	1.31	28.5	1900	2500	1.31
PURCHASING DIVISION	8	1600	2100	1.31	8	1600	2100	1.31	8.5	1600	2100	1.31
ENGINEERING DIVISION	4	400	600	1.5	6	600	800	1.33	6	600	800	1.33
PUBLIC WORKS	8.5	1300	1700	1.3	10	1500	2000	1.33	10	1500	2000	1.33
SURVEYOR	6	1600	2100	1.31	6	1600	2100	1.31	6	1600	2100	1.31
ZONING AND PLANNING	7.5	1200	1600	1.33	9.5	1500	2000	1.33	9.5	1500	2000	1.33
DATA PROCESSING DIVISION	30	4700	6200	1.32	35	5000	6500	1.33	33.5	5000	6500	1.3
COUNTY TREASURER DEPARTMENT	6	2300	3000	1.3	6	2300	3000	1.3	6	2300	3000	1.3
SUBTOTAL	157	24,200	31,800	1.31	175	25,400	33,300	1.31	179	26,300	34,500	1.31

	1980	1985	1990
DEPARTMENTAL GROSS TO BUILDING GROSS AREA CONVERSION FACTOR = 1.4	x1.4	x1.4	x1.4
TOTAL GROSS AREA	44,520 GROSS SF	46,620 GROSS SF	48,300 GROSS SF
MAXIMUM TOTAL GROSS S.F. AVAILABLE FOR EXPANSION [5]	48,000 GROSS SF	48,000 GROSS SF	48,000 GROSS SF
TOTAL UNASSIGNED AREA	+3,480 UNASSIGNED SQ FT	+1,380 UNASSIGNED SQ FT	-300 DEFICIT

REMARKS

(1) AREAS SHOWN REPRESENT ROUND-OFFS OF SPECIFIC DEPARTMENTS. FOR DETAILED BREAKDOWNS SEE INDIVIDUAL SPATIAL PROJECTIONS SHEETS.

(2) NET REPRESENTS NET ASSIGNABLE SQUARE FEET/DEPT. OR DIVISION.

(3) D.G. REPRESENTS DEPARTMENTAL OR DIVISIONAL GROSS INCLUDING DEPARTMENTAL INTERNAL CIRCULATIONS, CLEARANCES, ETC.

(4) E.R. REPRESENTS NET TO DEPARTMENTAL GROSS EFFICIENCY RATIO.

(5) SEE STRUCTURAL REPORT.

Figure 9.31

establishing the analysis procedure which structures this study have been:

1. Spatial analysis
2. Departmental relationships analysis
3. Spatial development analysis

For these purposes, none of the following items have been studied in depth:

1. Operational performance
2. Regulatory survey
3. Functional interfacing

In compliance with the expansion possibilities of the existing structure in which the previously mentioned functions are housed, the following evaluations have been conducted:

1. Structural analysis, evaluation and recommendations
2. Circulation patterns and exiting
3. Spatial change, growth and flexibilities
4. Environmental conditions

Objectives. The primary objective of this survey has been to determine which departmental functions are currently performed within acceptable spatial parameters and which are not and, in this last instance, to recommend possible solutions, alternatives and improvements regarding the spatial functionalism, area requirements, spatial interfacing and relationships, environment, engineering systems, circulation, spatial development and flexibilities of such functions, based on the results of a spatial analysis designed to provide the following information:

1. Problem identification
2. Operational functions definition
3. Spatial evaluation
4. Operational and functional relationships
5. Growth and change projections

Procedure. The procedure employed to determine the previously mentioned information has been based on a data collection approach and carried out by means of personal interviews with the heads of departments and divisions listed in the statement of the project's scope, studies of population projections based on the state demographic services center report (1977 edition) and the United States Census Bureau, an architectural evaluation and recommendation of the collected data, review and consultation with members of the administrative sector of the county official's office; plus a second review by the initially interviewed department heads, to which final architectural reviews, evaluations and recommendations were added. To further identify and analyze the spatial and environmental conditions outlined by this procedure, a comprehensive employee survey of fifty participants, distributed proportionately among the departments listed in this report's statement of scope, has been conducted, reviewed and evaluated.

Structure. Based on the objectives definition of this study, the basic structure of the spatial analysis employed can be outlined as follows:

1. Operational functions definition
2. Spatial analysis and evaluation
3. Functional relationships analysis and evaluation
4. Spatial development analysis
5. Problem identification
6. Spatial development projections
7. Spatial programming

For this purpose, the data collection forms shown in Figures 9.32, 9.33, 9.34 and 9.35 have been employed.

As illustrated in Figure 9.32, these spatial analysis data sheets list the functional components and their constitutive elements on a departmental or divisional basis, describe the function of each element and show which room and area they presently occupy. The sheets also indicate whether each particular element requires public access and they describe the actual occupancy, dividing it into males and females, if fixed, or visitors if transient. Such counts are placed in the upper rectangle for full-time employees or in the lower rectangle for part-time employees or visitors.

The equipment column refers directly to the equipment inventories as related to the floor plan layouts.

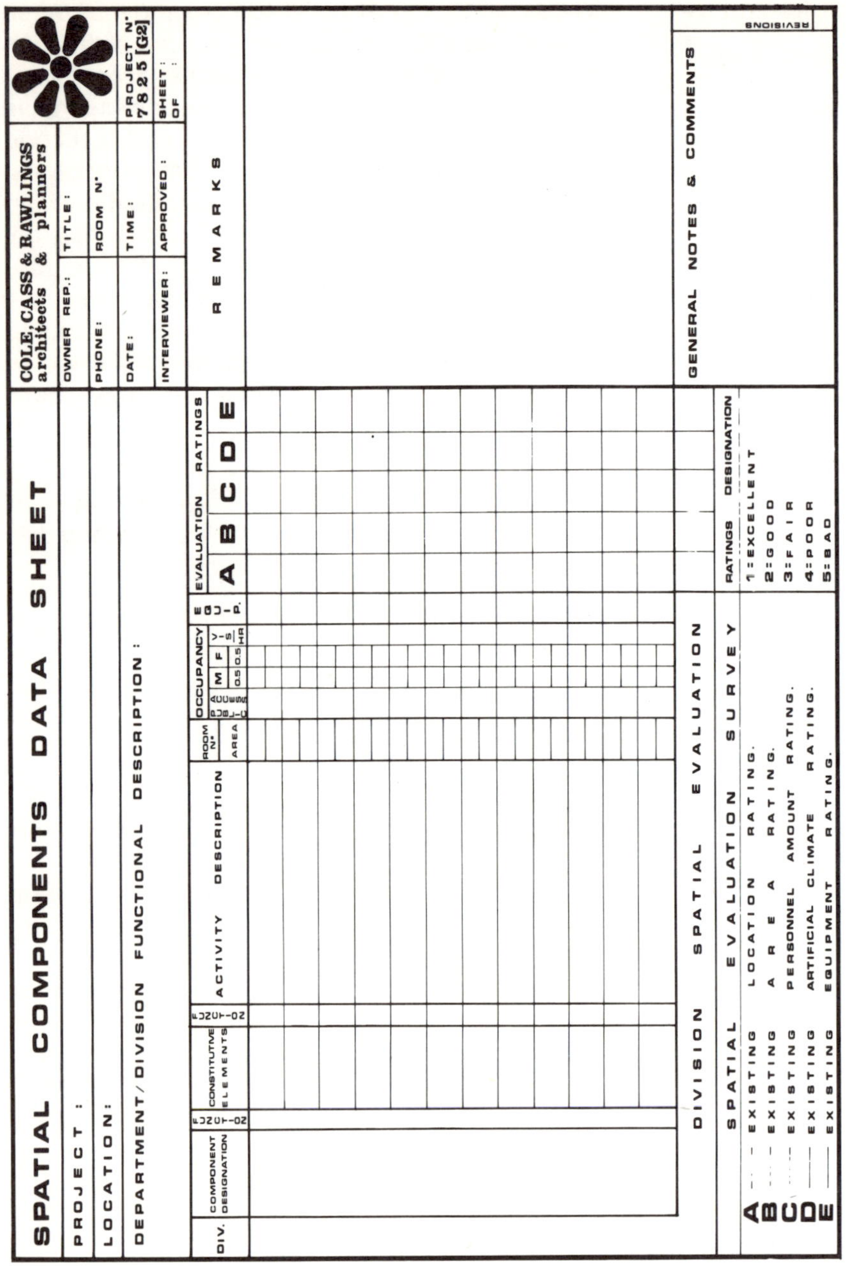

SPATIAL COMPONENTS DATA SHEET

COLE, CASS & RAWLINGS
architects & planners

OWNER REP.:   TITLE:
PHONE: ROOM N°
DATE: TIME:
INTERVIEWER: APPROVED:

PROJECT N° 7825 [G2]
SHEET: OF:

REMARKS

GENERAL NOTES & COMMENTS

REVISIONS

PROJECT:
LOCATION:
DEPARTMENT/DIVISION FUNCTIONAL DESCRIPTION:

DIV. | COMPONENT DESIGNATION | CONSTITUTIVE ELEMENTS | ACTIVITY DESCRIPTION | ROOM N° / AREA | OCCUPANCY | EVALUATION RATINGS: A B C D E

DIVISION SPATIAL EVALUATION

SPATIAL EVALUATION SURVEY

A — — EXISTING LOCATION RATING.
B — — EXISTING AREA RATING.
C — — EXISTING PERSONNEL AMOUNT RATING.
D — — EXISTING ARTIFICIAL CLIMATE RATING.
E — — EXISTING EQUIPMENT RATING.

RATINGS DESIGNATION
1:EXCELLENT
2:GOOD
3:FAIR
4:POOR
5:BAD

Figure 9.32

SPATIAL RELATIONSHIPS DATA SHEET

COLE, CASS & RAWLINGS
architects & planners

OWNER REP.:   TITLE:
PHONE: ROOM N°
DATE   TIME:
INTERVIEWER: APPROVED:

PROJECT N° 7825 [G2]
SHEET: OF:

REMARKS

GENERAL NOTES & COMMENTS

REVISIONS

PROJECT:
LOCATION:

INTRA DEPARTMENTAL RELATIONSHIPS ANALYSIS

DIV. | COMPONENT DESIGNATION | ELEMENT | RELATIONSHIPS DESCRIPTION | ROOM N° | TO: ELEMENT | V D T C F E

RELATIONSHIPS DESCRIPTIONS INDEXES

V: VALUE
1.ESSENTIAL
2.COMPLEMENTARY
3.NONESSENTIAL
4.UNDESIRABLE
5.UNACCEPTABLE

D: DYNAMICS
D·DIRECT
I·INDIRECT

T: TYPE
1.PHYSICAL
2.AUDIOVISUAL
3.AUDITIVE
4.VISUAL

C: CLASS
1·HUMAN/HUMAN
2·EQUIP/EQUIP
3·N.A.
4·HUMAN/EQUIP

F: FREQUENCY
1 CONSTANT
2 REPETITIVE
3 OCCASIONAL
4 SCARCE

E: PRESENT EVALUATION
1·EXCELLENT
2·GOOD
3·FAIR
4·POOR
5·BAD

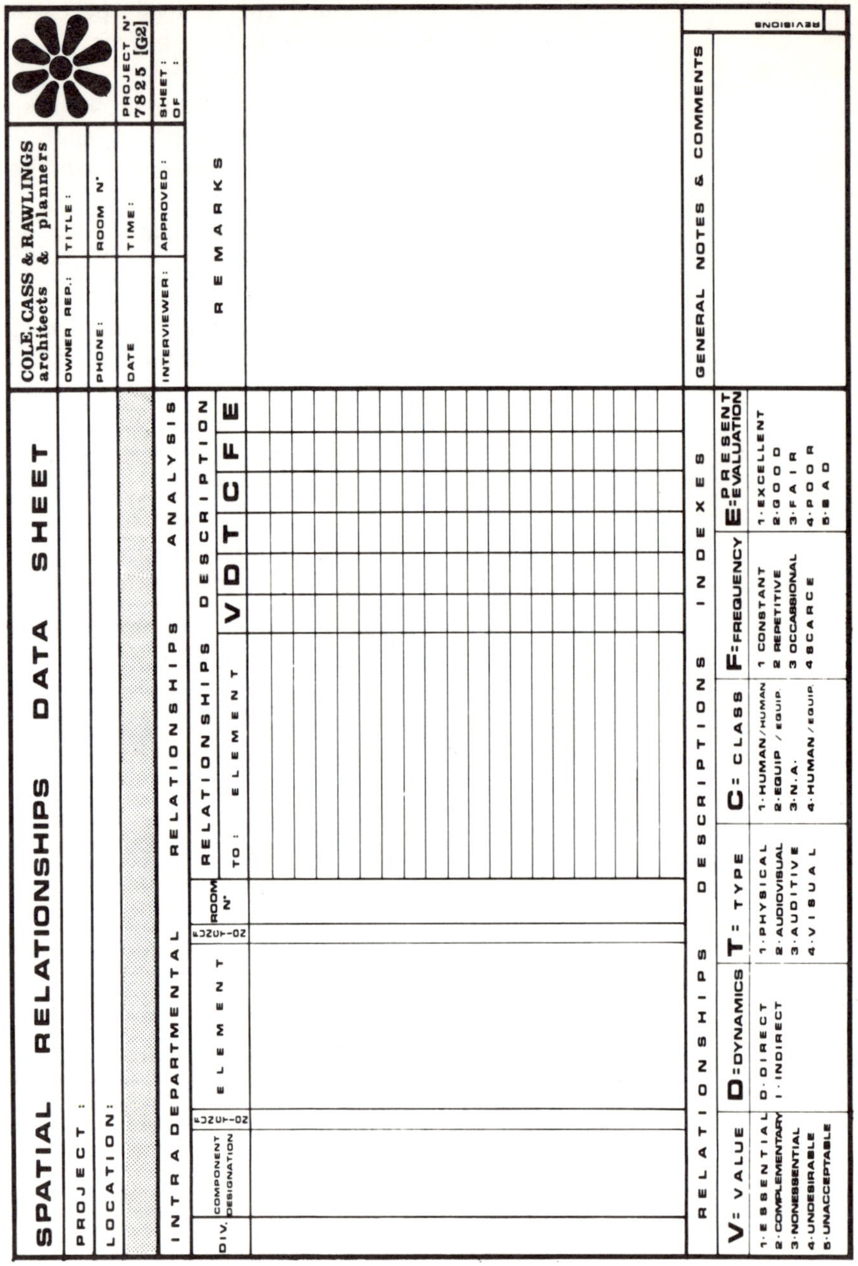

Figure 9.33

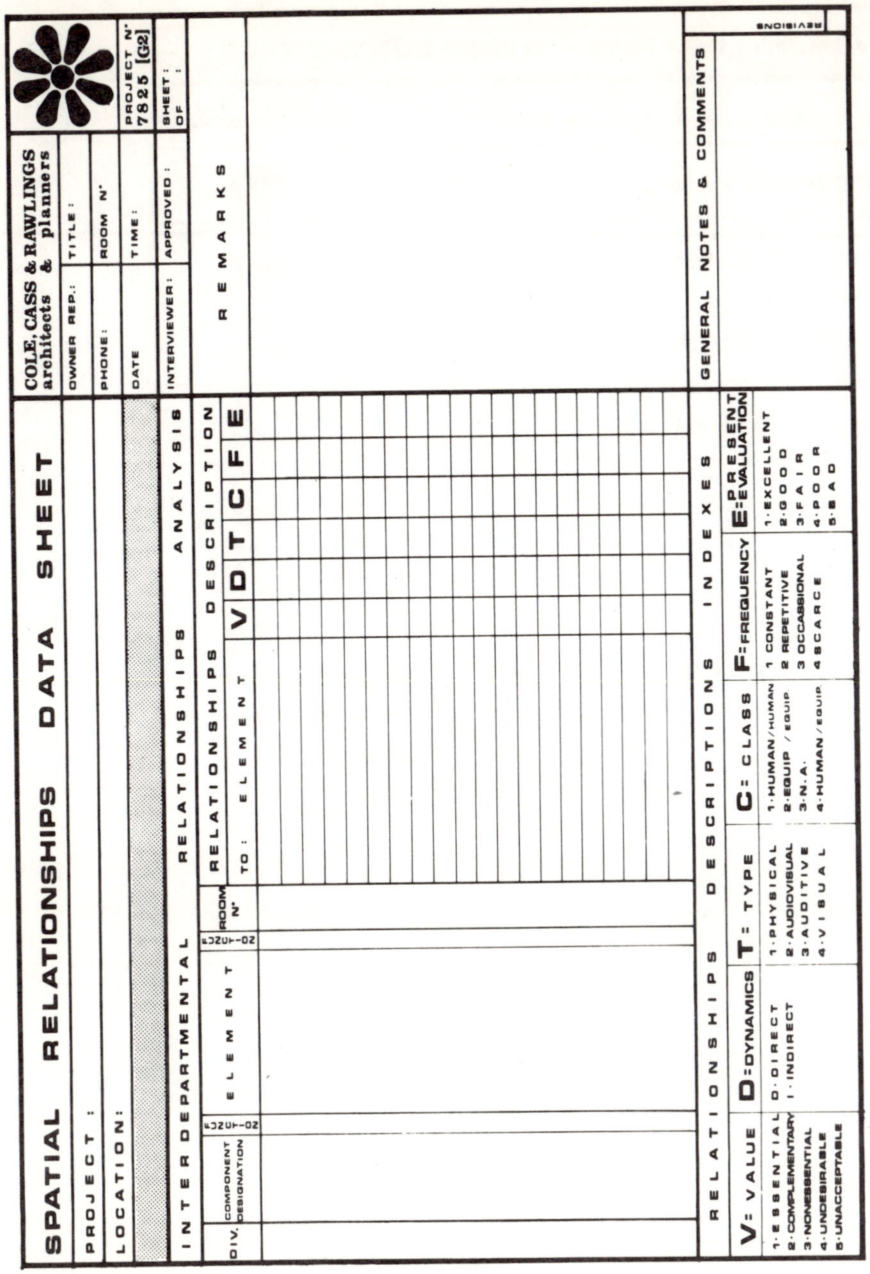

SPATIAL RELATIONSHIPS DATA SHEET

COLE, CASS & RAWLINGS
architects & planners

PROJECT:
LOCATION:

INTER DEPARTMENTAL RELATIONSHIPS DESCRIPTION ANALYSIS

OWNER REP.:   TITLE:
PHONE: ROOM N°:
DATE   TIME:
INTERVIEWER: APPROVED:

PROJECT N° 7825 [G2]
SHEET: OF:

REMARKS

RELATIONSHIPS DESCRIPTIONS INDEXES

V: VALUE
1 ESSENTIAL
2 COMPLEMENTARY
3 NONESSENTIAL
4 UNDESIRABLE
5 UNACCEPTABLE

D: DYNAMICS
D · DIRECT
I · INDIRECT

T: TYPE
1 · PHYSICAL
2 · AUDIOVISUAL
3 · AUDITIVE
4 · VISUAL

C: CLASS
1 · HUMAN / HUMAN
2 · EQUIP / EQUIP
3 · N.A.
4 · HUMAN / EQUIP

F: FREQUENCY
1 CONSTANT
2 REPETITIVE
3 OCCASIONAL
4 SCARCE

E: PRESENT EVALUATION
1 · EXCELLENT
2 · GOOD
3 · FAIR
4 · POOR
5 · BAD

DIV. | COMPONENT DESIGNATION | ELEMENT | ROOM N° | TO: ELEMENT | V D T C F E

GENERAL NOTES & COMMENTS

REVISIONS

Figure 9.34

SPATIAL PROJECTIONS DATA SHEET

COLE, CASS & RAWLINGS
architects & planners

PROJECT:
LOCATION:

GROWTH & CHANGE PROJECTIONS OBJECTIVES & BASIS DESCRIPTION

OWNER REP.:   TITLE:
PHONE: ROOM N°:
DATE   TIME:
INTERVIEWER: APPROVED:

PROJECT N° 7825 [G2]
SHEET: OF:

REMARKS

GROWTH CHANGE

L: LOCATION
1 — SITE
2 — NEIGHBOR'D
3 — CITY
4 — COUNTY

P: PLACEMENT &
D: DIRECTION
1 — ADJACENT
1.1 HORIZONTAL
1.2 VERTICAL
2 REMOTE

E: EXTENT (OR SCOPE)
F — FULL
P — PARTIAL

F: FUNCTION
T — TOTAL
P — PARTIAL

DIV. | COMPONENT DESIGNATION | CONSTITUTIVE ELEMENTS | ROOM N° | YEAR 1970 | YEAR 1975 | YEAR 1980 | YEAR 1985 | YEAR 1990 | GROWTH CHANGE L P/D E F

TOTALS
FULL TIME OCCUPANTS
SQUARE FEET PER OCCUPANT
TOTAL AREA IN SQUARE FEET
PART TIME OCCUPANTS

SF AREA
0 SF AREA 0
0.5 SF AREA 0

GENERAL NOTES & COMMENTS

REVISIONS

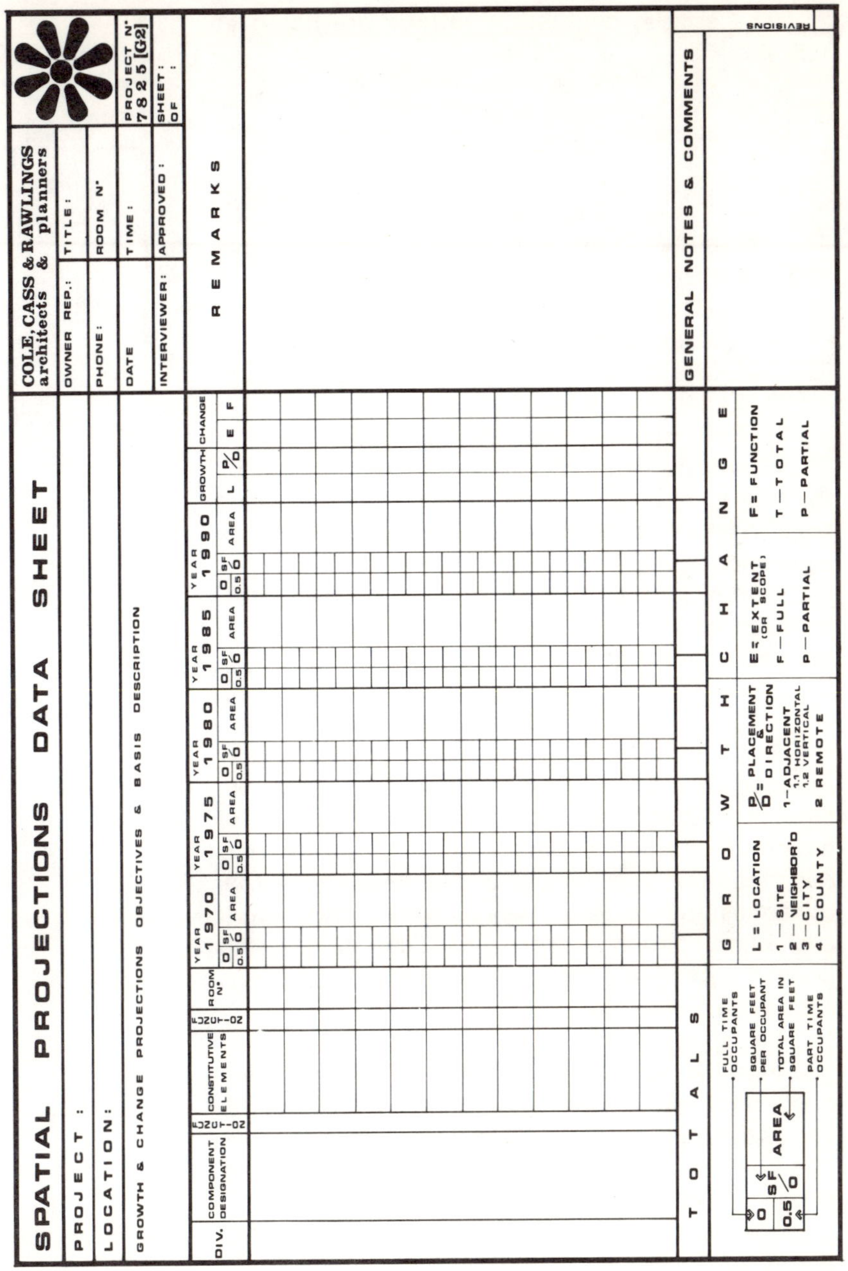

Figure 9.35

The spatial evaluation ratings are listed as follows:

Excellent — 1
Good — 2
Fair — 3
Poor — 4
Bad — 5

The issues these ratings relate to are listed on each individual sheet as follows:

A. Existing element location rating
B. Existing element area rating
C. Existing element personnel amount rating
D. Existing artificial climate rating
E. Existing equipment rating

The overall departmental space evaluation contains the ratings of these specific issues in a generalized fashion when applied to the department or division as a whole.

The graphic indications sometimes used throughout these charts represent the following temporal considerations:

----- Future function or element
+++++ Past function or element

The spatial relationships data sheets (either intradepartmental or interdepartmental, as shown in Figures 9.33 and 9.34) list the relationships among elements and the departments respectively and describe them with regard to their frequency, type, dynamics and value, as well as listing the present evaluation of such relationships. The ratings listed in each case follow the code listed at the bottom of each data collection form.

The spatial relationships themselves are further illustrated in spatial relationships diagrams which contain the basic information gathered through the use of the data collection forms. Value and description have been represented in the graphic manner shown in Figure 9.36.

The spatial development data sheets list the temporal development of the analyzed elements from the year 1970 through the year 1990 by means of the area and occupancy listed in each case. The occupancy square is divided into a top and bottom portion to signify full-time or part-time employees (see Figure 9.35).

RELATIONSHIPS GRAPHIC REPRESENTATIONS

ITEM/CODE		DESCRIPTIONS	GRAPHIC DESIGNATION
VALUES	1	ESSENTIAL	●
	2	COMPLEMENTARY	◑
	3	NON ESSENTIAL	○
	4	UNDESIRABLE	⊘
	5	UNACCEPTABLE	⊗
DYNAMICS	D	DIRECT	——
	I	INDIRECT	∿

Figure 9.36

The final analysis sheets total these results at the bottom of the page. The graphic symbol of Past (+++++) and Future (-----) are used to signify the specific temporality of listed elements.

To further illustrate the analysis and evaluation reports provided by these forms, a spatial and equipment layout has been made of each surveyed department or division. These spatial and equipment layouts illustrate only the conditions prevailing at the specific time the interviews were conducted.

In considering the application and use of the preceeding data collection forms on a general basis, it is important to take into account the following remarks:

1. All listed areas are given in square feet and represent either an approximation of present spatial parameters in the case of existing conditions, or the MINIMUM areas required for the optimum performance of the functions to be carried out within these limits in the case of spatial developments or projections.

2. Most of the functions statements listed in the spatial analysis data sheets have been obtained directly from the 1978 Westchester County Organization and Functions Report, but each has been verified with the corresponding department or division head.

3. The spatial evaluation responses and remarks columns contain not only comments and statements made by the respective department heads at the time of the interviews, but also footnotes and clarifications added by the interviewer during that period.

4. The spatial development areas listed as programmatic statements for the years 1980, 1985 and 1990, as well as the occupancy projections, have been established by approximations given by the specific department or division directors, and by adjustments based on previous spatial and occupancy developments and reviews by the county officials and the office of the administrative sector. The areas and occupancies listed represent, again, the MINIMUM requirements based on an analysis of past functional development plus current trends regarding operational performances, department function, etc. In some cases, footnotes indicate other factors taken into account when dealing specifically with foreseeable occurrences which might affect trends of development in one way or another.

5. The letters NA when used in the spatial analysis forms indicate "not available" or "non-applicable."

6. Basement storage areas, although recorded in the spatial analysis data sheet, have not been taken into account in the programming portion of this study; that is, they have not been included in either the spatial development portion or the tabulation of the general summaries. The reason for this omission is that such spaces, by reason of their functional role as well as their remote location with respect to the actual departmental functions, do not constitute a major determinant in the functional programming of the operations themselves, but rather a need which can be satisfied by means of several other alternatives. The subject of inactive storage is discussed in more detail in the Recommendations section of this report.

7. To facilitate the understanding of the final results of the departmental spatial analysis studies, a general summary of areas and occupancy developments has been provided in this report. This summary is intended to simplify a comprehensive view of all departments and divisions, but any clarifications regarding the spatial requirements or occupancy figures must be obtained by referring directly to the specific department or division study found in this document.

8. The actual occupancies recorded in these data collection forms might not coincide with those on the personnel files on a departmental basis, because part-time employees have been listed as 0.5 occupant and when added to others produce either fractions or units not physically measurable. Only those

employees who actually occupy workstations have been surveyed. Transient occupants such as field workers, etc., have not
been included as significant elements in the
spatial requirements considerations. Here
again, the recorded figures represent only the
occupancy at the specific time when the
spatial survey was conducted.

Regarding the employee survey structure, the
format in Figure 9.37 illustrates the questionnaire
distributed among the participants.

The results obtained can be found in the
Employee Survey section of this report.

PART II — DEPARTMENTAL ANALYSIS SUMMARIES

As described in the Spatial Analysis Structure
portion of this report, the following pages contain a
detailed study of the surveyed departments and
divisions in the form of the previously outlined Data
Collection Sheets, Spatial Analysis Layouts, Spatial
Relationships Diagrams and Equipment Inventories.

The order of these surveys is the same as that
established in the description of Part I of this
document. Conclusions, recommendations and
general comments regarding these studies can be
found in the Spatial Recommendations section of
this document. (See Figures 9.38 through 9.49)

Interdepartmental Relationships Diagram. In a
generalized fashion, the spatial interactions and
relationships of the surveyed departments can be
illustrated as shown in Figure 9.50.

SPATIAL STUDY QUESTIONNAIRE

DEPARTMENT: ______________________

DIVISION : ______________________

Name and Title of person completing this questionnaire:

DATE: ______________   PROJECT NO: **7825 [G2]**

EVALUATION RATINGS:

1. EXCELLENT
2. GOOD
3. FAIR
4. POOR
5. BAD

PLEASE ANSWER QUESTIONS NO. 1 & 2 BY RATING EACH ISSUE WITH THE EVALUATION RATINGS LISTED ABOVE, AND QUESTION NO. 3 BY PLACING AN "X" IN THE CORRESPONDING SQUARE.

DO NOT WRITE HERE

Reviewed ______________

CODE __________ RECORD __________

1. REGARDING YOUR DEPARTMENT AREA IN GENERAL; HOW WOULD YOU RATE THE PRESENT PERFORMANCE OF THE FOLLOWING?:   RATINGS   COMMENTS

   1.1 LOCATION WITHIN THE COURTHOUSE BUILDING . . . .
   1.2 AREA (SIZE) . . . .
   1.3 HEATING . . . .
   1.4 COOLING . . . .
   1.5 VENTILATION . . . .
   1.6 LIGHTING . . . .
   1.7 CIRCULATIONS . . . .
   1.8 NOISE AND SOUND BARRIERS . . . .
   1.9 EQUIPMENT . . . .
   1.10 SECURITY . . . .

2. REGARDING YOUR PERSONAL WORKSTATION; HOW WOULD YOU RATE THE FOLLOWING?:

   2.1 LOCATION WITHIN YOUR DEPARTMENT . . . .
   2.2 AREA (SIZE) . . . .
   2.3 HEATING . . . .
   2.4 COOLING . . . .
   2.5 VENTILATION . . . .
   2.6 LIGHTING . . . .
   2.7 VISUAL PRIVACY . . . .
   2.8 SOUND PRIVACY . . . .
   2.9 EQUIPMENT . . . .
   2.10 SECURITY . . . .

3. PLEASE INDICATE IN WHICH OF THE FOLLOWING AREAS YOU THINK THE COURTHOUSE BUILDING NEEDS IMPROVEMENT.

   3.1 EXTERIOR APPEARANCE . . . .
   3.2 INTERIOR APPEARANCE . . . .
   3.3 DEPARTMENTAL ZONING . . . .
   3.4 BUILDING SIZE . . . .
   3.5 MECHANICAL SYSTEMS (HEATING,
       VENTILATING & AIR CONDITIONING) . . . .
   3.6 ELECTRICAL SYSTEMS . . . .
   3.7 TOILET FACILITIES . . . .
   3.8 CIRCULATIONS . . . .
   3.9 ELEVATORS . . . .
   3.10 SECURITY . . . .

GENERAL COMMENTS

____
____
____
____
____

Figure 9.37

SPATIAL COMPONENTS DATA SHEET

COLE, CASS & RAWLINGS architects & planners

PROJECT: SPATIAL ANALYSIS SURVEY OF WESTCHESTER COUNTY GOVERNMENT OFFICES

LOCATION: 11073 JEFFERSON DRIVE, GREEN OAKS, MO.

DEPARTMENT/DIVISION FUNCTIONAL DESCRIPTION: ADMN. DEPT.-PERSONNEL DIV. FUNCTION: TO OFFER EQUAL EMPLOYMENT OPPORTUNITIES TO ALL APPLICANTS, SUBSEQUENTLY PROVIDING A QUALIFIED STAFF OF EMPLOYEES FOR COUNTY AGENCIES. PROGRAMS OPERATED BY THIS DIVISION INVOLVE EMPLOYEE RECRUITMENT, WAGE & SALARY ADMN., MANPOWER NEEDS ANALYSES, COLLECTIVE BARGAIN., UNION CONTRACT ADMN.& EMPLOY. TRAIN.

OWNER REP.:	TITLE:
KEN LARSON	MANAGER
PHONE: 444-0944	**ROOM N°** 104
DATE: 3-27-78	**TIME:** 10:00 A.M.
INTERVIEWER: JL	**APPROVED:**

PROJECT N° 7 8 2 5 [G2]

SHEET: 1 **OF:** 1

DIV.	COMPONENT DESIGNATION	FUNCTION	CONSTITUTIVE ELEMENTS	FUNCTION	ACTIVITY DESCRIPTION	ROOM N° / AREA	OCCUPANCY PUBLIC ACCESS	OCCUPANCY M 0.5	OCCUPANCY F 0.5	VIS/HR	EQUIP.	A	B	C	D	E
A D M I N I S T R A T I O N / P E R S O N N E L	MANAGEMENT (E)		MANAGER	E	GENERAL ADMINISTRATION UNION CONTRACTS NEGOTIATIONS	104 / 160	NO	1 / 0	0 / 0	6 / 2	(1)	4 (3)	3 (4)	NA	5 (5)	3 (4)
			CONFERENCE	C	ORAL BOARDS, CONFERENCES & INTERVIEWS	104A / 200	NO	0 / 0	0 / 0	20 / 10	(4)	2	4 (11)	NA	4 (13)	4
			ASSISTANT MANAGER	E	PERSONNEL ANALYSIS, MGMT ASSIST. EXAMS DEVELOP. RECRUIT	105 / 154	NO	1 / 0	0 / 0	6 / 1	(2)	4 (3)	2	2	4 (5)	2
			PERSONNEL TECHNICIAN	E	NON-PROFESSIONAL RECRUIT. & CLASSIFICATION	107 / 152	NO	1 / 0	0 / 0	6 / 1	(3)	(6) 4	2	2	(6) 4	2
			PERSONNEL ANALYST	C	HELP ASSISTANT MANAGER WITH PERSONNEL ANALYSIS FUNCTIONS	NA / NA	NO	— / —	— / —	NA	NA	NA →				
	CLERICAL (E)		RECEPTION	E	RECEPTION AND APPLICATION FORMS	107 / 250	YES	0 / 0	0 / 0	80 / 15	(5)	5 (10)	5 (11)	2	5 (10)	4 (12)
			APPLICA. FILL. (20)	E	EMPLOYMENT APPLICATION FORMS FILLING	NA / NA	YES	— / —	— / —	60 / 8	NA	NA →				
			GENERAL OFFICE	E	CLERICAL FUNCTIONS. INCLUDES SUPER. WORKSTATION & FILES	106 / 370	NO	0 / 0	4 / (1)	6 (2) / 1	(5)	2 (8)	5 (9)	2	5 (8)	3 (7)
	STORAGE (C)		RECORDS STORAGE	E	REMOTE/PRESENTLY (19) RECRUIT. FILES, REGISTERS, APPLICA. PRINT	G1-8 / 50	NO	0 / 0	0 / 0	NA	NA	5 (15)	5 (11)	NA	NA	5 (16)

REMARKS

(1) OCCASIONALLY, ONE WORKSTATION IS USED BY A LIMITED TIME EMPLOYEE OR A PART-TIME TYPIST DEPENDING ON SEASONAL DEMANDS.
(2) VISITORS LISTED ARE FOR SUPERVISOR ONLY.
(3) BAD TRAFFIC PATTERN IN FRONT OF OFFICE.
(4) COULD BE LARGER; NEEDS A CONFERENCE TABLE.
(5) NOISE TRANSMISSION IS HIGH. THIS IS BAD DUE TO THE CRITICAL/CONFIDENTIAL NATURE OF SOME CONFERENCES.
(6) VISUAL DISTRACTION AND NOISE FROM CORRIDOR.
(7) DEPARTMENT WOULD PREFER MAG. CARDS.
(8) BAD TRAFFIC PATTERN AND NOISE LEVEL.
(9) TOO SMALL. NEEDS MORE SPACE.
(10) RECEPTION/GENERAL OFFICE VISUAL CONTACT WITH VISITORS IS NOT DESIRABLE.
(11) TOO SMALL.
(12) COUNTER IS TOO HIGH FOR COMFORTABLE WRITING.
(13) TOO COLD.
(14) NEEDS OVERHEAD PROJECTOR, SCREEN, AND T-SHAPED CONFERENCE TABLE.
(15) TOO FAR AWAY.
(16) NEEDS MORE SHELVING.
(17) HVAC CONTROLS NEED IMPROVEMENT.
(18) PERSONNEL DIVISION SHOULD REMAIN ON GROUND FLOOR.
(19) SHOULD BE NEAR DEPARTMENT.
(20) LOCATE ADJACENT TO RECEPTION.

DIVISION SPATIAL EVALUATION

A	B	C	D	E
2 (18)	4 (11)	2	5 (17)	3

SPATIAL EVALUATION SURVEY

		RATINGS DESIGNATION
A	——— EXISTING LOCATION RATING.	1 = EXCELLENT
B	——— EXISTING AREA RATING.	2 = GOOD
C	——— EXISTING PERSONNEL AMOUNT RATING.	3 = FAIR
D	——— EXISTING ARTIFICIAL CLIMATE RATING.	4 = POOR
E	——— EXISTING EQUIPMENT RATING.	5 = BAD

GENERAL NOTES & COMMENTS

NOISE LEVEL: VERY BAD. DIVISION COMPLAINS ABOUT EXCESSIVE NOISE LEVELS DURING MORNING AND EARLY AFTERNOON HOURS.
PARKING: LACK OF CONVENIENT PARKING PRESENTS A SERIOUS PROBLEM FOR APPLICANTS AS WELL AS COUNTY EMPLOYEES FROM OTHER LOCATIONS.

REVISIONS 25

Figure 9.38

159

1 A DESK & CHAIR
  B REFERENCE TABLE
  C CHAIR
  D LATERAL FILES
  E SHELVING UNIT
  F FILE CABINET

2 A DESK & CHAIR
  B REFERENCE TABLE W/ 2 CHAIRS
  C CHAIR
  D FILE CABINET
  E WOOD SHELVING
  F STEEL SHELVING UNIT

3 A DESK & CHAIR
  B WOOD SHELVING
  C CHAIR

4 A CONFERENCE TABLE W/ 9 CHAIRS
  B STORAGE UNIT
  C BLACKBOARD
  D COAT RACK
  E COFFEE TABLE

5 A COUNTERTOP W/ FILE CABS. UNDER
  B WORK TABLE
  C CHAIRS
  D DESK & CHAIR
  E TABLET ARM CHAIRS
  F DESK W/ CHAIR
  G FILE CABINET
  H TABLE
  J FILE CABINETS
  K STORAGE UNIT

**ADMINISTRATION
PERSONNEL .**

EQUIPMENT LIST

3 · 28 · '78

Figure 9.39

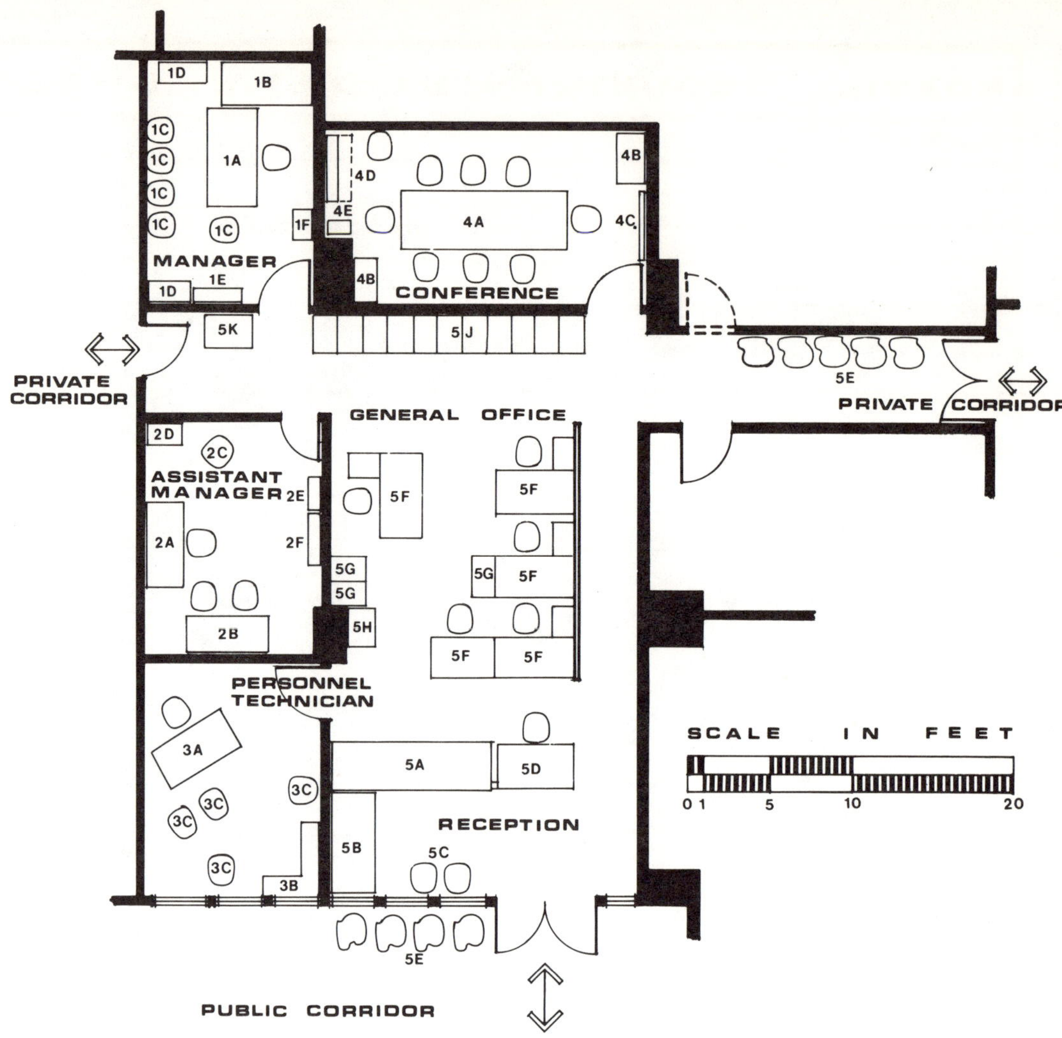

**ADMINISTRATION DEPARTMENT
PERSONNEL DIVISION**

EXISTING SPATIAL LAYOUT 3·27·'78

DIV.	COMPONENT DESIGNATION	FUNCTION	ELEMENT	FUNCTION	ROOM N°	RELATIONSHIPS DESCRIPTION — TO: ELEMENT	V	D	T	C	F	E
ADMINISTRATION / PERSONNEL	MANAGEMENT	E	MANAGER & CONFERENCE (2)	E	104 & 104A	ASSISTANT PERSONNEL MANAGER	1	D	1	1	1	2
						PERSONNEL TECHNICIAN	1	D	4		2	2
						GENERAL OFFICE	1	D	1		3	2
						RECEPTION	4	D/I	1-4		NA	2 (1)
						PERSONNEL ANALYST	3	D	1		4	NA
			ASSISTANT PERSONNEL MANAGER	E	105	PERSONNEL TECHNICIAN	1	D	1		1	4 (3)
						GENERAL OFFICE	1	D	1		1	2
						RECEPTION	4	D	1-4		4	2
						PERSONNEL ANALYST	1	D	1		1	NA
			PERSONNEL TECHNICIAN	E	107	GENERAL OFFICE	2	D	1		1	1
						RECEPTION	3	D	1		3	2
						PERSONNEL ANALYST	1	D	1		2	NA
			PERSONNEL ANALYST	C	NA	MANAGER AND ASSISTANT	5	NA	→		→	NA
						GENERAL OFFICE	2	D	1		3	NA
						RECEPTION & APPLICATION FILL.	3	D	1		4	NA
	CLERICAL	E	RECEPTION	E	107	CONFERENCE ROOM	2	I	3		3	2
						STORAGE	3	NA	→		→	NA
			GENERAL OFFICE	E	106	RECEPTION	1	D	1 (4)		1	2
						PERSONNEL ANALYST	2	D	1		2	NA
						STORAGE	2	D	1		4	5 (5)

REMARKS

(1) INTERCOM TO RECEPTIONIST WOULD HELP ANNOUNCEMENT OF VISITORS.

(2) DOOR BETWEEN OFFICE AND CONFERENCE ROOM WOULD AVOID HAVING TO CROSS GENERAL OFFICE FILING AREA WHEN ATTENDING MEETINGS, ETC.

(3) MUST CROSS GENERAL OFFICE TO ESTABLISH RELATIONSHIPS.

(4) GENERAL OFFICE EMPLOYEES FUNCTION AS OCCASIONAL BACK-UP FOR RECEPTIONIST.

(5) TOO FAR AWAY.

RELATIONSHIPS DESCRIPTIONS INDEXES

V = VALUE	**D** = DYNAMICS	**T** = TYPE	**C** = CLASS	**F** = FREQUENCY	**E** = PRESENT EVALUATION
1 · ESSENTIAL	D · DIRECT	1 · PHYSICAL	1 · HUMAN / HUMAN	1 CONSTANT	1 · EXCELLENT
2 · COMPLEMENTARY	I · INDIRECT	2 · AUDIOVISUAL	2 · EQUIP / EQUIP.	2 REPETITIVE	2 · GOOD
3 · NONESSENTIAL		3 · AUDITIVE	3 · N.A.	3 OCCASSIONAL	3 · FAIR
4 · UNDESIRABLE		4 · VISUAL	4 · HUMAN / EQUIP.	4 SCARCE	4 · POOR
5 · UNACCEPTABLE					5 · BAD

GENERAL NOTES & COMMENTS

UNLESS OTHERWISE TABULATED, ALL RELATIONSHIPS LISTED ARE RECIPROCAL.

REVISIONS

28

Figure 9.40

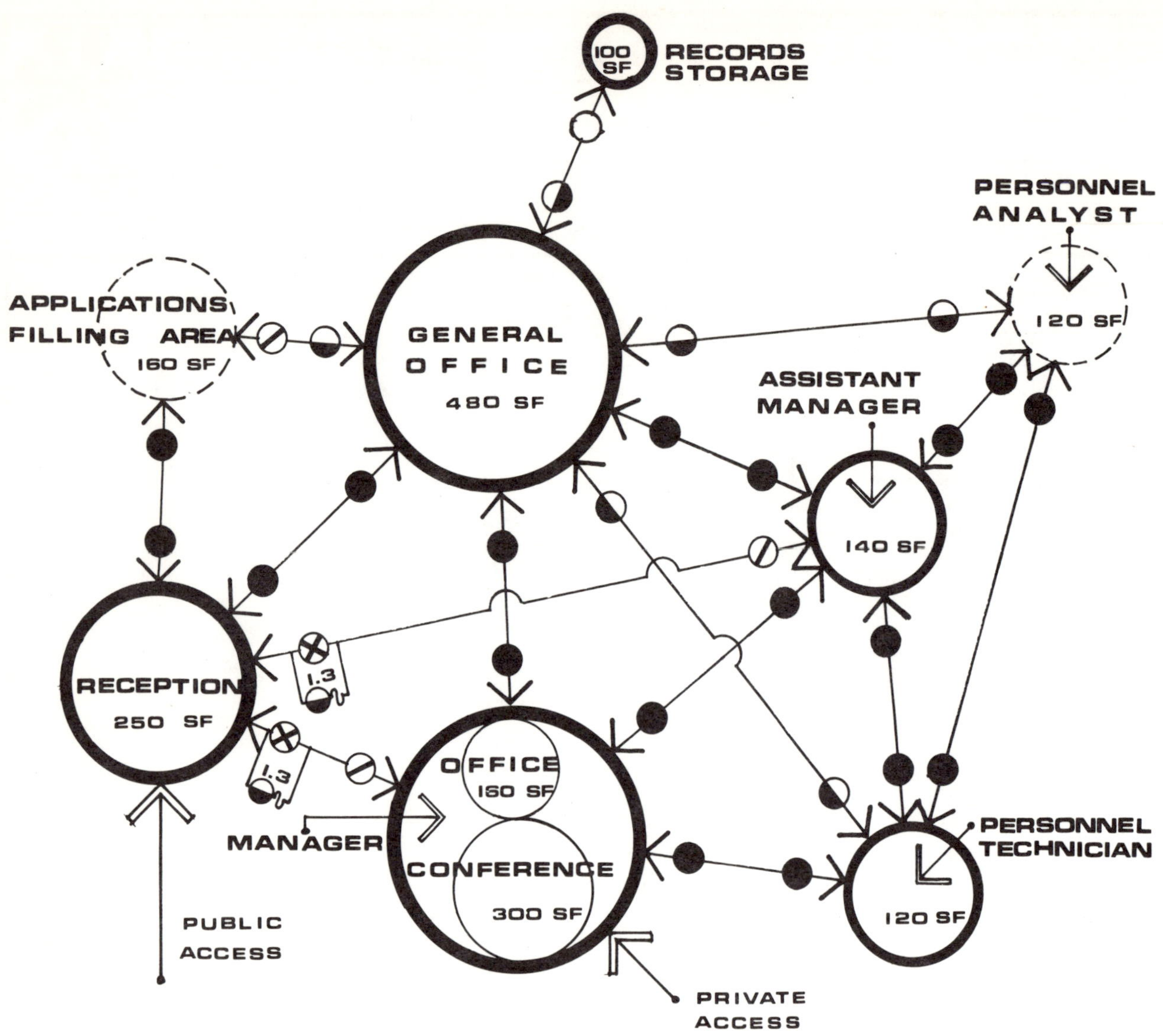

COUNTY OFFICIAL ADMINISTRATIVE SECTION

PERSONNEL DIVISION

INTRADEPARTMENTAL RELATIONSHIPS DIAGRAM

NO SCALE

3 - 27 - 78

 Figure 9.41

SPATIAL RELATIONSHIPS DATA SHEET

COLE, CASS & RAWLINGS architects & planners

PROJECT: SPATIAL ANALYSIS SURVEY OF WESTCHESTER COUNTY GOVERNMENT OFFICES

LOCATION: 11073 JEFFERSON DRIVE, GREEN OAKS, MO.

OWNER REP.: KEN LARSON
TITLE: MANAGER
PHONE: 444-0944
ROOM N° 104
DATE 3-27-78
TIME: 10:00 A.M.
PROJECT N° 7825 [G2]

INTER DEPARTMENTAL RELATIONSHIPS ANALYSIS

INTERVIEWER: JL
APPROVED:
SHEET: 1 **OF** : 1

DIV.	COMPONENT DESIGNATION	FUNCTION	ELEMENT	FUNCTION	ROOM N°	TO: ELEMENT	V	D	T	C	F	E
ADMINISTRATION / PERSONNEL	MANAGEMENT	E	MANAGER			CLERK OF COURTS	1	I	3	1	2	2
					104	CORPORATE COUNSEL	2	I	3	1	3	2
			ASSISTANT PERSONNEL MANAGER		105	COUNTY SHERIFF DEPARTMENT	2	I	3	1	2	3
				E		COUNTY REGISTER OF DEEDS	1	D	1	1	2	1
			PERSONNEL TECHNICIAN			COUNTY CLERK/ACCOUNTING	1	D	1	1	1	1
					107	ADMINISTRATION DEPARTMENT	1	D	1	1	1	1
	CLERICAL	E	GENERAL OFFICE	E	106	COUNTY CLERK/ACCOUNTING DIV.	1	D	1 (1)	1	2	2
						(TO ALL REMAINING DEPARTMENTS)	3	I	3 (2)	1	2	2

REMARKS

(1) BECAUSE OF PROXIMITY TO ACCOUNTING OFFICE, MOST CONTACT IS PHYSICAL, ALTHOUGH IT COULD BE AUDITIVE.

(2) PERSONNEL DIVISION DOES NOT NEED IMMEDIATE OR DIRECT PHYSICAL CONTACT WITH OTHER DEPARTMENTS.

RELATIONSHIPS DESCRIPTIONS INDEXES

V = VALUE	D = DYNAMICS	T = TYPE	C = CLASS	F = FREQUENCY	E = PRESENT EVALUATION
1 · ESSENTIAL	D · DIRECT	1 · PHYSICAL	1 · HUMAN / HUMAN	1 CONSTANT	1 · EXCELLENT
2 · COMPLEMENTARY	I · INDIRECT	2 · AUDIOVISUAL	2 · EQUIP / EQUIP.	2 REPETITIVE	2 · GOOD
3 · NONESSENTIAL		3 · AUDITIVE	3 · N.A.	3 OCCASSIONAL	3 · FAIR
4 · UNDESIRABLE		4 · VISUAL	4 · HUMAN / EQUIP.	4 SCARCE	4 · POOR
5 · UNACCEPTABLE					5 · BAD

GENERAL NOTES & COMMENTS

NOTE: LISTED RELATIONSHIPS ARE NOT RECIPROCAL; SEE INDIVIDUAL DEPARTMENT'S INTERDEPARTMENTAL RELATIONSHIPS DATA SHEETS FOR INFORMATION.

REVISIONS 29

Figure 9.42

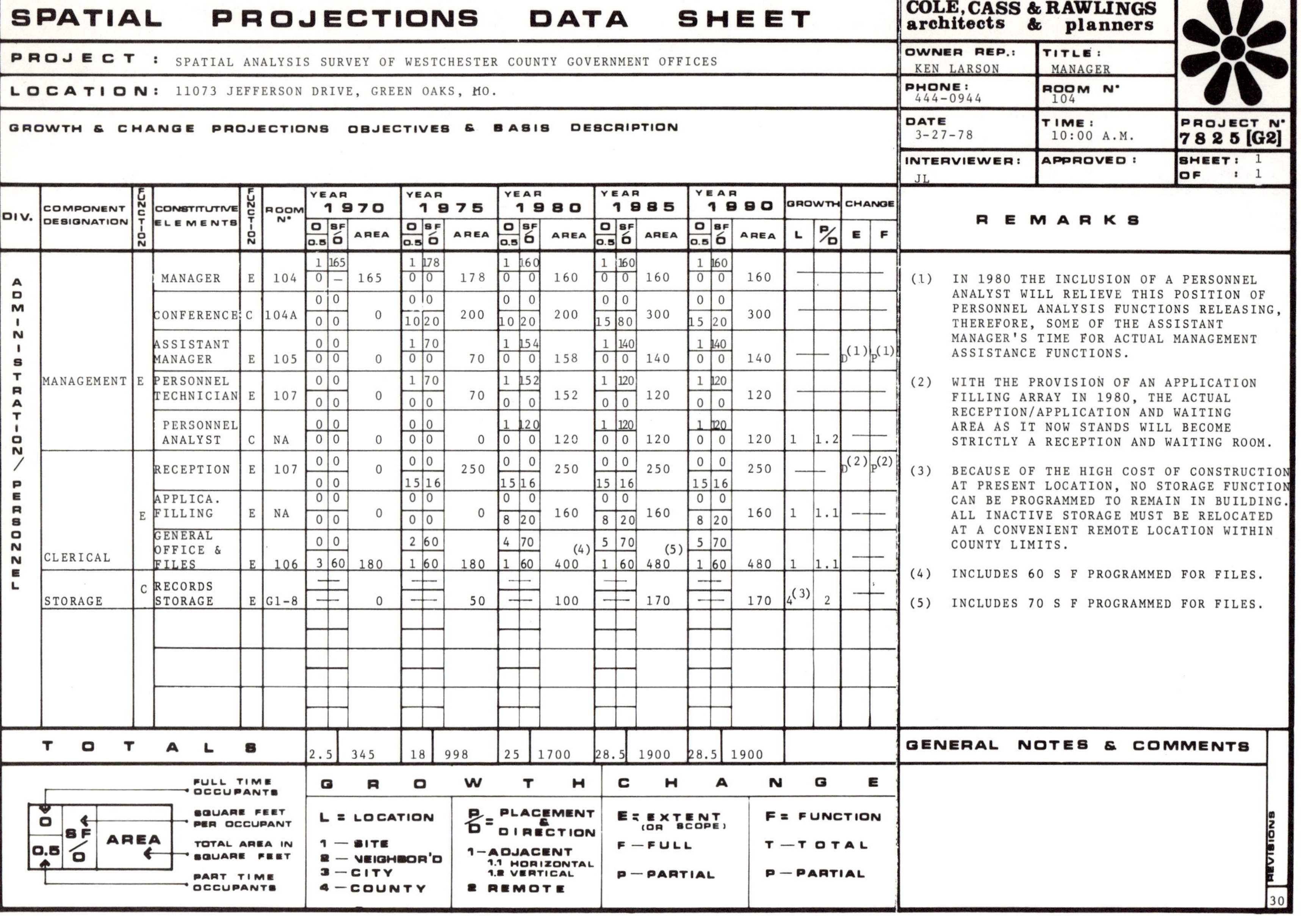

SPATIAL PROJECTIONS DATA SHEET

COLE, CASS & RAWLINGS architects & planners

PROJECT: SPATIAL ANALYSIS SURVEY OF WESTCHESTER COUNTY GOVERNMENT OFFICES

LOCATION: 11073 JEFFERSON DRIVE, GREEN OAKS, MO.

GROWTH & CHANGE PROJECTIONS OBJECTIVES & BASIS DESCRIPTION

OWNER REP.: KEN LARSON	TITLE: MANAGER
PHONE: 444-0944	ROOM N° 104
DATE 3-27-78	TIME: 10:00 A.M.
INTERVIEWER: JL	APPROVED:

PROJECT N° 7825 [G2] — SHEET: 1 OF: 1

Legend for occupant cells: **O** = full time occupants (top), **0.5** = part time occupants (bottom), **SF/O** = square feet per occupant, **AREA** = total area in square feet. Each year cell below is given as O(full/part), SF/O(top/bottom), AREA.

DIV.	COMPONENT DESIGNATION	FUNC	CONSTITUTIVE ELEMENTS	FUNC	ROOM N°	1970 O	1970 SF	1970 AREA	1975 O	1975 SF	1975 AREA	1980 O	1980 SF	1980 AREA	1985 O	1985 SF	1985 AREA	1990 O	1990 SF	1990 AREA	GROWTH L	GROWTH P/D	CHANGE E	CHANGE F
ADMINISTRATION / PERSONNEL	MANAGEMENT (E)		MANAGER	E	104	1/0	165/—	165	1/0	178/0	178	1/0	160/0	160	1/0	160/0	160	1/0	160/0	160				
			CONFERENCE	C	104A	0/0	0/0	0	0/10	0/20	200	0/10	0/20	200	0/15	0/80	300	0/15	0/20	300				
			ASSISTANT MANAGER	E	105	0/0	0/0	0	1/0	70/0	70	1/0	154/0	158	1/0	140/0	140	1/0	140/0	140			D(1)	P(1)
			PERSONNEL TECHNICIAN	E	107	0/0	0/0	0	1/0	70/0	70	1/0	152/0	152	1/0	120/0	120	1/0	120/0	120				
			PERSONNEL ANALYST	C	NA	0/0	0/0	0	0/0	0/0	0	1/0	120/0	120	1/0	120/0	120	1/0	120/0	120	1	1.2		
	CLERICAL (E)		RECEPTION	E	107	0/0	0/0	0	0/15	0/16	250	0/15	0/16	250	0/15	0/16	250	0/15	0/16	250			D(2)	P(2)
			APPLICA. FILLING	E	NA	0/0	0/0	0	0/0	0/0	0	0/8	0/20	160	0/8	0/20	160	0/8	0/20	160	1	1.1		
			GENERAL OFFICE & FILES	E	106	0/3	0/60	180	2/1	60/60	180	4/1	70/60	400 (4)	5/1	70/60	480 (5)	5/1	70/60	480	1	1.1		
	STORAGE (C)		RECORDS STORAGE	E	G1-8	—	—	0	—	—	50	—	—	100	—	—	170	—	—	170	4(3)	2		
TOTALS						2.5		345	18		998	25		1700	28.5		1900	28.5		1900				

REMARKS

(1) IN 1980 THE INCLUSION OF A PERSONNEL ANALYST WILL RELIEVE THIS POSITION OF PERSONNEL ANALYSIS FUNCTIONS RELEASING, THEREFORE, SOME OF THE ASSISTANT MANAGER'S TIME FOR ACTUAL MANAGEMENT ASSISTANCE FUNCTIONS.

(2) WITH THE PROVISION OF AN APPLICATION FILLING ARRAY IN 1980, THE ACTUAL RECEPTION/APPLICATION AND WAITING AREA AS IT NOW STANDS WILL BECOME STRICTLY A RECEPTION AND WAITING ROOM.

(3) BECAUSE OF THE HIGH COST OF CONSTRUCTION AT PRESENT LOCATION, NO STORAGE FUNCTION CAN BE PROGRAMMED TO REMAIN IN BUILDING. ALL INACTIVE STORAGE MUST BE RELOCATED AT A CONVENIENT REMOTE LOCATION WITHIN COUNTY LIMITS.

(4) INCLUDES 60 S F PROGRAMMED FOR FILES.

(5) INCLUDES 70 S F PROGRAMMED FOR FILES.

GENERAL NOTES & COMMENTS

Legend:
- O = FULL TIME OCCUPANTS
- SF/O = SQUARE FEET PER OCCUPANT
- AREA = TOTAL AREA IN SQUARE FEET
- 0.5 = PART TIME OCCUPANTS

GROWTH
- L = LOCATION
  - 1 — SITE
  - 2 — NEIGHBOR'D
  - 3 — CITY
  - 4 — COUNTY
- P/D = PLACEMENT & DIRECTION
  - 1 - ADJACENT
    - 1.1 HORIZONTAL
    - 1.2 VERTICAL
  - 2 REMOTE

CHANGE
- E = EXTENT (OR SCOPE)
  - F — FULL
  - P — PARTIAL
- F = FUNCTION
  - T — TOTAL
  - P — PARTIAL

REVISIONS — 30

Figure 9.43

SPATIAL COMPONENTS DATA SHEET

COLE, CASS & RAWLINGS
architects & planners

OWNER REP.: MARK HAYDEN
TITLE: COUNTY CLERK
PHONE: 444-2141
ROOM N°: 112

PROJECT: SPATIAL ANALYSIS SURVEY OF WESTCHESTER COUNTY GOVERNMENT OFFICES

LOCATION: 11073 JEFFERSON DRIVE, GREEN OAKS, MO.

DEPARTMENT/DIVISION FUNCTIONAL DESCRIPTION: THE COUNTY CLERK FUNCTIONS AS SECRETARY OF THE BOARD OF SUPERVISORS, ATTENDS ALL MEETINGS & KEEPS RECORDS OF ITS PROCEEDINGS. SUPERVISES ALL COUNTY-WIDE ELECTIONS, NOMINATION PAPERS, PREPARES ELECTION BALLOTS, ISSUES MARRIAGE LICENSES & SERVES AS STATE AGENT FOR ISSUANCE OF HUNTING & FISH. LICENSES. ALSO SERVES AS TAX LISTER.

DATE: 3-30-78
TIME: 10:00 A.M.
PROJECT N°: 7825 [G2]
INTERVIEWER: JL
APPROVED:
SHEET: 1
OF: 1

DIV.	COMPONENT DESIGNATION	FUNCTION	CONSTITUTIVE ELEMENTS	FUNCTION	ACTIVITY DESCRIPTION	ROOM N° / AREA	OCCUPANCY PUBLIC ACCESS	M 0.5	F 0.5	VIS/HR	EQUIP.	A	B	C	D	E	REMARKS
C O U N T Y C L E R K	COUNTY CLERK	E	CLERK'S OFFICE	C	USED MOSTLY FOR PRIVATE CONFERENCES (3)	112 / 340	NO	0 / 1	0 / 0	0	①1	2	1	NA	2	2	
			WORK-STATION (6)	E	SPENDS 90% OF TIME AT WORK-STATION IN GENERAL OFFICE	110 / 80	YES	1 / 0	0 / 0	4 / 0.5	③3	3(1)	2	NA	2	2	
			VAULT	E	ACCOUNTING RECORDS STORAGE	112A / 180	NO	0 / 0	0 / 0	0	NA	2	2	NA	2	2	
			MAIL ROOM	E	MAIL ROOM & WORKSTATION FOR COUNTY SUPERVISORS (2)	111 / 260	NO	0 / 0	0 / 0	10 / NA	②2	5(2)	2	NA	2	2	
	CLERICAL	E	RECEPTION	C	RECEPTION AND WAITING	110A / 180	YES	0 / 0	0 / 0	40 / 5	⑤5	1	1	NA	2	2	
			GENERAL OFFICE	E	CLERICAL & FILING ACTIVITIES	110 / 600	YES	0 / 0	4 / 0	24 / 3	③3	1	1	2	3(2)	1	
			MARRIAGE LICENSES	E	ISSUANCE OF MARRIAGE LICENSES	110B / 100	YES	0 / 0	1 / 0	16 / 2	④4	1	2	2	3(2)	1	
			FILES/ (5) STORAGE	C	FILING & STORAGE	110B / 120	NO	0 / 0	0 / 0	0	NA	NA →					

REMARKS

(1) COULD USE VISUAL SCREENING FROM RECEPTION AREA.

(2) COUNTY CLERK WOULD LIKE SUPERVISORS MAIL ROOM RELOCATED SO AS TO ALLOW CLOSING HIS DEPARTMENT AFTER OFFICE HOURS WHILE MAINTAINING THE MAIL ROOM OPEN, SINCE SUPERVISOR'S HOURS GENERALLY RUN PAST HIS OWN DEPARTMENT CLOSING TIME.

(3) IN GENERAL, THE BUILDING COULD USE MORE MEETING ROOMS FOR COUNTY BOARD COMMITTEES MEETINGS.

(4) HVAC ZONING NEEDS IMPROVEMENT IN THESE AREAS

(5) THIS AREA WOULD BE NEEDED FOR INCREASED FILES AND RECORDS STORAGE AND FOR FUTURE VOTERS REGISTRATION ON A COUNTY WIDE BASIS. PROGRAM FOR AT LEAST 100 S.F.

(6) LOCATED WITHIN GENERAL OFFICE (SEE SPATIAL LAYOUT).

DIVISION SPATIAL EVALUATION: 2 | 2 | 1 | 2 | 2

GENERAL NOTES & COMMENTS

NOISE LEVEL: TOLERABLE.

PARKING: DOES NOT APPEAR TO CAUSE SERIOUS PROBLEMS FOR THIS DEPARTMENT.

SPATIAL EVALUATION SURVEY

			RATINGS DESIGNATION
A	EXISTING	LOCATION RATING.	1: EXCELLENT
B	EXISTING	AREA RATING.	2: GOOD
C	EXISTING	PERSONNEL AMOUNT RATING.	3: FAIR
D	EXISTING	ARTIFICIAL CLIMATE RATING.	4: POOR
E	EXISTING	EQUIPMENT RATING.	5: BAD

REVISIONS

15

Figure 9.44

1
A   STORAGE UNIT
B   DESK W/CHAIR
C   CHAIR
D   FILE CABINETS
E   COFFEE TABLE
F   COAT RACK
G   STORAGE UNITS
H   CONFERENCE TABLE/6 CHAIRS

2
A   MAILBOXES UNIT
B   WORK TABLE
C   TYPEWRITER STAND
D   DESK W/ CHAIR
E   TABLE

3
A   DESK W/ CHAIR
B   DESK W/ CHAIR
C   DESK W/ CHAIR
D   FILES
E   DESK /CHAIR/ FILES
F   COUTERTOP W/ FILES UNDER
G   VAULT / SAFE
H   STORAGE UNIT
J   SHELVING UNIT
K   BOOKSHELVES
L   TACKBOARD
M   WORK TABLE
N   STORAGE UNIT
O   FILES
P   STORAGE UNIT
Q   SHELVING UNIT
R   DESK / CHAIR / FILE
S   COAT RACK

4
A   DESK W/ CHAIR
B   APPLICANTS' CHAIRS
C   TABLE
D   STORAGE UNIT
E   TABLE
F   PLANTER

5
A   WAITING CHAIRS (10)

COUNTY
CLERK

EQUIPMENT
LIST

3 · 30 · '78

SCALE   IN   FEET
0 1   5   10   20

COUNTY CLERK OFFICE
EXISTING SPATIAL LAYOUT
3 · 30 · 78

VAULT
1A
CLERK'S
OFFICE
1F
2A
MAIL ROOM
1C
1D
1B
1E
2E
2D
1G
2C
2B
3H
3Q
3S
3P
FILES / STORAGE
3R
3J
3O
3D
3M
3N
CLERK'S
WORKSTATION
3K
3C
GENERAL   OFFICE
3N
4D
3
4C
3B
MARRIAGE
LICENSES
3G
4A
3A
3F
4B
3L
4F
RECEPTION
5A
5A
4E
PUBLIC CORRIDOR

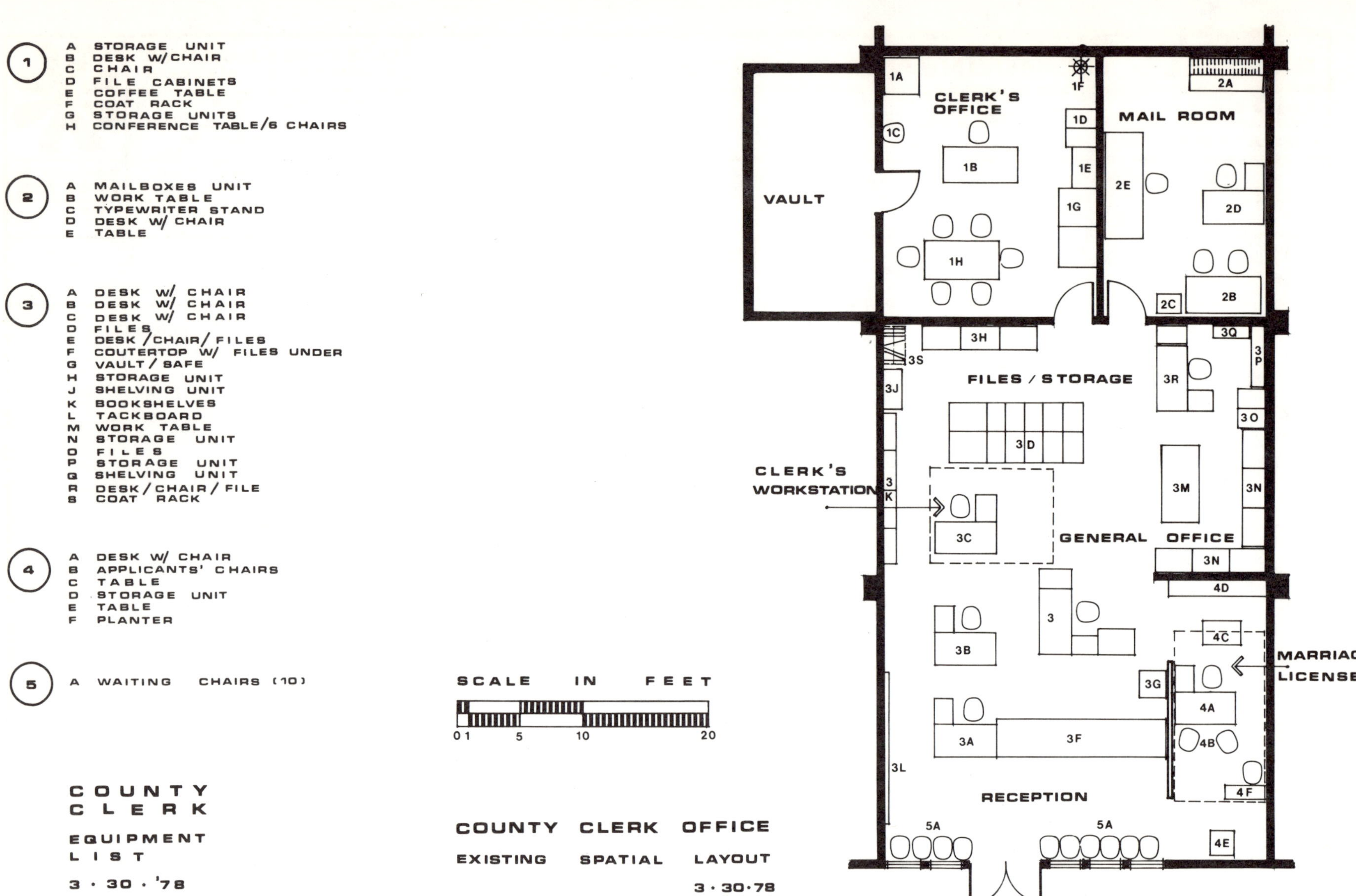

Figure 9.45

SPATIAL RELATIONSHIPS DATA SHEET

COLE, CASS & RAWLINGS architects & planners

PROJECT: SPATIAL ANALYSIS SURVEY OF WESTCHESTER COUNTY GOVERNMENT OFFICES

LOCATION: 11073 JEFFERSON DRIVE, GREEN OAKS, MO.

OWNER REP.: MARK HAYDEN	TITLE: COUNTY CLERK
PHONE: 444-2141	ROOM N°: 112
DATE: 3-30-78	TIME: 10:00 A.M.

PROJECT N°: 7825 [G2]

INTERVIEWER: JL	APPROVED:	SHEET: 1 OF : 1

INTRA DEPARTMENTAL RELATIONSHIPS ANALYSIS

DIV.	COMPONENT DESIGNATION	FUNCTION	ELEMENT	FUNCTION	ROOM N°	TO: ELEMENT	V	D	T	C	F	E
COUNTY CLERK	COUNTY CLERK	E	CLERK'S OFFICE WORKSTATION VAULT	E	112 & 110	GENERAL OFFICE	1	D	1	1	1	1
						SUPERVISORS MAIL ROOM	4	D	1	NA	4	5 (1)
						MARRIAGE LICENSES	3	D	1	1	3	2
						RECEPTION	4	D	1	1	2	2
			SUPERVISOR'S MAIL ROOM	E	111	GENERAL OFFICE	2	I	3	1	3	2
						CLERK'S WORKSTATION	2	I	3	1	3	2
	CLERICAL	E	GENERAL OFFICE	E	110	RECEPTION	2	D	1	1	1	1
						MARRIAGE LICENSES	2	D	1	1	3	2
						SUPERVISORS MAIL ROOM	4	D	1	NA	3	5 (1)
			MARRIAGE LICENSES	E	110B	GENERAL OFFICE	1	D	1	1	1	2
						RECEPTION	2	D	1	1	2	1
			RECEPTION	C	110A	GENERAL OFFICE	1	D	1	1	2	1
						MARRIAGE LICENSES	1	D	1	1	1	1
						COUNTY CLERK	3	D	1	1	4	2
			FILES/STORAGE	C	NA	GENERAL OFFICE	2	D	1	1	3	NA

REMARKS

(1) TOO CLOSE. DEPARTMENT MUST BE CROSSED TO REACH THIS AREA. TELEPHONE OR INTERCOM CONTACT IS SUFFICIENT.

RELATIONSHIPS DESCRIPTIONS INDEXES

V: VALUE	D: DYNAMICS	T: TYPE	C: CLASS	F: FREQUENCY	E: PRESENT EVALUATION
1·ESSENTIAL	D·DIRECT	1·PHYSICAL	1·HUMAN/HUMAN	1 CONSTANT	1·EXCELLENT
2·COMPLEMENTARY	I·INDIRECT	2·AUDIOVISUAL	2·EQUIP/EQUIP.	2 REPETITIVE	2·GOOD
3·NONESSENTIAL		3·AUDITIVE	3·N.A.	3 OCCASSIONAL	3·FAIR
4·UNDESIRABLE		4·VISUAL	4·HUMAN/EQUIP.	4 SCARCE	4·POOR
5·UNACCEPTABLE					5·BAD

GENERAL NOTES & COMMENTS

UNLESS OTHERWISE TABULATED, ALL RELATIONSHIPS ARE RECIPROCAL.

REVISIONS

Figure 9.46

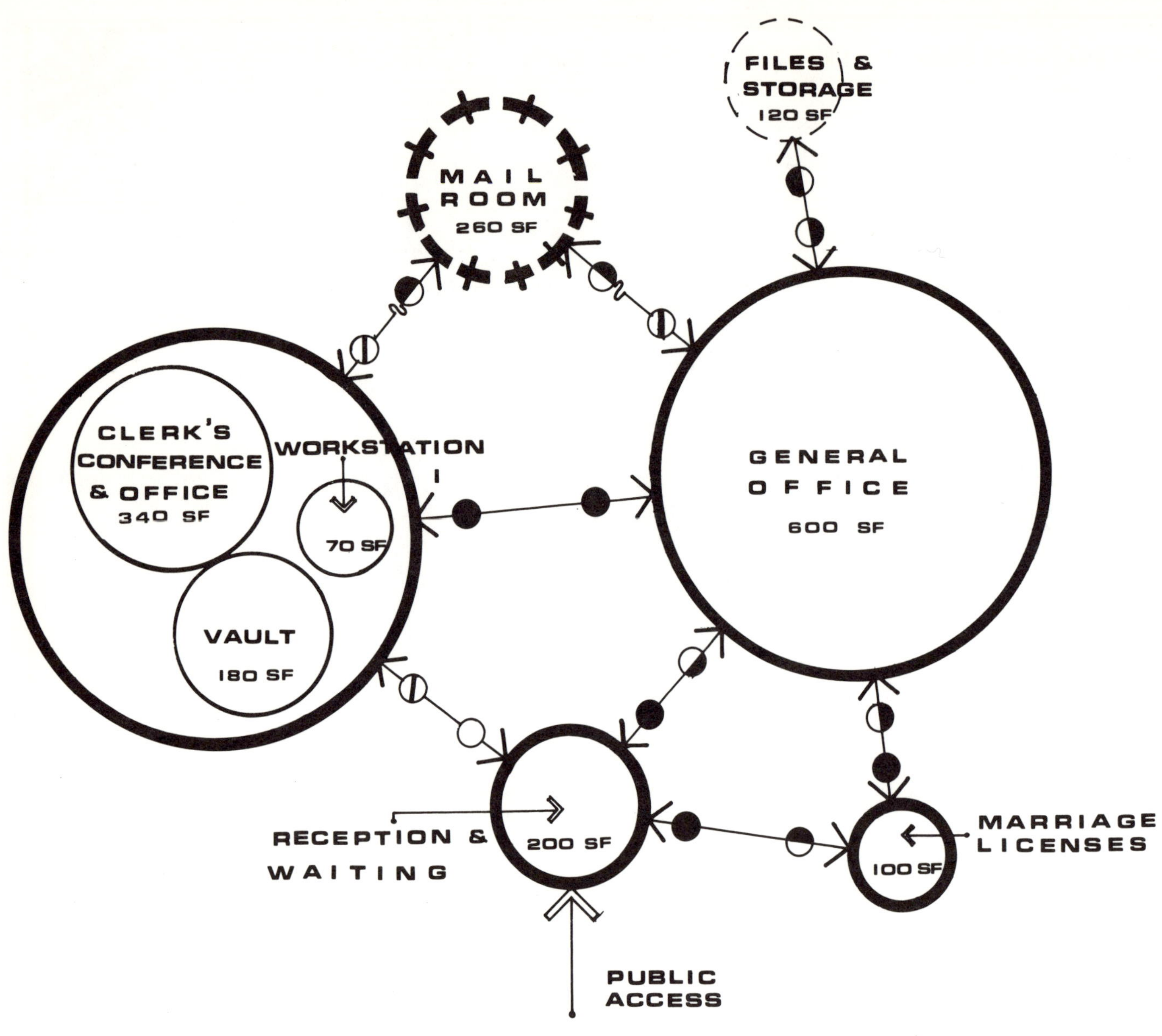

COUNTY CLERK INTRADEPARTMENTAL RELATIONSHIPS DIAGRAM

NO SCALE
3 - 30 - 78

SPATIAL RELATIONSHIPS DATA SHEET

COLE, CASS & RAWLINGS — architects & planners

PROJECT:	SPATIAL ANALYSIS SURVEY OF WESTCHESTER COUNTY GOVERNMENT OFFICES
LOCATION:	11073 JEFFERSON DRIVE, GREEN OAKS, MO.

OWNER REP.: MARK HAYDEN	TITLE: COUNTY CLERK
PHONE: 444-2141	ROOM N° 112
DATE: 3-30-78	TIME: 10:00 A.M.

PROJECT N° 7825 [G2]

INTERDEPARTMENTAL RELATIONSHIPS ANALYSIS

INTERVIEWER: JL	APPROVED:	SHEET: 1 OF : 1

DIV.	COMPONENT DESIGNATION	FUNCTION	ELEMENT	FUNCTION	ROOM N°	RELATIONSHIPS DESCRIPTION TO: ELEMENT	V	D	T	C	F	E
C O U N T Y C L E R K	COUNTY CLERK	E	CLERK'S OFFICE WORKSTATION	E	110 & 112	ACCOUNTING DIVISION	1	D	1	1	2	1
						COUNTY TREASURER	2	I	3	1	2	1
						PURCHASING DIVISION	2	D	1	1	3	1
		E	SUPERVISOR'S MAIL ROOM	E	111	SUPERVISOR'S CHAIRMAN OFFICE	2	I	3	1	3	3
						COUNTY OFFICIAL	2	D	1	1	4	2
	CLERICAL	E	GENERAL OFFICE	E	110	COUNTY REGISTER OF DEEDS	1	D	1	1	2	1
						ENGINEERING DIVISION/SURVEYOR	2	D	1	1	2	3
						PURCHASING	2	D	1	1	2	2
						ACCOUNTING DIVISION	1	D	1	1	2	3
						CORPORATE COUNSEL	2	I	3	1	3	4 (1)

REMARKS

(1) CORPORATION COUNSEL IS PRESENTLY LOCATED ON THIRD FLOOR LEVEL. DISTANT – TIME/TRAVEL.

RELATIONSHIPS DESCRIPTIONS INDEXES

V: VALUE	**D: DYNAMICS**	**T: TYPE**	**C: CLASS**	**F: FREQUENCY**	**E: PRESENT EVALUATION**
1·ESSENTIAL	D·DIRECT	1·PHYSICAL	1·HUMAN/HUMAN	1 CONSTANT	1·EXCELLENT
2·COMPLEMENTARY	I·INDIRECT	2·AUDIOVISUAL	2·EQUIP/EQUIP.	2 REPETITIVE	2·GOOD
3·NONESSENTIAL		3·AUDITIVE	3·N.A.	3 OCCASSIONAL	3·FAIR
4·UNDESIRABLE		4·VISUAL	4·HUMAN/EQUIP.	4 SCARCE	4·POOR
5·UNACCEPTABLE					5·BAD

GENERAL NOTES & COMMENTS

THESE RELATIONSHIPS ARE <u>NOT</u> RECIPROCAL. SEE INDIVIDUAL SURVEY FOR LISTED DEPARTMENTS' RELATIONSHIPS DESCRIPTIONS AND EVALUATIONS.

REVISIONS: 18

Figure 9.48

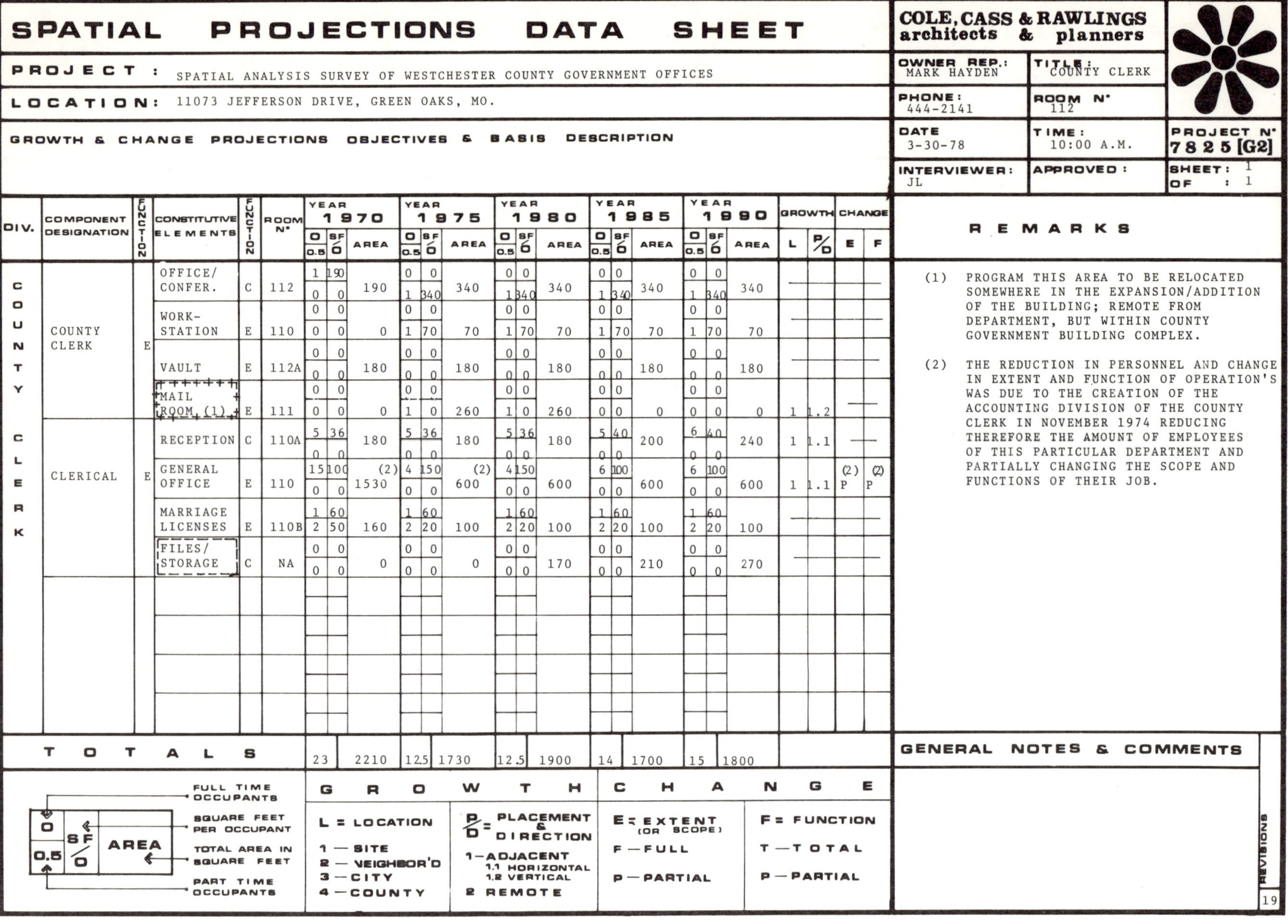

SPATIAL PROJECTIONS DATA SHEET

COLE, CASS & RAWLINGS architects & planners

PROJECT: SPATIAL ANALYSIS SURVEY OF WESTCHESTER COUNTY GOVERNMENT OFFICES

LOCATION: 11073 JEFFERSON DRIVE, GREEN OAKS, MO.

GROWTH & CHANGE PROJECTIONS OBJECTIVES & BASIS DESCRIPTION

OWNER REP.: MARK HAYDEN	TITLE: COUNTY CLERK
PHONE: 444-2141	ROOM N°: 112
DATE: 3-30-78	TIME: 10:00 A.M. — PROJECT N°: 7825 [G2]
INTERVIEWER: JL	APPROVED: — SHEET: 1 OF: 1

Occupant cells are given as FT (full-time occupants, top-left O), SF (square feet per occupant, top-right), PT (part-time occupants, bottom-left 0.5), O (bottom-right), AREA (total area in sq. ft.).

DIV.	COMPONENT DESIGNATION	FUNCTION	CONSTITUTIVE ELEMENTS	FUNCTION	ROOM N°	1970 FT	SF	PT	O	AREA	1975 FT	SF	PT	O	AREA	1980 FT	SF	PT	O	AREA
COUNTY CLERK	COUNTY CLERK	E	OFFICE/ CONFER.	C	112	1	190	0	0	190	0	0	1	340	340	0	0	1	340	340
			WORK-STATION	E	110	0	0	0	0	0	0	0	1	70	70	0	0	1	70	70
			VAULT	E	112A	0	0	0	0	180	0	0	0	0	180	0	0	0	0	180
			MAIL ROOM (1)	E	111	0	0	0	0	0	0	0	1	0	260	0	0	1	0	260
	CLERICAL	E	RECEPTION	C	110A	5	36	0	0	180	5	36	0	0	180	5	36	0	0	180
			GENERAL OFFICE	E	110	15	100	0	0	(2) 1530	4	150	0	0	(2) 600	4	150	0	0	600
			MARRIAGE LICENSES	E	110B	1	60	2	50	160	1	60	2	20	100	1	60	2	20	100
			FILES/ STORAGE	C	NA	0	0	0	0	0	0	0	0	0	0	0	0	0	0	170

CONSTITUTIVE ELEMENTS	1985 FT	SF	PT	O	AREA	1990 FT	SF	PT	O	AREA	GROWTH L	P/D	CHANGE E	F
OFFICE/ CONFER.	0	0	1	340	340	0	0	1	340	340	—	—	—	—
WORK-STATION	0	0	1	70	70	0	0	1	70	70	—	—	—	—
VAULT	0	0	0	0	180	0	0	0	0	180	—	—	—	—
MAIL ROOM (1)	0	0	0	0	0	0	0	0	0	0	1	1.2	—	—
RECEPTION	5	40	0	0	200	6	40	0	0	240	1	1.1	—	—
GENERAL OFFICE	6	100	0	0	600	6	100	0	0	600	1	1.1	(2) P	(2) P
MARRIAGE LICENSES	1	60	2	20	100	1	60	2	20	100	—	—	—	—
FILES/ STORAGE	0	0	0	0	210	0	0	0	0	270	—	—	—	—

TOTALS	1970		1975		1980		1985		1990	
	23	2210	12.5	1730	12.5	1900	14	1700	15	1800

REMARKS

(1) PROGRAM THIS AREA TO BE RELOCATED SOMEWHERE IN THE EXPANSION/ADDITION OF THE BUILDING; REMOTE FROM DEPARTMENT, BUT WITHIN COUNTY GOVERNMENT BUILDING COMPLEX.

(2) THE REDUCTION IN PERSONNEL AND CHANGE IN EXTENT AND FUNCTION OF OPERATION'S WAS DUE TO THE CREATION OF THE ACCOUNTING DIVISION OF THE COUNTY CLERK IN NOVEMBER 1974 REDUCING THEREFORE THE AMOUNT OF EMPLOYEES OF THIS PARTICULAR DEPARTMENT AND PARTIALLY CHANGING THE SCOPE AND FUNCTIONS OF THEIR JOB.

GROWTH

L = LOCATION
1 — SITE
2 — NEIGHBOR'D
3 — CITY
4 — COUNTY

P/D = PLACEMENT & DIRECTION
1 - ADJACENT
1.1 HORIZONTAL
1.2 VERTICAL
2 REMOTE

CHANGE

E = EXTENT (OR SCOPE)
F — FULL
P — PARTIAL

F = FUNCTION
T — TOTAL
P — PARTIAL

Legend (area box): O = FULL TIME OCCUPANTS; SF = SQUARE FEET PER OCCUPANT; AREA = TOTAL AREA IN SQUARE FEET; 0.5 = PART TIME OCCUPANTS.

GENERAL NOTES & COMMENTS

REVISIONS 19

Figure 9.49

170

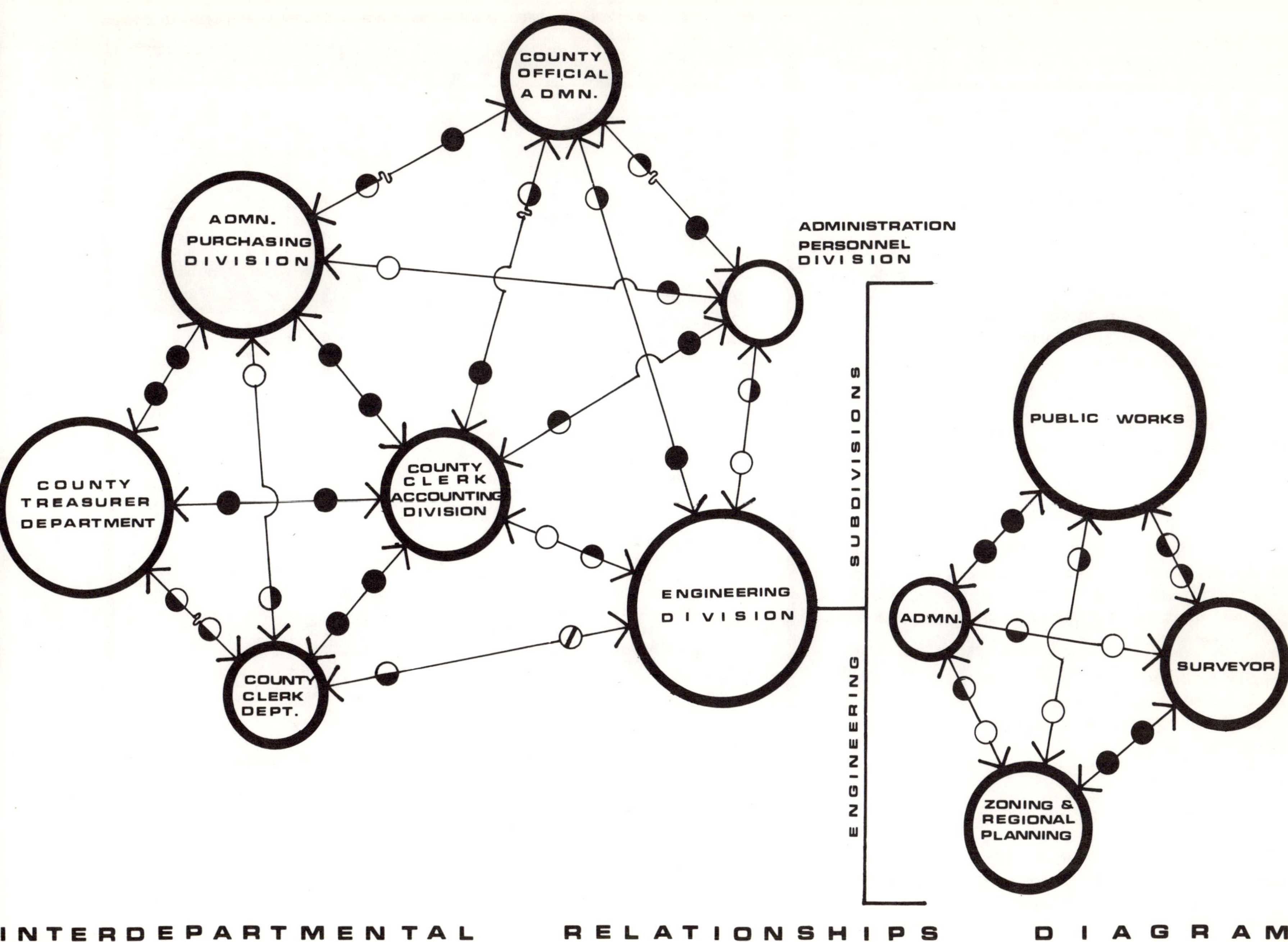

INTERDEPARTMENTAL RELATIONSHIPS DIAGRAM

Figure 9.50

Employee Survey. Based on the size and personnel of each department or division, the following number of survey questionnaires were distributed at random among the employees:

- County Clerk Department — 2
  - Accounting Division — 4
- County Official Department — 2
  - Administrative Sector — 2
    - Personnel Division — 6
    - Purchasing Division — 6
    - Engineering Division — 2
      - Public Works Subdivision — 6
      - Surveyor Subdivision — 2
      - Zoning and Planning
        Subdivision — 4
    - Data Processing Division — 10
- County Treasurer Department — 4

The results obtained by means of the questionnaire are as shown in Figure 9.51.

PART III — SPATIAL RECOMMENDATIONS

When considering the spatial needs for the analyzed functions, several essential factors become apparent immediately with respect to the problem areas listed in the preceeding pages. These factors can be summarized as:

1. Flexibility/Growth and Change
2. Noise/Noise Control
3. Privacy (Auditive and Visual)
4. Storage
5. Grouping and Departmental Zoning
6. Circulations and Accesses
7. Areas and Spatial Interfacing

The pertinency of these factors varies in each case as the different functions executed within the departmental or divisional parameters dictate; but in spite of these differences, the general summary of these factors represents a comprehensive spatial diagnoses of conditions which prevail (with a greater or lesser degree of significance as the individual cases may indicate) throughout the surveyed areas.

1. REGARDING YOUR DEPARTMENT AREA IN GENERAL; HOW WOULD YOU RATE THE PRESENT PERFORMANCE OF THE FOLLOWING?:

	EXELLENT/ GOOD %	FAIR %	POOR/ BAD %
1.1 LOCATION WITHIN THE COURTHOUSE BUILDING	76	20	4
1.2 AREA (SIZE)	30	22	48
1.3 HEATING	34	32	34
1.4 COOLING	38	28	34
1.5 VENTILATION	30	24	46
1.6 LIGHTING	68	20	12
1.7 CIRCULATIONS	32	38	30
1.8 NOISE AND SOUND BARRIERS	16	38	46
1.9 EQUIPMENT	56	34	10
1.10 SECURITY	52	28	20

2. REGARDING YOUR PERSONAL WORKSTATION; HOW WOULD YOU RATE THE FOLLOWING?:

	EXELLENT/ GOOD %	FAIR %	POOR/ BAD %
2.1 LOCATION WITHIN YOUR DEPARTMENT	66	16	18
2.2 AREA (SIZE)	40	30	30
2.3 HEATING	30	34	36
2.4 COOLING	30	36	34
2.5 VENTILATION	25	20	55
2.6 LIGHTING	66	20	14
2.7 VISUAL PRIVACY	30	20	50
2.8 SOUND PRIVACY	16	18	66
2.9 EQUIPMENT	56	28	16
2.10 SECURITY	48	34	18

3. PLEASE INDICATE IN WHICH OF THE FOLLOWING AREAS YOU THINK THE COURTHOUSE BUILDING NEEDS IMPROVEMENT.

	%
3.1 EXTERIOR APPEARANCE	2 %
3.2 INTERIOR APPEARANCE	58 %
3.3 DEPARTMENTAL ZONING	20 %
3.4 BUILDING SIZE	30 %
3.5 MECHANICAL SYSTEMS (HEATING, VENTILATING & AIR CONDITIONING)	76 %
3.6 ELECTRICAL SYSTEMS	18 %
3.7 TOILET FACILITIES	34 %
3.8 CIRCULATIONS	22 %
3.9 ELEVATORS	16 %
3.10 SECURITY	22 %

Figure 9.51

Flexibility. Currently, the flexibility of the surveyed spaces is moderately acceptable if the heating, ventilating and air conditioning issues noted throughout the spatial analysis are considered of a secondary nature. However, if we are to fully assess the impact of such environmental factors as key determinants in the optimum performance of the surveyed functional elements, then the flexibility evaluation proves inadequate with respect to those issues dealing with mechanical ventilation, heating, etc. Therefore, any remodeling or expansion of the existing facilities MUST involve the functional and environmental flexibility factor as a major design determinant.

Furthermore, the inevitable growth and change concepts related to this facility can only be accomodated through a careful design approach which fully considers the importance of these issues. The functionalism of any spatial allocation for these administrative offices depends greatly on the ability of the area and environment provided to change or exceed the predetermined functional parameters.

Therefore, flexibility is more than a recommendation, it is a mandate for any conceptual design that attempts to effectively satisfy the needs of these functions.

Noise. One negative factor which appears to upset almost every operation surveyed is noise.

In a building of this nature, a certain amount of noise is always present, and in most cases it is inevitable. Still, the sound transmission and complaints about the noise levels exceeded the ratings of normal expectancy for operations of this category and importance.

In several cases, the complaints and remarks regarding noise went beyond the simple "distractive effect" statement and zeroed in on the actual sound transmission problem, which has become critical in many departments and divisions where meetings and discussions regarding confidential matters have to be conducted. In addition, the environmental factors (such as poor ventilation, heating, etc.) of some offices force the occupants to leave their doors open during conferences, which aggravates the noise and sound transmission problems. This interfacing of negative factors must also be noted and prevented.

As a final evaluation, the noise control and sound transmission in the present facilities leave much to be desired. Improvement in this area is definitely needed.

Privacy. Directly related to the sound transmission and noise control factors is the need for privacy (auditive or visual) expressed by many surveyed elements.

Auditive privacy is directly related to the previous section and its aspects have already been outlined. Visual privacy, however, while not readily available throughout the facilities, can be (and in many cases HAS BEEN) easily provided by merely rearranging movable partitions, equipment, etc.

Although many surveyed elements expressed their dissatisfaction regarding visual privacy, a general evaluation at this point cannot underline this need as a factor of sufficient magnitude to be considered a major programmatic statement.

Storage. Storage is a major problem in the Westchester County Courthouse. Almost 80% of the surveyed areas indicated some lack of storage, either active or inactive. The problem lies in the fact that departmental growth has increased so rapidly within the past ten years that areas originally allocated for inactive storage, office supplies, etc. have been overtaken by new workstations, filing cabinets, etc. Because of the spatial limitations with respect to the expansion of this building, the storage allocated in the spatial analysis summaries of this document has been programmed to contain only the essential equipment and supplies necessary for the bi-weekly or monthly support of the analyzed functions.

Due to the high cost of construction on a facility of this nature and the spatial limitations, the provision of extensive storage areas has been considered impractical and nearly impossible to realize.

Regarding the need for departmental inactive storage, it seems more advisable to provide space for storage expansion at another location where the cost and spatial limitation factors would not be of such significance.

Grouping. Although theoretically improvement could be made in this area, it appears that few of the surveyed departments or divisions evaluated the existing conditions as bad or even poor. The surveyed elements agree when the occupants of one department leave their area, the distance or proximity to another department or division located within the building is not a major cause of concern or functional inadequacy. In spite of this, we recommend that careful attention be given to the grouping and departmental zoning of these functions. Complaints should never be the primary force behind improvements.

When determining these groupings, aside from the logical interdepartmental relationships, extreme care must be paid to the projected growth of each department and its possibility of expansion.

The interdepartmental relationships and spatial development projections of each area are fully documented in the corresponding analysis data sheets of the individual summaries, as well as in the interdepartmental relationships diagram.

Circulation. Circulation within some departments is a major source of spatial inadequacy. This is one issue which, in the majority of cases, is directly related to the available area. Intradepartmental circulation and access are in definite need of improvement. Functional zoning and realistic circulation factors could help tremendously in this regard.

On the other hand, interdepartmental circulation appears to be adequate throughout the building. Elevators in the southwestern portion of the building could further improve the overall circulation pattern if their installation is economically feasible.

Area. A general evaluation of the surveyed departments MUST underline the need for spatial expansion. In the majority of cases, this expansion is more than a recommendation, it is a clear necessity.

Of the surveyed areas, only two departments have been found to occupy a spatial parameter which fully satisfies the functions performed within its boundaries. These two departments are the County Clerk and the County Treasurer.

The remaining departments and divisions have been found to have either insufficient area for their functional tasks or inadequate spatial layouts from which to operate properly.

Among the many problems caused by these factors is the obvious spatial interfacing of several operations, such as doubled storage and breakroom areas (as in the case of the Data Processing Division), conference rooms used for breakrooms, workstations or file rooms, etc.

The only viable solution to these problems is the very serious consideration of spatial expansion. The functions presently contained within the county government sector of the Westchester County Courthouse need more space. This is not just a suggestion, it is an indisputable fact.

9.3 Comments

The case studies illustrated represent only two variations of the innumerable layouts which can be used in the development of program formats. The viable possibilities in this respect may vary as often as building type and size, depending on the extent and complexity of the building's contents.

In the two cases outlined above the programming phase was clearly defined in a contractual agreement. This constituted a definite advantage when developing the programmatic formats, since their layouts must inevitably respond to the requirements outlined in the contract. In cases where no such contractual agreement is available, it is always advisable to confirm by means of a letter which areas and procedures will be covered and which will not.

One of the reasons the two projects discussed were selected as illustrative examples for this text is that in both cases the sequence, Objectives Definition - Analysis - Programming - Requirements, was followed without variation or modification. However, this is not always the sequence which MUST be followed, since often the analysis phase becomes the supportive procedure of a programmatic activity which overwhelmingly determines the requirements of a building project. This transposition will become more accentuated when the initial statements of purpose and intent lean more towards the immediate creation or adaptation of new systems, approaches and operational procedures rather than follow a gradual transition towards their implementation, as the operational and functional requirements of a building project literally venture into new methodologies, procedures and techniques instead of gradually developing into them.

Another characteristic which should be underlined in the two case studies is that the programmatic layout was essentially designed to be a working tool for the design phase, in spite of the fact that its contents also represent the documented result of an investigative procedure. In Case Study No. 2, on the other hand, the report had to be designed to be a self-supporting document from beginning to end, since its contents would inevitably become a matter of public record. One very special characteristic of institutional projects is most of the contents of documents such as the one illustrated in the second case study will be read and interpreted by so many different special interests that clarification and notes clearly stating the extent, development and scope of the data and conclusions are essential.

For the most part, both projects were executed in total compliance with the contractual requirements and programming systems. Although several limitations regarding accurate employee counts and areas definition were encountered when developing the spatial history of each operation, this is a very common occurrence and should not become a matter of great concern, since the effect of such limitations is negligible.

Finally, the first case study took a total of:
Principals' Time — 32 man/hr
Programmers' Time — 320 man/hr
Clerical Time — 45 man/hr
Reproduction/Format — $450,000 approx. cost

Reimbursable Expenses — $50.00 approx. cost
Case Study No. 2 totaled the following expenses:
Principals' Time — 45 man/hr
Programmers' Time — 430 man/hr
Clerical Time — 80 man/hr
Reproduction/Format — $1800.00 approx. cost
Reimbursable Expenses — $220.00 approx. cost

The issue of the actual cost of architectural programming functions is a complex subject which can only be determined by the specific design firm providing such services. There is no accurate "rule of thumb" based on square footage of surveyed area, building type, occupancy or total professional design fee percentage, which can be recommended at this point without risking erroneous results. It is better to base the costs of architectural programming on budgeted man hours and reproduction/format costs as previously illustrated. Here again, experience is the best guide.

Ultimately, at any reasonable rate and whatever the relative costs, this figure might mean to the professional designer and his client the difference between failure and success when related to the operational performance of a building project. Actually, that is what makes programming valuable.

---

[22]For the sake of simplification, only one portion of this survey is illustrated in this text.

[23]The term "departmental gross" was used in this report in lieu of the more common "Standard Net Assignable" to signify the result of the net area requirement plus a previously established circulation factor of 1.3 (77% E.R.); the different terminology was used at the specific request of the county building commission.

10
Ramifications and Alternatives

10.1 Simplifications and Developments

Before we attempt to define the areas in which some of the previously outlined methodologies could be simplified or developed, and the procedures for doing so, we must determine the effect of such simplification or developments on the final program. In that regard, the simplification of the different phases involved in architectural programming does not necessarily represent a reduction in the scope of the final document produced. In the following pages we present possible "short-cuts" which may result in a simplification of tasks in specific cases, but which do not constitute a list of possible deletions that would reduce the scope of the final program layout, since steps of the latter nature can only be based on the individual characteristics and requirements of each project.

The developments outlined here do not necessarily mean increases in the programming scope, but rather indicate courses of action to follow in cases where the need for more detailed information becomes apparent.

SITE ANALYSIS. Perhaps the greatest simplification procedure which can be achieved in definition of natural characteristics is the proper recording of the environmental data for the region in which most of the projects of a specific professional design firm are based. Such information should be recorded in a reproducible form, since the items included in such a report, like solar and wind analysis, precipitation graphs, etc., will probably have to be done only once. On the other hand, since the remaining factors involved in the site analysis phase will probably be executed by specialized consultants, any attempt to simplify their execution is, for the most part, useless. These items, which can be referred to as supporting data should be made part of the program appendix whenever possible so they will be available for immediate reference when needed. By the same token, the artificial conditions portion of a site analysis is so greatly determined by the site itself that any attempt at generalized simplifications would produce negative rather than positive results.

In site planning, it is convenient to limit the extent of any spatial relationships diagram to a simple divisional interactions level only and to avoid overdetailing the charts anymore than is essentially necessary.

SPATIAL ANALYSIS. The proper establishment of divisions is probably the most important step in simplifying any spatial analysis process, since it will determine, to a great extent, the subsequent fragmentations which must take place prior to any full investigation or programming of functions. Without a sensible divisional fragmentation the subsequent analytical steps can become so confusing that instead of clarifying a functional network they obscure it, resulting in a data-clogged document without sense or value. For this reason it is advisable at this point to list only those functional relationships which hold any value (either positive or negative), and not those which would merely represent reciprocal non-essential values.

By the same token, if during the functions analysis all the constitutive elements of operational components are considered essential, it is advisable to eliminate the corresponding "function" column in the final data collection report, since it will serve no purpose. The listing of operational functions is needed so functional priorities will be readily understood by the design phase, but it is also the best means of documenting the client's sense of values, hierarchies and programmatic input. More often than not, we find ourselves referring back to "minutes of the meeting" and letters written as a result of conferences confirming unclear statements and priorities established during brainstorming sessions, which priorities may or may not constitute abiding program requirements. In any case, the elimination of functional descriptions must be done with caution. A program that includes the proper functional designation becomes, not only the detailed spatial requirements summary for a definite building structure, but also, the pattern of values from which those requirements are derived. There lies the importance of these functional descriptions; and accordingly, while the actual relationships description (type, class, volume, etc.) can be omitted from a specific functional relationships diagram, the value (essential, non-desirable, etc.) and the dynamics (direct or indirect) should ALWAYS be recorded.

Moreover, whenever an item is not applicable to a particular aspect or circumstance, it should be listed accordingly. In data collection and general programming, documentation is a key word which must NEVER be forgotten. This rule invariably applies to documents or data labeled "by owner."

Another way of simplifying the analysis and programming procedures is through the use of proper analysis questionnaires in lieu of personal interviews and work sessions. This procedure, when properly executed, can effectively reduce man hours and extensive conferences. On the other hand, it can also produce extremely inaccurate results. For example, during the review of the employee survey which served as model for the one illustrated in Chapter 9 of this book, we observed by mere chance that in one particular response the location of a specific workstation was rated as "poor" by the typist who occupied it while the manager of her department rated it as "good" during his personal interview. When contacted by telephone, the typist's response was that she had rated the location of her workstation as "poor" because a few months back several filing cabinets had been placed in front of her desk and she could no longer maintain the necessary visual contact with the reception area.

As we can see, a surveyed element's interpretation of "location" can be erroneous enough to invalidate the results of a summary conducted by means of a simple questionnaire. The fact is that in the previous example, clearly the "poor" location was of the filing cabinets, NOT the workstation. Yet, if one *specific* remark of one *specific* manager about one *specific* workstation had not been accidentally recalled during the review of that one *specific* questionnaire, chances are that the misinterpretation would have gone unnoticed.

Even after correcting such a mistake, one will always wonder: "How many other responses of this nature are there in this survey?"

Several feedback procedures also help simplify programmatic assumptions and tabulations through the use of established or recommended minimum/maximum parameters of spatial densities ratios for definite functions such as those illustrated in Figure 10.1, circulation patterns or even building types.

Beyond the previously mentioned simplifications, several developments can further aid the clear determination and outline of programmatic statements. Just as any fragmentation process can isolate the functions and relationships of operational constituents within a project, this same procedure can be employed in the determination and detailing of specific activities within a complex. For example, although the general office component of an operation might contain fifteen secretaries and fifteen workstations within the same spatial envelope, the description and relationships of these elements can be defined by the simple use of a micro-analysis technique in exactly the same manner in which we established the definition of components characteristics and their interrelationships. If necessary, even the relationships of several of those elements to the whole can be defined by means of simple explanatory notes within the charts. This approach, which might seem useful only in a few special cases, can be the best tool available to a designer for the development of furnishing layouts, placement of doors and windows, establishment of points of access, movement patterns, (see Figure 10.2), and even circulation diagrams.

On the other hand, while the micro-analysis is useful in some specific cases, the opposite type of approach is always advisable in urban planning and general site development; that is, the detailed description of a project site, neighborhood, or even region as it relates to its neighboring structures, roads, surroundings, etc. This macro-analysis, which can be carried out by following the same general guidelines illustrated previously, can be of extreme significance in clarifying and programming

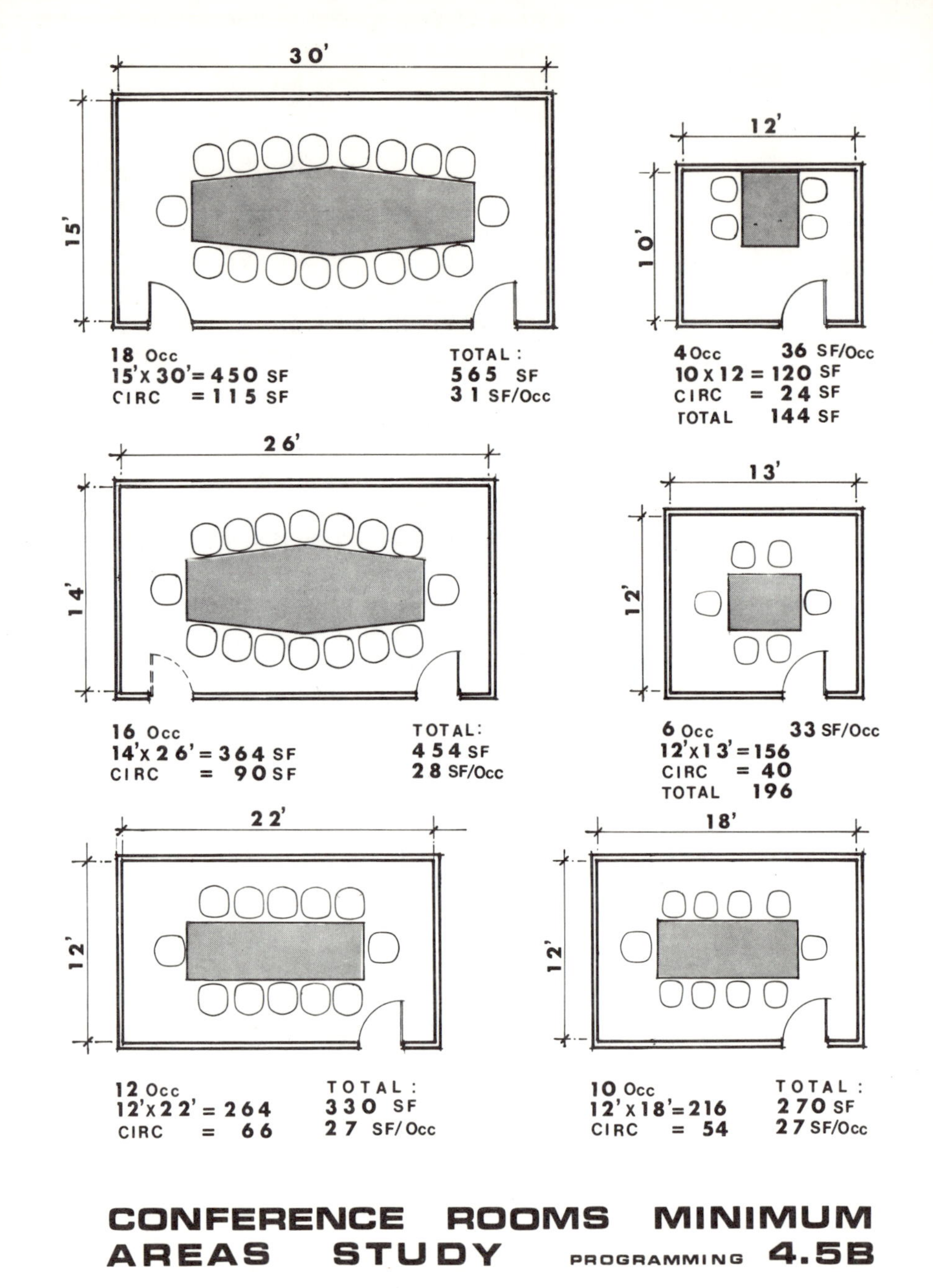

Figure 10.1

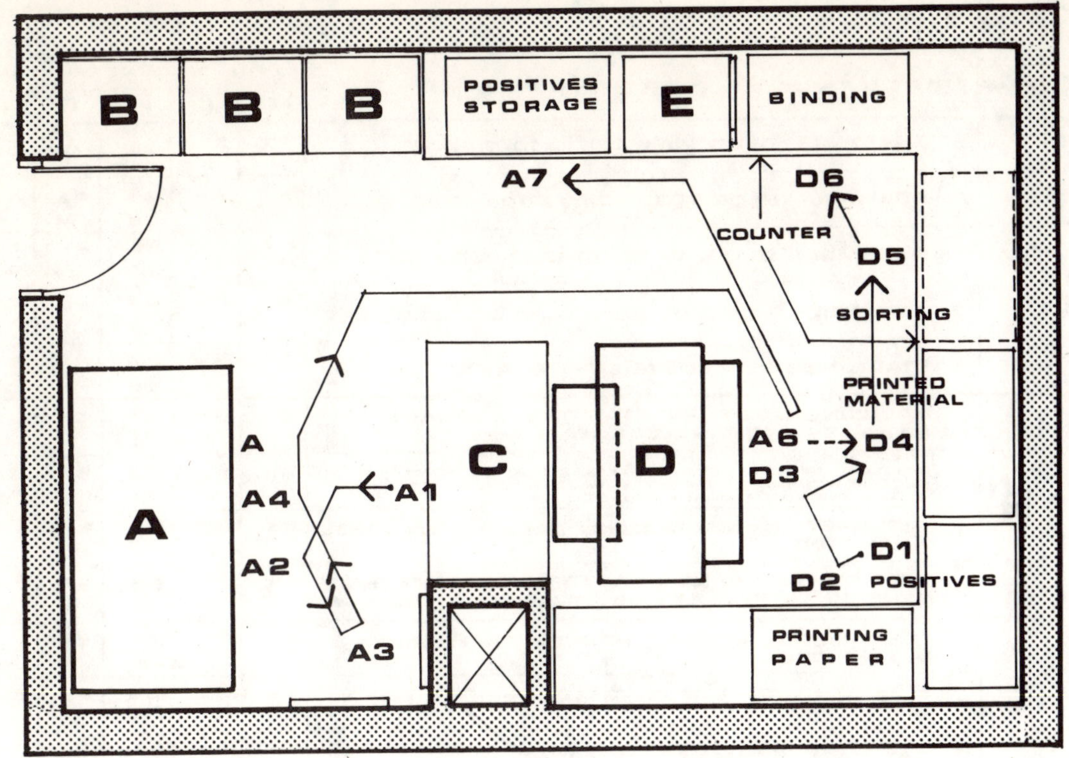

Figure 10.2

the impact of a project with regard to location, and the possible consequences of placement decisions for the project itself.

ECONOMIC ANALYSIS. Simplifications and developments in this sector depend entirely on the degree of accuracy as well as the extent of the analysis performed when establishing the budget, costs control and economic factors affecting a project.

To either simplify or develop the cost analysis studies based on building functions, type and quality of construction, project size, etc., the best resources are specialized references, computer analysis or reliable consultants who can readily and accurately produce proper results.

REGULATORY SURVEYS. The great similarity between climatological investigations and regulatory surveys is once the first detailed survey has been executed for a determined project, its layout can become a reliable reference for other projects of the same nature with relatively few entries needed to upgrade its contents. One way in which the regulatory surveying process can be greatly simplified during the programming stage is by the use of the regulatory file index illustrated on Figures 10.3 and 10.4.

Since the regulations of all building codes are essentially based on either the occupancy or class of construction of a building, by merely listing all occupancy classifications and types of construction, a table can be developed which shows whether or not a specific occupancy type is permitted with a certain type of construction and where in the file the requirements which affect and regulate this combination are explained in detail, as shown in Figure 10.4. A simple occupancy interactions matrix like the one developed in the left hand portion of Figure 10.3 also shows at a glance the fire ratings requirements for separation between adjacent occupancies.

The possible developments of regulatory surveys include interviews with the proper agency officials, computer-aided code reviews, specialized reports and even variances and waivers documenta-

(*) DATA BASED ON UNIFORM BUILDING CODE 1979 EDITION

NP = NOT PERMITTED

SEPARATION OF MIXED OCCUPANCY (*)

(1) REDUCE TO 2HR IF B-1 IS LIMITED TO MOTOR VEHICLE STORAGE < 9 OCC.

REGULATORY SURVEY GENERAL INDEX (*)

Group		Occupancy Description	I	II	III	IV	V
A	1	1 ASSEMBLY BLDG. W/STAGE > 1000 OCC.	A_1 I	A_1 II	NP	NP	NP
A	2	2 ASSEMBLY BLDG. W/STAGE > 300 < 1000 OCC.	A_2 I	A_2 II	A_2 III	A_2 IV	A_2 V
A	2.1	2.1 ASSEMBLY BLDG. W/ NO STAGE > 300 OCC.	$A_{2.1}$ I	$A_{2.1}$ II	$A_{2.1}$ III	$A_{2.1}$ IV	$A_{2.1}$ V
A	3	3 ASSEMBLY BLDG. W/ NO STAGE < 300 OCC.	A_3 I	A_3 II	A_3 III	A_3 IV	A_3 V
A	4	4 STADIUMS & AMUSEMENT PARKS	A_4 I	A_4 II	A_4 III	A_4 IV	A_4 V
B	1	1 GASOLINE & SERVICE STATIONS, STORAGE GARAGES / NON HAZARDOUS	B_1 I	B_1 II	B_1 III	B_1 IV	B_1 V
B	2	2 WHOLESALE/RETAIL STORES, OFFICE BLDGS. DRINKING/DINING W/OCC < 50, POLICE/FIRE STATS., FACTORIES & EDUCATIONAL BLDGS. OVER 12TH GRADE < 50 OCC.	B_2 I	B_2 II	B_2 III	B_2 IV	B_2 V
B	3	3 AIRCRAFT HANGARS, OPEN PARKING GARAGES & HELIPORTS	B_3 I	B_3 II	B_3 III	B_3 IV	B_3 V
B	4	4 ICE & POWER PLANTS, FACTORIES & WORKSHOPS W/NO COMBUSTIBLES/EXPLOSIVES, STOR./SALES RMS	B_4 I	B_4 II	B_4 III	B_4 IV	B_4 V
E	1	1 EDUCATIONAL BLDGS. THROUGH 12TH GRADE W >50 OCC.	E_1 I	E_1 II	E_1 III	E_1 IV	E_1 V
E	2	2 EDUCATIONAL BLDGS. THROUGH 12TH GRADE W< 50 OCC.	E_2 I	E_2 II	E_2 III	E_2 IV	E_2 V
E	3	3 DAY CARE CENTERS W/ > 6 CHILDREN	E_3 I	E_3 II	E_3 III	E_3 IV	E_3 V
H	1	1 STORAGE & HANDLING OF HAZARDOUS MATERIALS	H_1 I	H_1 II	H_1 III	H_1 IV	H_1 V
H	2	2 STORAGE & HANDLING OF CLASSES I, II & III-A LIQUIDS, PAINT STORES, SHOPS & SPRAY BOOTHS	H_2 I	H_2 II	H_2 III	H_2 IV	H_2 V
H	3	3 WOODWORKING PLACES, BOX FACTORIES, SHOPS & FACTORIES WHERE COMBUSTIBLE FIBERS/DUST ARE MANFR'D	H_3 I	H_3 II	H_3 III	H_3 IV	H_3 V
H	4	4 REPAIR GARAGES	H_4 I	H_4 II	H_4 III	H_4 IV	H_4 V
H	5	5 AIRCRAFT REPAIR HANGARS	H_5 I	H_5 II	H_5 III	H_5 IV	H_5 V
I	1	1 NON AMBULATORY NURSING HOMES/SANITARIUMS, NURSERIES < 6 YRS. OLD & HOSPITALS	I_1 I	I_1 II	I_1 III	I_1 IV	I_1 V
I	2	2 AMBULATORY NURSING HOMES, NURSERIES W/ 5 OCC. OR MORE & >6 YRS OLD	I_2 I	I_2 II	I_2 III	I_2 IV	I_2 V
I	3	3 MENTAL HOSPITALS, SANITARIUMS, JAILS, PRISONS & REFORMATORIES	I_3 I	I_3 II	NP	NP	NP
M	1	1 PARKING GARAGES, CARPORTS, SHEDS & AGRICULTURAL BLDGS.	SEE "M" GENERAL				
M	2	2 FENCES > 6'H, TANKS & TOWERS	SEE "M" GENERAL				
R	1	1 HOTELS, APARTMENT HOUSES, CONVENTS & MONASTERIES > 10 OCC.	R_1 I	R_1 II	R_1 III	R_1 IV	R_1 V
R	3	3 DWELLING & LODGING HOUSES	R_3 I	R_3 II	R_3 III	R_3 IV	R_3 V

  Figure 10.3

HEIGHT:

NO LIMIT Stories

NO LIMIT Feet

AREA/FLOOR:

NO LIMIT Square Feet

FIRE RESISTANCE CHARACTERISTICS:

Exterior Walls

    Bearing   **4**

    Non-Bearing   **3**

Interior Walls

    Bearing   **3**

    Non-Bearing   **0**

Structural Frame   **4 > 8 STORIES** / **3 < 8 STORIES**

Floors   **3**

Roof   **2**

TOILET FIXTURE COUNT:

N° OF OCC / EMPLOYEES OF EA. SEX	WC M	WC F	UR	LAV M	LAV F	SH [1]	DF [2]
1 – 200	1	1	0	1	1	(1)	1
0 – 15	1	1	0	1	1	(1)	1
201 – 400	1	2	1	1	1	0	2
16 – 35	1	2	1	1	1	NA	1
401 – 600	1	3	2	2	2	0	2
36 – 55	2	3	1	2	2	NA	1
601 – 800	2	4	2	2	2	0	3
56 – 80	2	4	2	2	2	NA	2
801 – 1000	2	5	3	3	3	0	3
81 – 110	3	5	2	3	3	NA	2
1001 – 1200	2	6	4	3	3	0	3
111 – 150	4	6	2	4	4	NA	3
OVER 1200	1/600	1/275	1/500	1/2 WC/UR		0	(2)
OVER 150	1/40	1/40	1/40	1/2 WC/UR		NA	1/100

Note: Common facilities accepted if accessible to occ. & employees.

(1) Provide showers where hazardous/ poisonous materials are present.

(2) 1 DF/150 occ. up to 600
    1 add./ea. 3000 occ.

Figure 10.4

EXIT REQUIREMENTS:

Fire Enclosure   **2 HRS**

Number of Exits:

OCCUPANCY	MIN 2 EXITS IF OCC ARE OVER ↓ (A)	HANDIC'PD ACCESS
OFFICES	30	YES
STORES	50 MAIN FL / 10 UPPER FL	YES
FACTORIES	30	NO
ALL OTHERS	50	YES

(A) 100 % FIRST FL.
    50 % ADJ / FIRST FL.
    25 % ADJ / ADJ / FIRST FL

Width of Exits:

$$\frac{\text{Total Occupancy}}{50}$$

Smokeproof Enclosure:

    Over 75' Height

Distance to Exits:

    150' Unsprinklered

    200' Sprinklered

Dead Ends: **20'** Max.

Corridors:

    Width   **44"**

    Height   **7'-0"**

Stairs:

    Width   **44"**

Doorways:

    Width   **36"**

---

tion. Extensive developments of this process should always be avoided. The extent of a regulatory survey at the programming stage should be limited only to those basic areas in which the functional structure of spaces will be affected and never extended to those sections which concern design procedures, such as riser and thread sizes, parapet walls, floor finishes, etc.

TEMPORAL ANALYSIS. The most effective way of simplifying the growth and change development of operations is by limiting the number of years the development will extend into the future. Most growth or change projections which exceed ten years become dangerously speculative and should only be executed when specifically requested by the client with a clear understanding of their inherent limitations.

One aspect that should never be overlooked is the study of past operational developments, since their analysis will inevitably aid understanding of functional requirements and hierarchies of growth and directions, as well as provide a concrete foundation on which to build a theory of future developments. The determination of spatial projections, when based on viable possibilities and not speculation, suggests the use of reliable sources and arguments which, at the same time, constitute the possible developments of this phase, such as population projections, market analysis, regional characteristics, urban developments, economic outlook, and an entire gamut of extremely specialized data which should only be derived from very reliable sources and, for the most part, executed by specialists.

Temporal projections should never be established by the programming design professional alone, but only reviewed by him/her as related to the overall structure of the program and the manner in which such projections might affect the project. The actual determination of growth and change projections, unless a specialized consultant has been retained for that purpose, is a team effort in which client, users, consultants and design professionals join in an attempt to determine, with the greatest possible degree of accuracy, HOW and TO WHAT EXTENT tomorrow should affect today.

10.2 Computer-Aided Programming

Bearing in mind present computer technology capabilities and limitations, the effective use of computers as an aid to architectural programming today must be considered of an interactive nature, as a system which allows a human operator to interact with a computer in a continuous way. The use of proper data bases and computerized systems in this interactive manner can be extremely helpful in the recording and processing of all feedback data regarding spatial densities, areas and volumes studies, allocation and entries with respect to building types, plan efficiency ratios, economic analyses and even regulatory surveys.[24]

Actually, computer capabilities for numerical computations and data processing make the use of computers extremely helpful in any area related to data analysis. Furthermore, several of our present systems and procedures rely entirely on the use of computers for their application, since their usefulness is based strictly on a computer's capability to perform certain tasks at a rate which manual processing could NEVER achieve.

Today, there are several specialized titles which deal with the use of computers in the design professions and which cover subjects ranging from the simple use of computers for accounting and business operations to the more specialized computer graphics and computer-aided design. For more information regarding these publications, refer to the bibliography at the end of this text.

Access to computer facilities can be easily achieved by simply procuring the temporary use of an existing computer center, by employing an in-house terminal connected to a larger computer network, by acquiring a complete in-house computer system, or through the use of a mini-computer system which functions as an actual terminal by means of time-sharing devices, as well as several other possible combinations of these alternatives.

Since most of the data contained in reference manuals and project follow-ups and feedback forms can be stored in a computer memory or in tape or disc libraries, the possibilities involving the use of computer terminal systems as an aid to the programming functions of professional design practices are unlimited. Furthermore, with the development of computer graphics, most diagrams which concern the analysis and programming functions, whether involving two or even three dimensional graphics, can be easily and readily executed at a speed and accuracy otherwise unattainable.

In addition, many man-hour costs and a lot of related expenses can be saved by translating specific spatial investigations descriptions and functional interactions analysis into a computer's program. The savings are due to the flexibility gained by converting the spatial analysis and programming system into data which can be readily processed by the computer.

In architectural programming, the basic areas in which the use of automated systems has been found to be of significant help are:

Site Analysis
    Climatological Data Storage
    Slopes and Drainage Analysis
    Cut/Fill Analysis
    Passive Solar Analysis
    Views, Accessibility and Orientation Analysis
    Maps, Circulation, Flow and Overlay Analysis

Spatial Analysis
    Data Storage and Analysis
    Circulation Analysis/Flow Diagram
    Spatial Relationships Diagram Generation

Economic Analysis
    Quantities, Determination
    Costs Analyses and Projections

Temporal Analysis
    Scheduling/CPM Diagrams
    Temporal and Statistical Analyses

Feedback
    Project Description, Data and Evaluation

One word of caution, although the implementation of automated systems can improve the effectiveness and performance of analysis and programming procedures within a professional design organization, such implementation should NEVER be based solely on the advantages realized by this programming process, since it is very unlikely that the specific advantages realized during the architectural programming phase will justify expenditures of the nature computerized systems represent. Ultimately, the implementation of systems such as those mentioned above will have a positive effect on the analysis and programming phases of professional design services; however, it is unlikely that such positive effects alone could, in the majority of cases, justify the acquisition of the necessary computer hardware and software.

10.3 Follow-Ups and Feedbacks

The word "performance" when related to the qualitative evaluation of systems and functions generally becomes a point of dissention among those parties involved in its definition.

Just as subjective inputs can benefit or damage the procedures and results of any professional design discipline, the performance evaluation, when determined by subjective approaches, can provide either the most valuable information concerning a project's excellence, or the most damaging misconception regarding its functional values, and since future decisions may be based on this evaluative process, it is essential that a conscientious method of feedback data collection for completed projects be established.

These data collection procedures might not necessarily be executed by a programmer, but whenever possible they should be, since the establishment of systems, groupings, functional relationships and performances as well as areas, volumes and spatial parameters during this professional design phase, enables the programmer to reuse the input received from this investigative process more effectively than any other discipline.

In analyzing the placement of a door leading from one room to another, the production department in any professional design office will probably be interested in the door's mechanics; in other words, "Does it work properly?" The designer will probably analyze whether the door was placed in the right location; that is, "Aesthetically and functionally, is it placed in the best possible location?" The programmer, on the other hand, will probably concern himself less with these questions than with the door's operational performance; that is, "Is it supporting the function it was programmed to support?" And is so, "How?" or if not, "Why?" This question, which by nature of its comprehensiveness includes the other two within its basic structure, indicates the value of a broad rather than a specific approach in the design of evaluation procedures, since invariably the comprehensive outlook carries much more weight and determines more clearly the direction to be taken in any significant follow-up method.

Different specializations within the same professional design organization are interested in different feedback information. That is why the establishment of an overall system of detailed information for rating the operational performance of a project can lead to nothing but confusion.

For example, the management sector of a professional design firm might be interested in the following facts:

1. Was the client satisfied?
2. Did the professional design services rendered by our firm meet his expectations?
3. How well is the building performing?
4. Are the users satisfied with the facilities provided?

These questions, as we can clearly see, define only general performances from different angles without detailing problem areas. At the management level this is the way it should be, but not in the programming sector. The satisfaction or dissatisfaction of a client might or might not depend on the spatial performance of the building, and the excellence of professional design services as a whole is a matter beyond the control of one particular specialist.

Therefore, any follow-up method must be designed to provide a comprehensive summary of pertinent and basic data related to specific projects, which data can be used by each and every phase involved in the spatial development of those projects. The method must not concern itself with extensive evaluations and studies of individual subdisciplines or facts useful only to specific phases of professional design.

Furthermore, since each of these phases originated in the program, it provides the logical starting point and frame of reference that will enable us to establish a checklist of value judgment for analyzing the functional performance of each phase. This is another reason why this task should fall within the programming scope.

The evaluation sheet illustrated in Figure 10.5 provides data at all the levels mentioned above, outlining, at the same time, an overall view of the client and user's input as they relate to each component. Assigned ratings of excellent, good, fair, poor or bad should be established by a joint evaluation when judging each component's functional performance.

The primary advantage of this particular table is that it offers at a glance all the essential data necessary for a project's follow-up evaluation, while allowing for selectivity of information. For example, the cost data and general evaluation of professional design services might be used while the specific elements evaluation might be omitted, etc.

Another advantage of this breakdown is that problem areas can be easily identified, and errors either corrected or avoided in the future, if the detailed evaluative breakdown of areas, shown in the lower half of the table, is executed.

The form includes number of occupants, spatial densities and area columns just as a verification of programmed functions for future use in the programming library.

Since numerical factors can vary tremendously in subjective interpretations and judgments, we should try to offer as few judgment alternatives as possible, thereby forcing an average evaluation of each case instead of an extensive list of possible qualifications, like very bad, bad, not so bad, poor, not so poor, fair. With the five general categories —

bad, poor, fair, good and excellent — all the possible remarks can be synthesized in a definite pattern which will cover everything from the direct criticism of a "bad," to the praise of an "excellent."

Once again, these designations should be decided upon jointly by user and design professional. Some clients, because of their own personalities, occupations, etc., tend to be more critical than others. It is impossible to determine the actual performance factor for a project that is evaluated on the basis of ONE opinion only, without modifications or re-evaluations. A client might label as "poor" the functional performance of a room because of a feature determined by factors beyond the scope of a specific spatial function, as a window being too high to be seen through by someone sitting down at a desk, when the window sill had to be raised because of an increase in the size of a unit heater below it, which requirement resulted from the client's request during construction for sliding glass doors across the room.

Situations like this happen all the time, and since follow-up and feedback data methodologies are essentially designed to serve the professional design practice, filtering these comments is not only advisable but essential if such methodologies are to illustrate a true picture of spatial conditions and effects. The primary task of feedback systems is to provide our professions with the invaluable benefit of actually knowing whether the processes which commenced with a programming action and ended with a feedback analysis of functional evaluation and operational performance have effectively provided the spatial creation in question with adequate functional characteristics and capabilities to fulfill the essential goal of the professionals' joint efforts.

10.4 The Programming Library
Most professional design firms have within their general or specialized reference library sections dedicated to the analysis and programming of buildings. The problem, however, is that these sections are generally mixed with data pertaining to other phases and are usually filed within a general

PROJECT EVALUATION DATA SHEET

ARCHITECT

project no	date
contract date	SHEET
completion date	OF

PROJECT

LOCATION

project arch	interior design
program	contract documents
design	const. admn.

BUILDING TYPE: ____________________ S. FT: _______ CU. FT: _______

CONTRACT	NAME	CONTRACT $	S.F. $	C.F. $	% of TOTAL
GENERAL CONST.					
PLUMBING					
H V A C					
ELECTRICAL					
OTHER					
TYPE OF CONST: N° OF STORIES :	TOTALS				

C O N S U L T A N T S

structural	electrical
plumbing	landscaping
hvac	other
comments	

OVERALL PROFESSIONAL DESIGN SERVICES EVALUATION

PROFESSIONAL	DESIGN	SERVICES				ENGINEERING		SYSTEMS		CONSTRUCTION		
function	aesthetics	materials	const. admn.	interiors	landscaping	plumbing	h v a c	electrical	other	quality	cost	time

COMPONENT	ELEMENTS	ROOM N° AREA	OCC 0.5	SF O	ARCH.		I.D.			ENG.			REMARKS
					function	form	fin.	color	furn.	plbg	hvac	elec	

185

Figure 10.5

outline of building types. While such building type designations do classify structures according to their generalized functions, they are too general to provide a clear definition of the particular programming considerations inherent in definite functional variations within specific building types. Many professional design publications may have devoted themselves to the subject of "schools," but does that mean that within their pages we will find the necessary data to aid us in the programming process of a military academy? A building of this nature is, after all, a school. However, we will probably find only a few pages related to our problem after spending an entire day checking our schools file. This is why a logical fragmentation of building types is not only useful but imperative in the conscientious organization of ANY reference library, just as it was essential in the primary stages of the analytical process. For the purposes of this text, we will only concern ourselves with the programming section of such a library, although the procedures outlined in this sub-chapter may also be applied to reference sections, design, production, etc.

One way of organizing a Programming Library can be outlined as follows:

SECTION I. PROGRAMMING MANUALS, TECHNIQUES, PROCEDURES, SYSTEMS, POLICIES, ETC. This is a generalized section which contains ALL the available data that relates to the actual PROGRAMMING activities. Theoretical in nature, it is not concerned with building types of any kind. It contains programming references, program formats, booklets, etc.

SECTION II. BUILDING TYPE REFERENCES. This section contains programming information regarding specific building types, and all the available data related to them from spatial parameters and cost data to environmental requirements, follow-ups and feedback investigations.

A viable way of subdividing this section can be developed as follows:

Division 1. Residential.
  1.1   Housing Site Planning/General
  1.2   Single Family/Detached
  1.3   Apartments
  1.4   Condominiums
  1.5   Housing for the Aged
  1.6   Public Housing
  1.7   Housing for the Handicapped
  1.8   Mobile Homes and Parks
  1.9   Youth Hostels
  1.10 Fraternity Housing
  1.11 Dormitories
  1.12 Townhouses
  1.13 Duplexes
  1.14 Cluster Housing
  1.15 Planned Unit Developments (P.U.D.)

Division 2. Educational.
  2.1   Nursery Schools
  2.2   Elementary Schools
  2.3   Secondary Schools
  2.4   Colleges and Universities
  2.5   Technical Schools
  2.6   Vocational and Art Schools
  2.7   Military Academies
  2.8   Special Education Facilities
  2.9   Research Centers
  2.10 Sports Instruction Centers

Division 3. Health and Hygiene.
  3.1   Hospitals
  3.2   Nursing Homes
  3.3   Clinics
  3.4   Rehabilitation Centers
  3.5   Mental Health Centers
  3.6   Medical and Dental Schools

Each sub-index should be divided into categories like the following:

Building Type

 Division 4. Commercial.
 4.5 Offices/General
  4.5A Site analysis and planning data/references (See Figure 10.6)
  4.5B Spatial analysis data and references (Figure 10.7)
  4.5C Economic data
  4.5D Related references (such as design approaches, construction systems, etc.)

SECTION III. PROGRAMMING FORMS AND AIDS. This section should contain those standard forms, survey formats, etc., which are used repeatedly throughout the analysis and programming procedures. (See Figure 10.8)

In many cases, programming references are contained within manuals and publications which are more directly related to the design or production sections of professional design libraries. In these cases, it is advisable to file in the programming section an index card referring to the title, date, author and page number of the book or publication in which these programming references are found.

An easy way to develop this reference library is to establish the general format and then fill in each section as the need arises. That is, when a particular project is researched, all the pertinent data should be recorded in the corresponding section for future use. If during the course of this research a

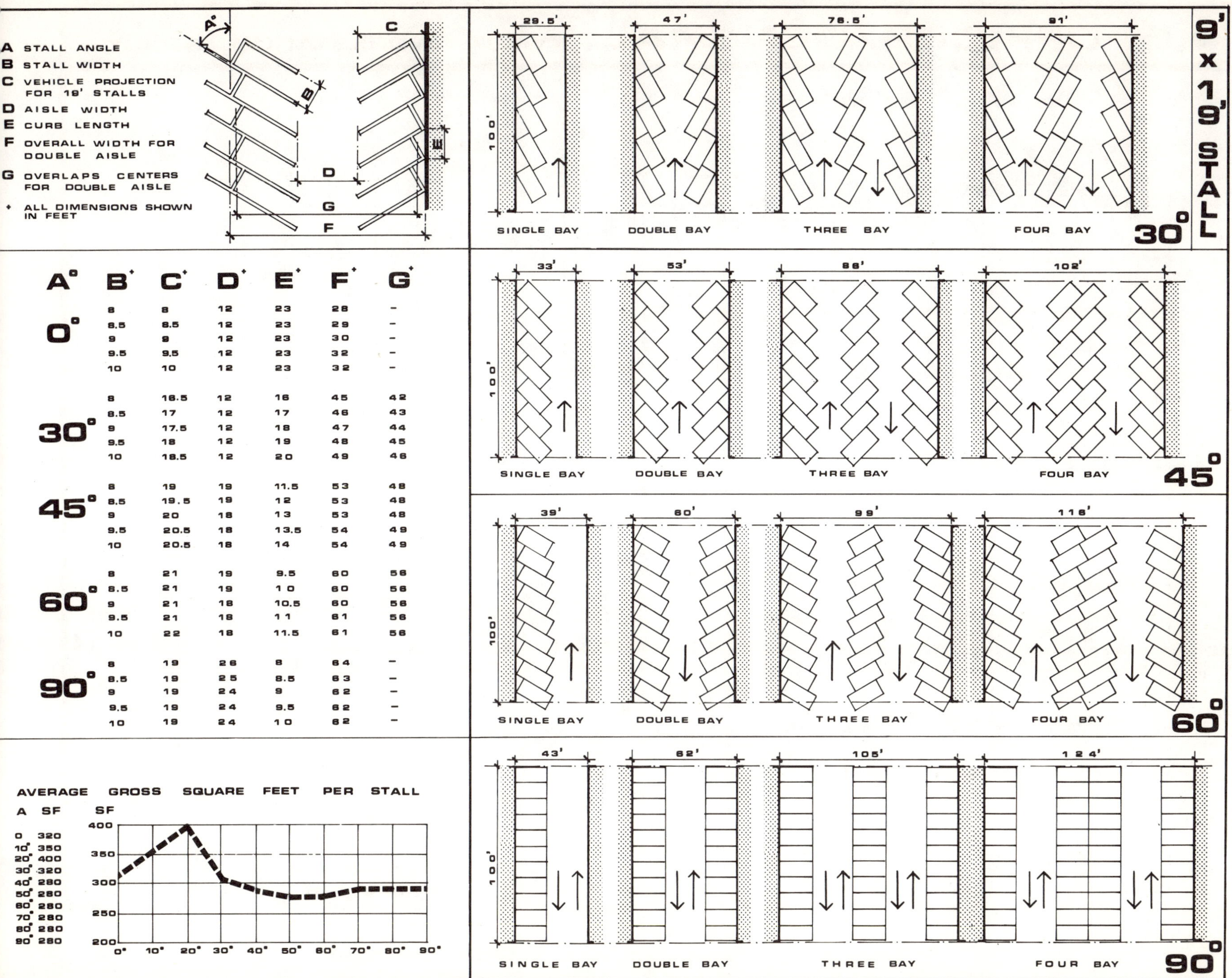

A STALL ANGLE
B STALL WIDTH
C VEHICLE PROJECTION FOR 19' STALLS
D AISLE WIDTH
E CURB LENGTH
F OVERALL WIDTH FOR DOUBLE AISLE
G OVERLAPS CENTERS FOR DOUBLE AISLE
• ALL DIMENSIONS SHOWN IN FEET

9' X 19' STALL

A°   B•   C•   D•   E•   F•   G•
0°   8   8   12   23   28   —
     8.5   8.5 12   23   29   —
     9   9   12   23   30   —
     9.5   9.5 12   23   32   —
     10   10   12   23   32   —
30° 8   16.5 12   16   45   42
     8.5   17   12   17   46   43
     9   17.5 12   18   47   44
     9.5   18   12   19   48   45
     10   18.5 12   20   49   46
45° 8   19   19   11.5 53   48
     8.5   19.5 19   12   53   48
     9   20   18   13   53   48
     9.5   20.5 18   13.5 54   49
     10   20.5 18   14   54   49
60° 8   21   19   9.5 60   56
     8.5   21   19   10   60   56
     9   21   18   10.5 60   56
     9.5   21   18   11   61   56
     10   22   18   11.5 61   56
90° 8   19   26   8   64   —
     8.5   19   25   8.5 63   —
     9   19   24   9   62   —
     9.5   19   24   9.5 62   —
     10   19   24   10   62   —

AVERAGE GROSS SQUARE FEET PER STALL
A   SF
0°   320
10° 350
20° 400
30° 320
40° 280
50° 280
60° 280
70° 280
80° 280
90° 280

SINGLE BAY   DOUBLE BAY   THREE BAY   FOUR BAY
30°   29.5'   47'   78.5'   91'
45°   33'   53'   86'   102'
60°   39'   60'   99'   118'
90°   43'   62'   105'   124'
100'

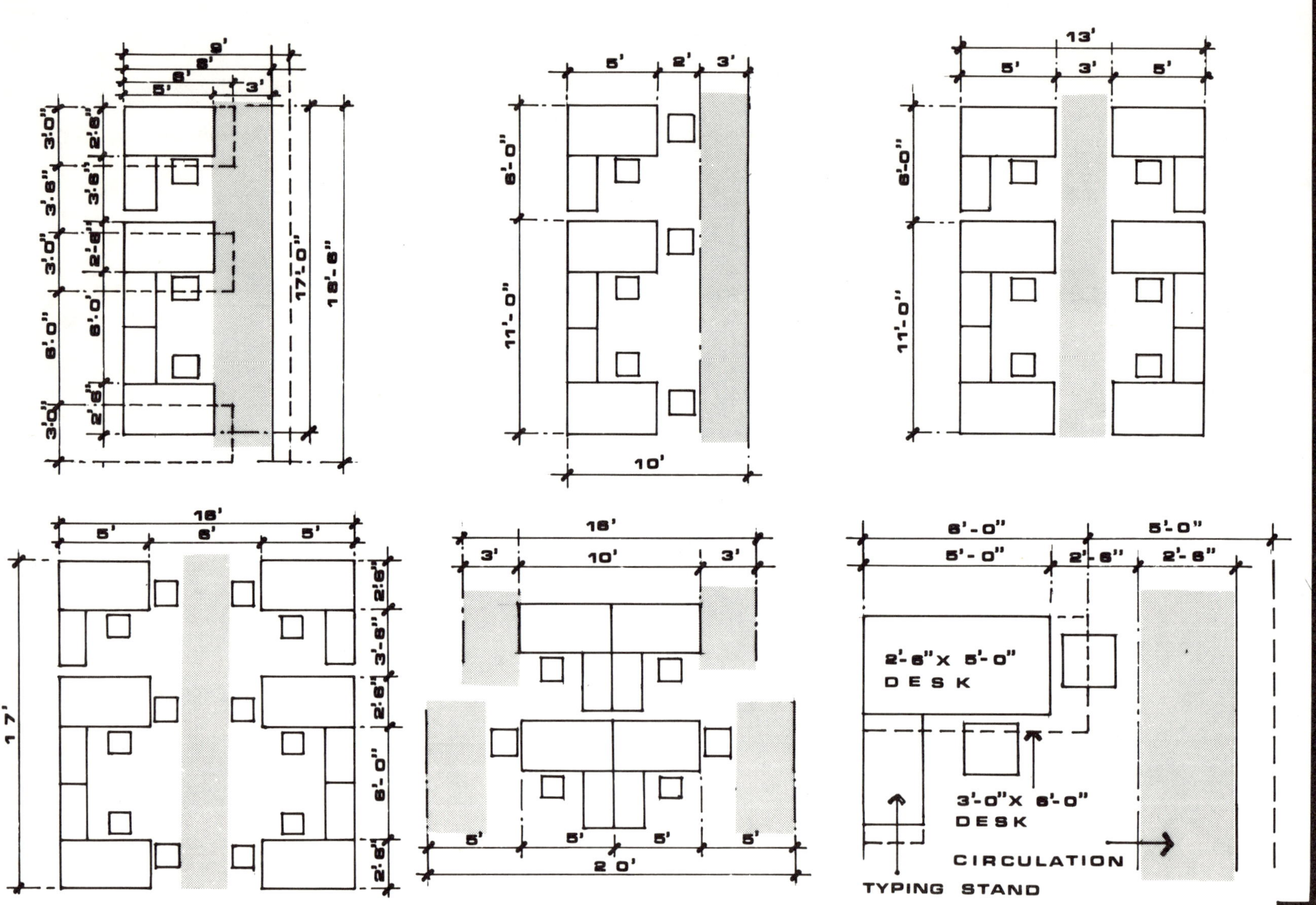

Figure 10.7

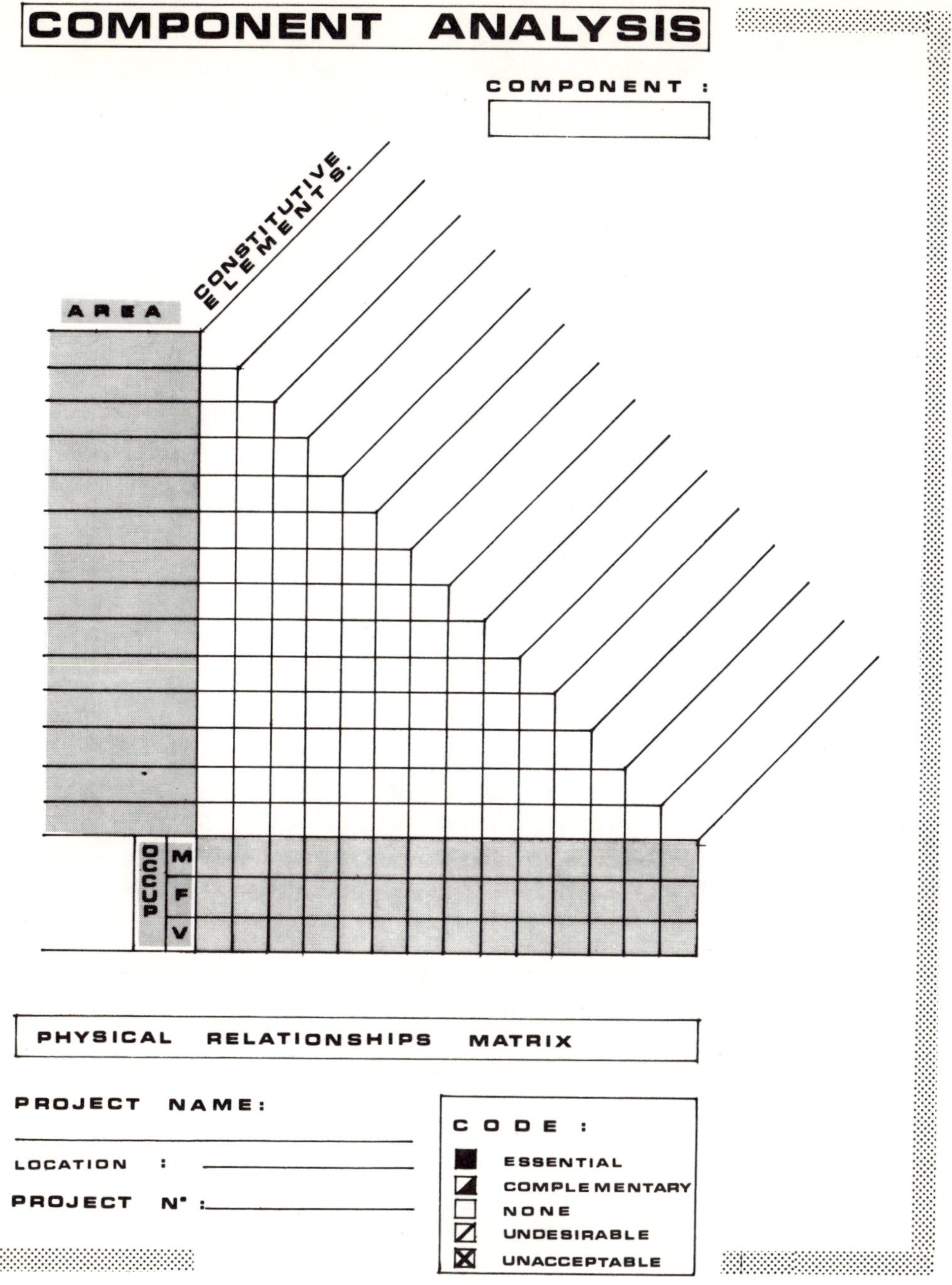

Figure 10.8

worthwhile reference is found which applies to another building type, this reference should also be recorded in its corresponding file.

If a firm discovers in the course of its programming activities that different forms and aids are necessary for different building types, then it is convenient to subdivide Section III into the major categories outlined in Section II and file such forms accordingly.

An uncomplicated and prompt retrieval technique which has proven successful for professional design libraries utilizes color coded formats.[25] In many cases, these formats can be also easily adapted to encompass specialized libraries, such as the one defined here, within their overall layout.

No matter which filing or retrieval technique is used, the cost of maintaining an updated reference library for each professional design phase is more than justified when compared to time wasted in researching dozens of books and magazines guided only by a series of titles that are usually too broad or misleading to accurately define the particular contents of their texts.

What actually defines "working experience" is the recorded information of an event for future use. This can only be accomplished through a systematic approach to data collection and processing; it makes the difference between organization and chaos and determines, to a great extent, the final degree of productivity, performance and achievement of any professional design activity.

---

[24] It is not altogether improbable that in the not too distant future the regulatory agencies will become part of the same network connected to the computer terminals of professional design firms or terminal "centers," and codes surveying, processing and updating will be executed entirely in this manner.

[25] Refer to "Color Coding the Firm's Library for Fast, Simple Retrieval" by Kathleen L. Kalt. *Architectural Record*, Jan. 1980.

Notes on Interfaces

In professional design, the only positive trends and activities are those which fertilize and facilitate creativity. Obviously the concept of creativity is such an essential element of the design professions' intrinsic structure that any trend which imperils or sets back the creative action will be in direct conflict with the universally accepted definitions of purpose and accomplishment of professional design tasks. The conceptual approach to programming methodologies is also subject to this condition. If a programming methodology works AGAINST, instead of facilitating, the creative essence of professional design, it is bound to fail, and what is worse, it is bound to cause a lot of damage before it does.

On the other hand — and precisely because creativity is an extremely individualistic concept in itself — the same programming system that complements and aids the creative task of a specific individual or professional design organization may limit and even damage the creative developments of others.

The question has been raised whether the contents of a building program should be structured and developed differently when done for a specific designer or design organization, or whether the program should be prepared as a format for the documentation of statement of purpose, intents and specific requirements for a building project prior to the selection of any professional to oversee its conceptual execution. Basically, the question is "Should the actual needs and requirements of a project vary depending on the designer in charge?"

One simple fact which comes to mind when trying to find a viable answer to the real question behind a program structure is that buildings can be looked upon as structures created BY human beings FOR human beings; but buildings can also be defined as structures created and executed by and for FUNCTIONS. If at first glance the distinction seems useless, let us not forget that in this discrepancy of interpretation lies the concept of "architectural humanization," without which there cannot be any creative purpose or justification whatsoever. Yet, the real issue seems to lie more in the balanced acceptance of this discrepancy rather than in absolute one-sided conclusions. Buildings are created by and for human beings AND the functions they perform; therefore buildings must serve BOTH.

That is why the main characteristic defining the TRUE comprehensiveness of a spatial programming methodology has to be the clarity and accuracy of its layout of flexibilities, controllable variables and programmatic statements.

To assume that the owners or users of spatial developments can be made subject to a design trend or even aesthetic values is wrong. Such an attitude will inevitably produce a "stand-still" in the long struggle that the design professions have been engaged in for years to achieve social significance and recognition. The innate aesthetic sensitivity of a designer might affect his sense of values so that certain commodities or issues are considered unimportant or even undesirable when viewed in the light of aesthetic achievement, but that does not mean the people or the functions he is designing for will have the same value pattern.

However, if the designer's formal and spatial sensitivities are ignored, the ultimate consequences could be more disastrous than operational sacrifices in exchange for aesthetic achievements or functional misinterpretations, since function, comfort and convenience are terms related not only to strict operational performance factors, but to the need for physically and psychologically desirable and acceptable environments.

That is why, to properly illustrate this conceptual duality, any programmatic system must concern itself with the delicate balance of form and function at the outset, not as a predesign activity which commences by closing or discarding alternatives but as an investigative and planning process which tries to OPEN possibilities or discover roads to follow in the successful execution of professional design achievements.

For this reason, programmers should try to avoid stylistic considerations in their aesthetic statements. While limiting certain stylistic trends at the outset can be extremely helpful to the design phase, ESTABLISHING STYLISTIC COURSES OF ACTION FOR THE DESIGN PROCEDURE IS HERETICAL, since the key to the successful execution of a programming activity is the realization that a building program is not meant to be an inflexible edict or list of mandates regarding a construction project, but rather, the conglomerate of identified needs, variables and flexibilities under the same "programmatic requirements" heading. This conglomerate is meant to facilitate the creation of a spatial envelope, never to limit or shackle it. Undoubtedly, there is a certain degree of subjectivity involved in programming activities. Indeed every human activity contains a certain degree of subjectivity — yet, in judging design or in simply analyzing

formal propositions for design solutions, the balance between objectivity and subjectivity can only be attained through a true understanding of the design problem on the part of those making the judgment.

There is no question that during programming, some issues, functions or elements are given more importance than others not on the basis of their strict operational hierarchy, but on the basis of compatibility of the function (or the people who execute that function) with the programmer or because such issues, functions or elements inherently offer more professional design advantages of one type or another. For this reason, the value system employed initially in developing programmatic priorities must be based on operational hierarchies ONLY. Aesthetic or design-related requirements should never be ignored, but until such time as operational functions are solely based on the spatial form which envelopes them, we must realize that the essential force behind functional design is the effective creation of formal developments FOR functional behaviors and not the opposite.

Another issue involving the degree of comprehensiveness of programmatic statements in the area of stylistic trends or conceptual approaches to design solutions is the conflict between the pursuit of prototypes or stereotyped accomplishments. A prototype is an achievement independent of formal repetitions and can only be designated through programmatic judgments. It serves as a model on which later stages are based or judged; for example, the Egyptian pyramid is a prototype (actually, each Egyptian pyramid is a prototype in itself). A stereotype, on the other hand, typifies or conforms to an unvarying pattern or manner and lacks any individuality. Unfortunately, stereotypes are the inevitable product of inflexible prototype-programmed issues; and this problem, which is directly related to that of programmatic activities overstepping their NATURAL boundaries, emerges as one of the most damaging results of improperly developed spatial programming formats, outlines or even methodologies.

For this reason, whenever the programmatic requirements outline becomes so stringent that it defines or patterns itself after a prototype, the result will probably be nothing but a stereotyped solution.

Furthermore, this rule can also be applied to an entire programming system. Indisputably, conceptions and developments in design can be directly affected by the development of the program which outlines the basis from which the design must spring. For example, just as the scale of design in relation to the designer can involve a conceptual development that acts as if "sculpturing a figurette," it can also take the form of "carving a palace"; and the results produced by the two approaches will probably be totally different from one another, even if executed by the same person, since the proportion or disproportion of each approach almost "redirects" the specific subjective pattern which will structure the design procedure and its directions ultimately affecting its achievements as well.

Because of this, and because programmatic statements objectives and requirements can effectively alter the conceptual development scale of design activities, extreme care must be exercised when dealing with areas which often cross these boundaries, such as statements and requirements involving qualitative issues like image, character, appeal, etc. The overdefinition or predetermination of these factors can have such a profound effect on the subsequent professional design procedures that the end result will be patterned after pre-established formats instead of being tailored after its own needs and stereotyped rather than properly and conscientiously designed.

Today, the necessity for proper spatial programming is accentuated by the enormous occupational fragmentation of the last few decades. This fragmentation has produced a variety of sociological elements with a high degree of specialization and individuals who know nothing about construction, professional design or related disciplines are more the rule than the exception. No longer do we find the family who built their own house or even partially helped to construct it. We live in the era of the specialist; and although this can be extremely advantageous, it also has serious disadvantages where understanding and assimilation are concerned. In addition, foreseeable crises and the depletion of natural resources in the near future seem to suggest a limit to the subjective or unjustified use of FORM as the essential ingredient of professional design formulas. This does not mean that designs must invariably justify themselves to every element of society, but it does mean that a design must ALWAYS justify itself to those elements of society which it serves.

Faced with the developments, uncertainties and crises of the world today, the proper definition of cultural and socio-economic adjustments, flexibilities and compromises is not just an alternative but a mandate, if we are to understand, define and resolve those issues which will undoubtedly shape our future. No longer can we maintain unilateral attitudes where strict functionalism or aesthetic value alone defines the primary causes or justifies effects of professional design tasks. In building projects, "character," "image" or "appeal" can become matters of such subjective interpretation that their overall sociological implications are not only left undefined, but also often completely unheard of. Only the properly balanced outline of conditions can satisfactorily determine and develop the conception and execution of spatial designs. The true social significance of the design professions lies beyond the simplistic "appeal" recognition by a few selected experts or a handful of self-appointed critics; it lies in the actual achievement of their essential objective: providing society with the proper spatial scenario in which to perform, produce and develop.

Throughout history men have left in their buildings the unequivocal mark of their era. No matter what the period or pattern of events, the buildings of Mankind have always represented a continuing tribute to the past and a record for the future.

No matter which time, event or consequence, Architecture has always told the story.

What will it tell about today?

Glossary of Terms

The following pages contain definitions of terms employed or directly related to this text and whose special usage during programming activities may need explanation or clarification.

Appeal: Any quality which arouses sympathetic responses or attraction.

Architectural Programming: The process which sets forth the requirements and procedures necessary for the architectural realization of a building project.

Area Diagram: A scale drawing which by using simple shapes shows the areas needed by units of an organization.

Bubble Diagram: A drawing showing relationships of functions, individuals or groups within an organization in freely drawn shapes which may be overlapped or contained one within another.

Building Programming: The process which sets forth the spatial and functional requirements of a determined structure or building.

Character: Symbolic statement about the type, qualities and characteristics of a structure or operation.

Cluster: A grouping of closely associated components, functions, elements or work units.

Compliance Condition: The restriction or prerequisite of accordance with pre-established rules, objectives, requests or universal factors.

Computer Hardware: The fundamental computing elements and associated devices of a computer system.

Computer Software: Collective designation for the various types of programs utilized in a computer system.

Constitutive Elements: The physical constituents or modules (either spatial or human) of operational components.

Construction Documents: The working drawings

and specifications for a building project.

Controllable Variable: Subjectively determined, influenced or regulated programmatic issues subject to unpredictable, inconstant, changing or fluctuating developments.

Decentralization: The act of breaking the concentration of several parts within a whole.

Design Professions: Those professions collectively responsible for the creative design or modifications of Mankinds' physical environment.

Economic Analysis: The study and evaluation of the economic constituents of a building project. The result of such study.

Environmental Criteria: Standards, rules or tests on which environmental judgments or decisions can be made.

Formality: Careful attention to order, conventionality, regularity or precision.

Functional Interactions: Reciprocal actions between functions.

Functional Relationships: The logical or relevant association between functions.

Geometry: The conformance characteristics of a body with geometric forms.

Gross Area: The sum of all floor areas included within the outside faces of exterior walls for all stories or areas which have floor surfaces.

Grouping: The act of arranging, composing or combining spatial elements or functions. The result of such act.

Hierarchy: A group of functions, elements or categories arranged in order of rank, grade, significance, class, etc.

Humanism: Quality of being human. Of human nature.

Image: A conceptualization or representation of something as a copy or derivative of another.

 Integration: To bring together into a whole. To unify.

Land Survey: The mapping of a specific site, which may include topography, boundaries, existing structures, etc.

Matrix: A chart in which all the relationships between elements or components are displayed. The two lists may be the same or different.

Net Area: The sum of all areas on all floors of a building assigned, or available for assignment, to programmed operational functions.

Network (based) Planning: Schedule or process that charts all or many step relationships and dependencies.

Open Plan: A common term for any spatial layout which uses little or no fixed partitioning.

Operational Analysis: The fragmentation of an operation into constituents for individual study and evaluation. The result of such study.

Operational Component: The functional constituents of each operational division structured by an organized hierarchy of constitutive elements.

Operational Division: The major functional constituents essential to an operation. A defined hierarchical grouping of operational components.

Operational Fragmentation: The separation or division of an operation into functional constituents for individual analysis.

Operational Interfaces: Functions or activities of common boundaries among operations.

Plan Efficiency Ratio: The relation or proportion between the net assignable and gross area of a structure.

Priority: The state of precedence of one concept, element or function over another.

Privacy: Personal interest of withdrawal from public attention, view or company. Seclusion.

Professional Design Practice: The practice of the design professions within the limits of ethics, professional standards and legal regulations applicable to the scope and execution of their functions, responsibilities and performance requirements.

Programmatic Concept: Thought, notion, idea or understanding which determines the intellectual approach to programming.

Program Format: The material form, outline or layout of an architectural program.

Program Objective: Goal of a course of action by the programmatic activities.

Program Requirements: Final layout of prerequisites, needs or wants for a project.

Project Budget: Established sum for the overall cost of a construction project including construction, land, equipment and contingency costs, financing costs, regulatory fees, legal and professional services compensations, etc.

Project Programming: The process which outlines the complete scope and programming procedures for a building project.

Quality: An essential characteristic which belongs to something and makes or helps to make it what it is. The degree of excellence or superiority which a thing possesses.

Regulatory Survey: The analysis, or recorded analysis, or applicable rules and regulations affecting a building project.

Scale: The proportion that a design bears to the subject it is designed for. The proportion between a graphic representation and the object it represents.

Sectoring: Dividing into sectors, portions or districts.

Site Analysis: The investigative process which deals with the study and evaluation of the natural or artificial characteristics of a specific site. The result of such process.

Space Criteria: A spatial standard, rule or test on which a judgment decision could be made.

Spatial Analysis: The fragmentation of a specific area into defined constituents for individual study and evaluation. The result of such study.

Spatial Density: The concentration of occupants within a space, usually designated in square feet per person.

Spatial Flexibility: The capabilities of change and growth of a space.

Spatial Parameter: A constant which determines, restricts or represents the area requirement for a specific function or element.

Spatial Relationship: The logical, natural or relevant association or connection between spaces.

Spatial Survey: The study or analysis of an operation or building in order to investigate, determine and evaluate space needs, characteristics and usage.

Status: The visible expression of level or importance of rank or position through physical arrangements.

Symmetry: Similarity of arrangement or form on either side of a dividing line or plane. Opposite-asymmetry.

Synthesis: The combining or agglutination of separate constituents to form a coherent whole. The whole so formed.

Temporal Analysis: The fragmentation of time and classification of occurrences into past, present and future for individual consideration. The development and study of operational histories and growth/change projections. The result of such studies.

Temporality: The limitations and characteristics imposed by time. Pertaining to, concerned with time.

Territoriality: A concern with identification of space for personally controlled use.

Value: The estimated or appraised worth or cost of an item. The ratio between worth and cost. That degree of quality of a thing according to which it is thought as being more or less desirable, important, useful, etc.

Appendix A

SOIL BORING

SOILS INVESTIGATIONS SERVICES, INC.

CONSULTING ENGINEERS

1845 FAIRVIEW DRIVE MONROE, WISCONSIN   PH: (608) *341·6166

November 17, 1978

Gray and Associates, Inc.
1990 Lincoln Road
Clearwater, GA

Subject: Foundation Investigation
       Sandy Creek Subdivision 8, Plat No. 18C

Gentlemen:

As requested, we have performed two standard soil borings for the subject project. Soil boring locations are presented on Drawing 8097-K1. Soil boring records are presented on Drawing 8097-K2. Prints of each of these drawings are a part of this report.

Ground water was encountered at 12 feet below existing ground surface at the time of the boring. The ground water rose to depths of 7 and 8 feet below existing ground surface 10 minutes after the completion of the borings. This indicates that the layer of organic silt forms a "seal" that prevents the ground water contained in the water bearing strata of sand to reach higher elevations. We expect this water level to reach higher elevations during other seasons of the year.

In general, soils encountered consist of fill materials over a layer of black organic peaty topsoil, over a layer of gray organic silt underlain by layers of fine to medium sand and brown fine to medium sand with some gravel in Boring #1 and underlain by layers of clayey silt and sand with some gravel in Boring #2. Please refer to the soil boring records for a complete description of soils. (See Figure A-1)

We understand that a single-story slab-on-grade building, to serve as a restaurant, is contemplated to be built at this site. Foundations for the building will probably consist of spread footings, although other types of deep foundations (piles) could be considered.

Based on the borings, we offer the following comments and recommendations to be implemented during the site preparations, construction, and design of this building:

1. Site preparations for this building should include the removal of all the fill materials, topsoil, and the organic silt from under the building area. The fill materials vary widely in nature, are relatively loose (low blow counts), and were probably placed under uncontrolled conditions. The layer of black peaty topsoil is completely unsuitable for the support of any type of structure. The layer of organic silt was very soft and highly susceptible to consolidation under load. This material is also susceptible to consolidation due to the decay of its organic content and to volume changes with changes in its moisture content. The extent of excavation is approximately 13½ feet in the Boring #1 area and 19½ feet in the Boring #2 area. Excavation to these depths will undoubtedly require the use of a dewatering system. Because of the proximity of Sandy Creek, we estimate that at least two large pumps and sumps will be needed to effectively dewater the site. We estimate that a "working mat" will be needed at the bottom of excavation consisting of 12 to 18 inch minimum layer of #2 (1½-inch to 2½-inch size) stone. This working mat will provide a base upon which the compacted granular fill could be placed and also provide a very "porous" material through which the ground water could flow easily and be pumped elsewhere.

2. Following the excavation and the placement of the working mat, the area could be built-up as needed using compacted granular fill. The compacted granular fill should consist of sand or sand and gravel placed in 6 to 8-inch layers with each layer being thoroughly compacted to at least 95% of the maximum dry density as determined for the materials used in accordance with A.S.T.M. Test Designation 1557-70, Method D. Some of the sand and gravel materials encountered in the fill could be used as compacted granular fill. However, silt and clay free materials are recommended in order to achieve the desired compaction. This could be accomplished through careful selection of the materials excavated. It may be advisable to restrict the granular fill to materials having not more than 5% passing the #200 mesh sieve. The compacted granular fill should extend down and out from the outer edge of footings by a 1:1 slope or less. Excavation limits should account for this. The dewatering operation should continue through the placement of the compacted granular fill. Special attention should be given to the moisture content of the materials in order to achieve the desired compaction.

3. An allowable bearing pressure of 3000 pounds per square foot could be utilized when the steps outlined in Sections 1 and 2 above are followed. This allowable bearing pressure assumes that footings will be placed at depths of approximately 4 to 5 feet and that more than 36 inches of compacted granular fill will be placed below the bottom of the footings. In order to minimize the potential settlement of the foundation, we recommend an increase in the amount of reinforcing steel to include two #5 bars in the footings.

4. Other types of foundations, such as deep pile foundations, could be considered for the construction of the proposed building on this site. If this type of foundation is considered further, we recommend that added deep borings be performed to determine if this type of alternative is feasible.

5. For parking, driving, and sidewalks, we estimate that the surface soils have been stabilized for this type of structure. However, due to the soft soils some settlement might still be experienced. If some settlement occurs, the affected areas could easily be repaired with a pavement overlay.

6. The excavated soft soils could only be used in landscaped areas. The site should be graded to provide positive drainage and avoid any surface water problems which in many instances can be as troublesome as ground water problems. The slab-on-grade should be underlaid with an effective moisture barrier.

7. Safety precautions such as the ones required by OSHA should be enforced throughout the entire project. This might include the proper sloping or bracing of excavated walls.

8. We recommend that a soils engineer be present at the time of excavation to implement or modify our recommendations given here on the basis of the borings performed and also to insure that the compacted granular fill has been placed properly. It is extremely important that the specified compaction be achieved.

Please contact us if you have any questions regarding this submittal, or if we could be of further assistance.

Respectfully submitted,

John D. Hick, P.E.

Triplicate

SANDY CREEK LANE ℄
108'
108'
279'
PARKING
PROPOSED BUILDING.
65'
83'
BORING Nº 1
BORING Nº 2
EXISTING TREES
EAST ROW OF HIGHWAY 97
NORTH

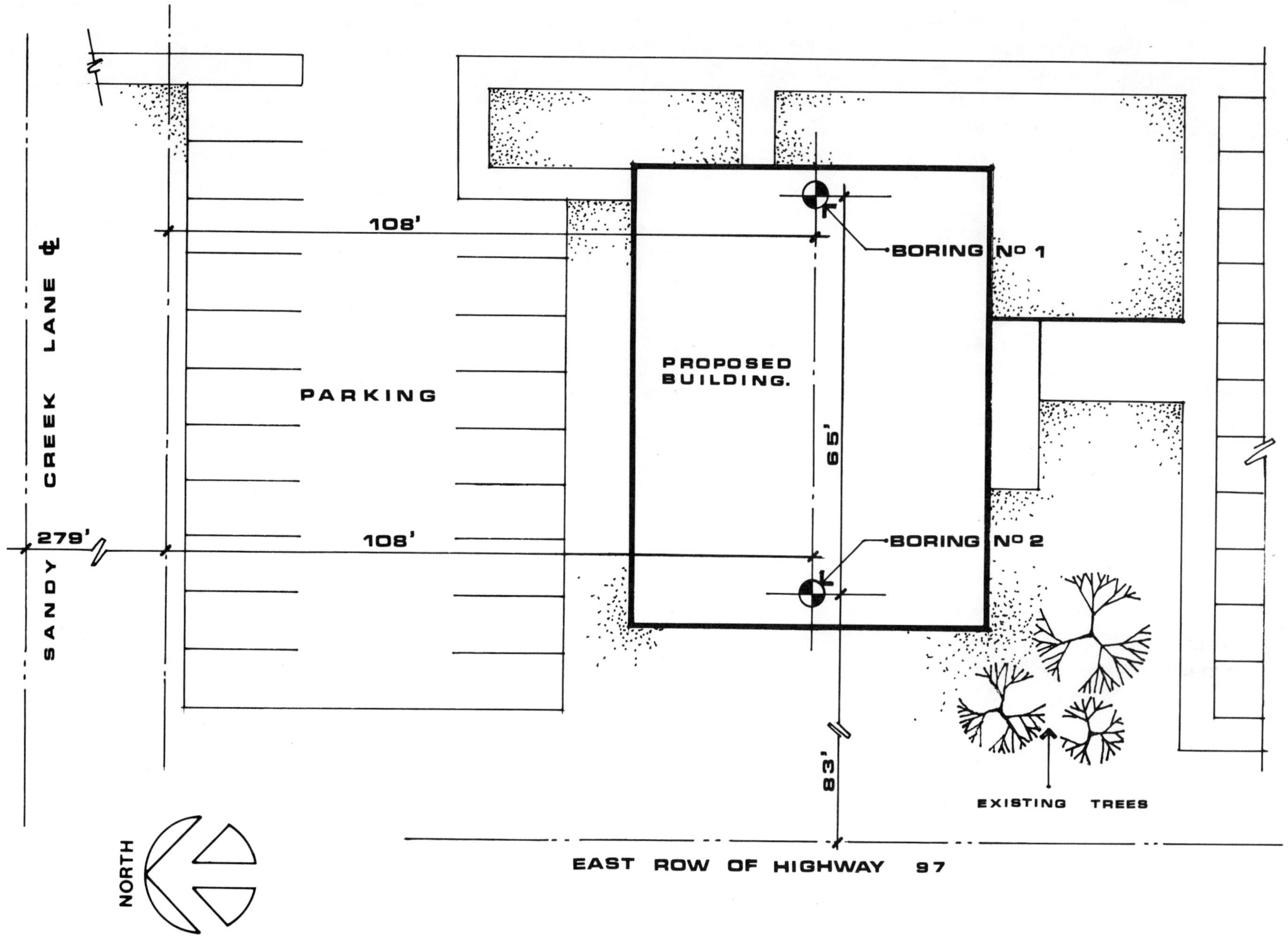

Figure A-1

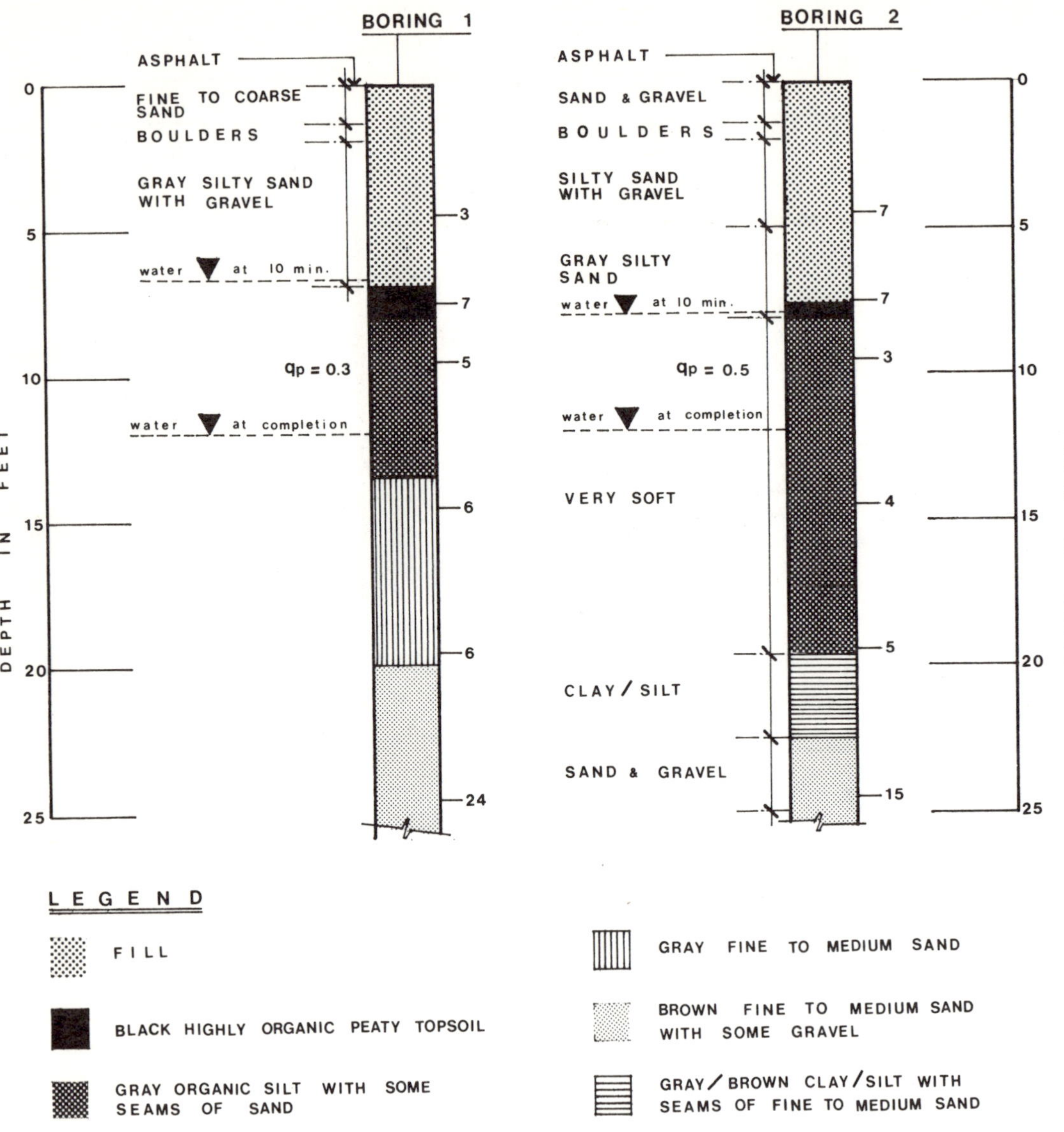

N O T E S

1   BORINGS BY ASTM TEST DESIGNATION D1586-67

2   THE NUMBER OF BLOWS REQUIRED TO DRIVE THE 2 INCH O.D. SPLIT SPOON SAMPLER 12 INCHES WITH A 140 lb. WEIGHT FALLING 30 INCHES IS RECORDED ON THE RIGHT HAND EDGE OF EACH BORING LOG. THIS IS THE STANDARD PENETRATION TEST.

3   BORINGS PERFORMED NOV. 19 1978

4   HOLES FILLED IN AFTER WATER LEVEL CHECK

5   q_p = PENETROMETER READING ; TONS / SQ. FT.

Figure A-2

Appendix B

SOLAR ANALYSIS

To make use of the solar charts included in this section, the following definitions must be clearly understood:

The respective position of the sun to any point on the surface of the earth is defined by two angles: azimuth and altitude. The different values of these angles are determined by the factors which affect the two essential elements involved; that is, the point in question by its latitude and the position of the sun by the date and hour (see Figure B-1).

The following table provides the values for the azimuth and altitude angles for the north latitudes[B1] 25°, 30°, 35°, 40°, 45°, and 50°, which cover the entire continental United States. If an accurate value for conditions which fall between the specific latitude illustrated is desired a simple interpolation of figures will produce the result. For example: If the azimuth and altitude of a 37° 30′ latitude north at 2:00 P.M. in winter is desired, from the table shown at Figure B-2 we can obtain the values of 150° for the azimuth and 22° 45′ for the altitude. The use of the

data provided by this table[B2] is essential in the development of graphic analyses for passive solar effects. However, neither azimuth nor altitude angles can be used as the direct elevation projection of the sun's angle for any graphic study development. The actual procedure to determine the shadows and angles in graphics is a matter of greater detail which does not necessarily correspond to the programmatic phase. An actual detailed explanation of such procedure can be found in *Time-Saver Standards, A Handbook of Architectural Design* by John Callender, fourth edition, page 80.

An excellent source and reference for the development and gathering of solar analysis data can be found in *Design With Climate* by Victor Olgyay.

[B1]Equivalent south latitudes can be easily developed by merely transposing winter and summer dates as well as fall and spring dates.

[B2]Figures based on data provided by Hendrik P. Maas in "Short Cuts to Solar Angles" of the *Time-Saver Standards,* page 39.

Sun path
W SUNSET 6:00 PM
N
Sun
h°
Horizon
12:00M
a°
S
11:00 AM
SUNRISE 6:00 AM
E
North : NORTH MERIDIAN (TRUE NORTH)
a°=Azimuth : ANGLE MEASURED HORIZONTALLY FROM NORTH MERIDIAN EASTERLY FOR A.M. TIME WESTERLY FOR P.M. TIME
h°=Altitude : ANGLE MEASURED VERTICALLY BETWEEN THE SUN AND THE HORIZONTAL PLANE OF THE HORIZON

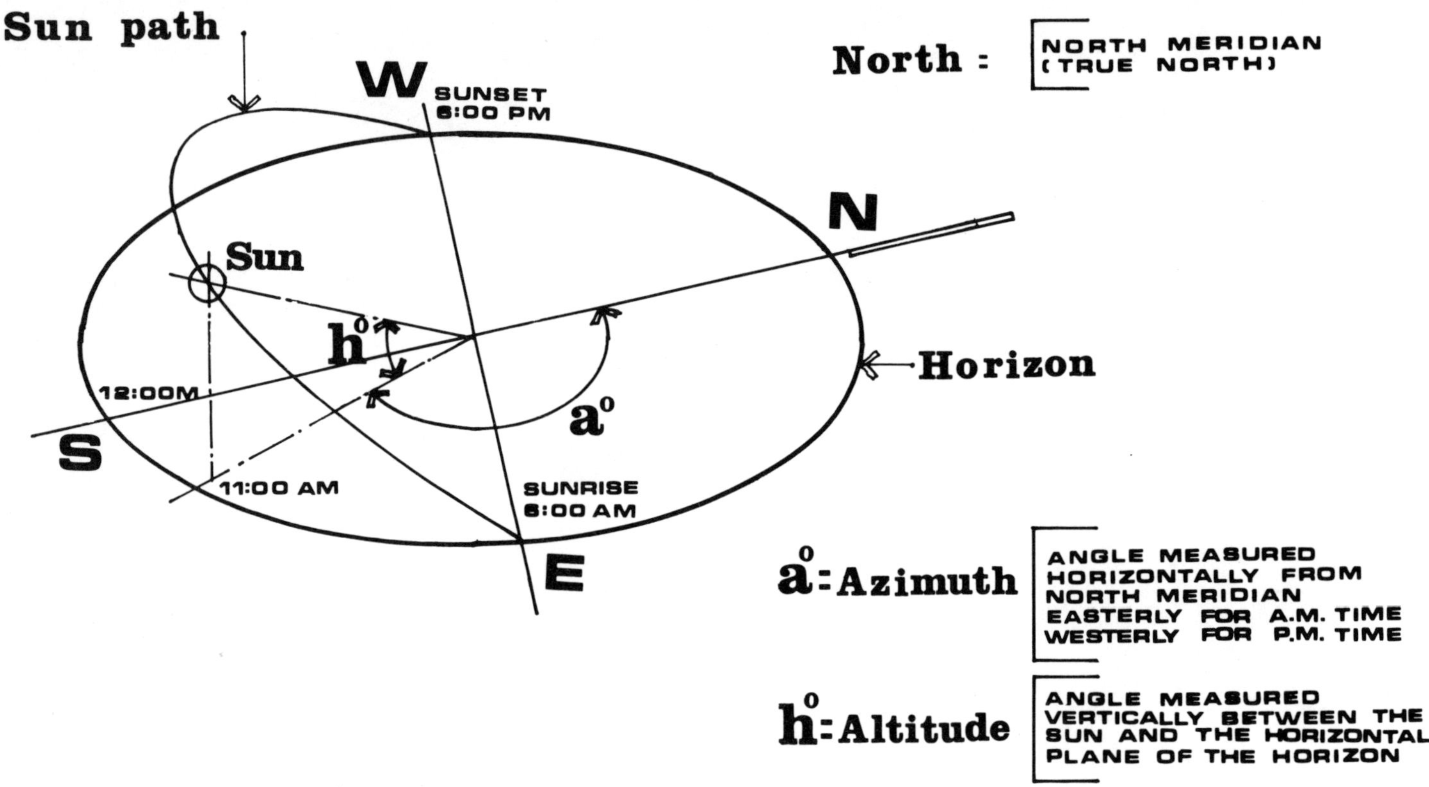

Figure B-1

SOLAR ANALYSIS DIAGRAMS

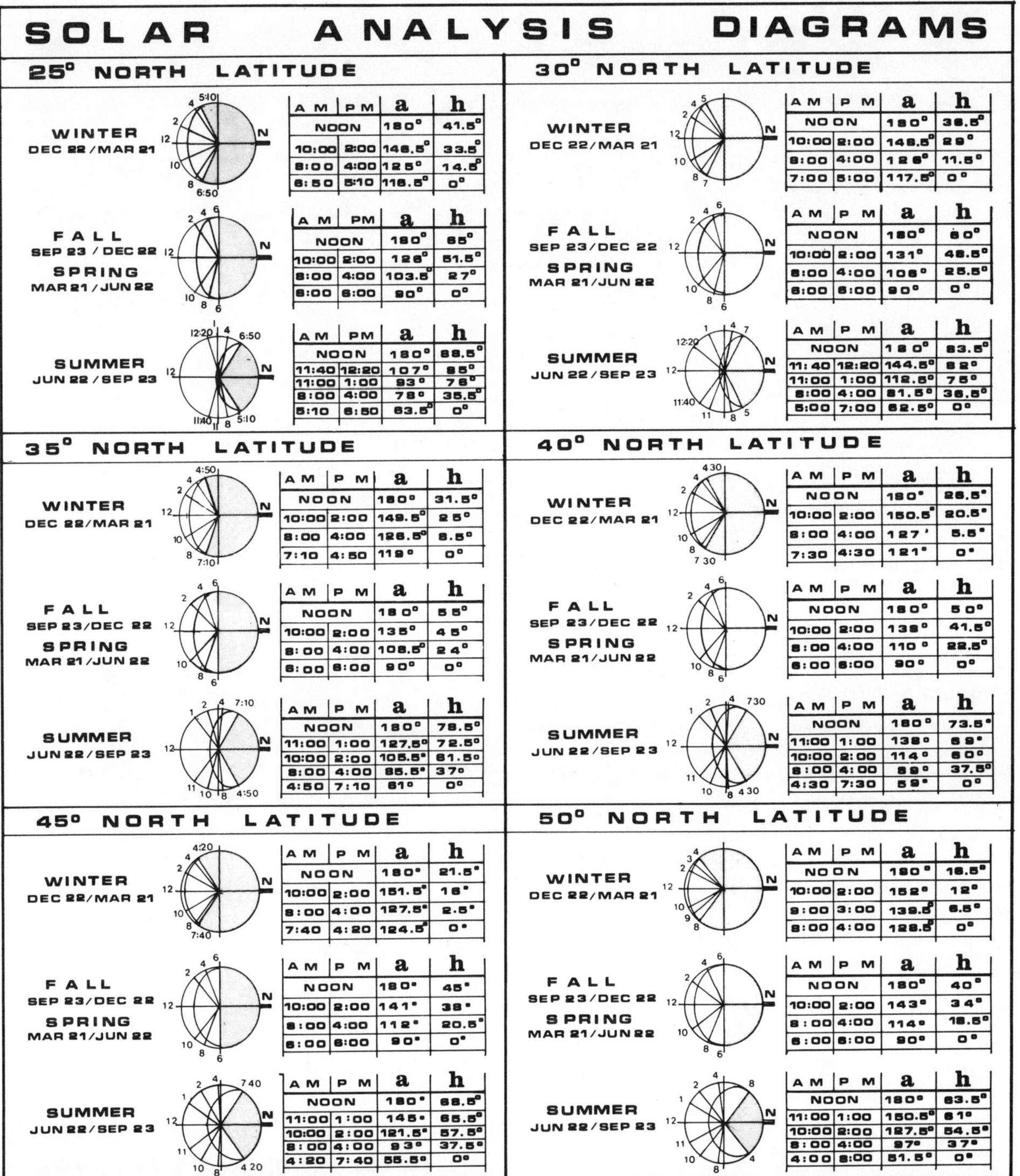

25° NORTH LATITUDE

WINTER DEC 22/MAR 21

AM	PM	a	h
NOON		180°	41.5°
10:00	2:00	146.5°	33.5°
8:00	4:00	125°	14.5°
6:50	5:10	116.5°	0°

FALL SEP 23/DEC 22 **SPRING** MAR 21/JUN 22

AM	PM	a	h
NOON		180°	65°
10:00	2:00	126°	51.5°
8:00	4:00	103.5°	27°
6:00	6:00	90°	0°

SUMMER JUN 22/SEP 23

AM	PM	a	h
NOON		180°	88.5°
11:40	12:20	107°	85°
11:00	1:00	93°	76°
8:00	4:00	78°	35.5°
5:10	6:50	63.5°	0°

30° NORTH LATITUDE

WINTER DEC 22/MAR 21

AM	PM	a	h
NOON		180°	36.5°
10:00	2:00	148.5°	29°
8:00	4:00	128°	11.5°
7:00	5:00	117.5°	0°

FALL SEP 23/DEC 22 **SPRING** MAR 21/JUN 22

AM	PM	a	h
NOON		180°	60°
10:00	2:00	131°	48.5°
8:00	4:00	108°	25.5°
6:00	6:00	90°	0°

SUMMER JUN 22/SEP 23

AM	PM	a	h
NOON		180°	83.5°
11:40	12:20	144.5°	82°
11:00	1:00	112.5°	75°
8:00	4:00	81.5°	36.5°
5:00	7:00	62.5°	0°

35° NORTH LATITUDE

WINTER DEC 22/MAR 21

AM	PM	a	h
NOON		180°	31.5°
10:00	2:00	149.5°	25°
8:00	4:00	126.5°	8.5°
7:10	4:50	119°	0°

FALL SEP 23/DEC 22 **SPRING** MAR 21/JUN 22

AM	PM	a	h
NOON		180°	55°
10:00	2:00	135°	45°
8:00	4:00	108.5°	24°
6:00	6:00	90°	0°

SUMMER JUN 22/SEP 23

AM	PM	a	h
NOON		180°	78.5°
11:00	1:00	127.5°	72.5°
10:00	2:00	105.5°	61.5°
8:00	4:00	85.5°	37°
4:50	7:10	61°	0°

40° NORTH LATITUDE

WINTER DEC 22/MAR 21

AM	PM	a	h
NOON		180°	26.5°
10:00	2:00	150.5°	20.5°
8:00	4:00	127'	5.5°
7:30	4:30	121°	0°

FALL SEP 23/DEC 22 **SPRING** MAR 21/JUN 22

AM	PM	a	h
NOON		180°	50°
10:00	2:00	138°	41.5°
8:00	4:00	110°	22.5°
6:00	6:00	90°	0°

SUMMER JUN 22/SEP 23

AM	PM	a	h
NOON		180°	73.5°
11:00	1:00	138°	69°
10:00	2:00	114°	60°
8:00	4:00	89°	37.5°
4:30	7:30	59°	0°

45° NORTH LATITUDE

WINTER DEC 22/MAR 21

AM	PM	a	h
NOON		180°	21.5°
10:00	2:00	151.5°	16°
8:00	4:00	127.5°	2.5°
7:40	4:20	124.5°	0°

FALL SEP 23/DEC 22 **SPRING** MAR 21/JUN 22

AM	PM	a	h
NOON		180°	45°
10:00	2:00	141°	38°
8:00	4:00	112°	20.5°
6:00	6:00	90°	0°

SUMMER JUN 22/SEP 23

AM	PM	a	h
NOON		180°	68.5°
11:00	1:00	145°	65.5°
10:00	2:00	121.5°	57.5°
8:00	4:00	93°	37.5°
4:20	7:40	55.5°	0°

50° NORTH LATITUDE

WINTER DEC 22/MAR 21

AM	PM	a	h
NOON		180°	16.5°
10:00	2:00	152°	12°
9:00	3:00	139.5°	6.5°
8:00	4:00	128.5°	0°

FALL SEP 23/DEC 22 **SPRING** MAR 21/JUN 22

AM	PM	a	h
NOON		180°	40°
10:00	2:00	143°	34°
8:00	4:00	114°	18.5°
6:00	6:00	90°	0°

SUMMER JUN 22/SEP 23

AM	PM	a	h
NOON		180°	63.5°
11:00	1:00	150.5°	61°
10:00	2:00	127.5°	54.5°
8:00	4:00	97°	37°
4:00	8:00	51.5°	0°

Figure B-2

Appendix C

WIND ROSES

A "Wind Rose" is a diagrammatic representation of the wind direction (see Figure C.1) experienced at a specific location for a definite period of time (days, months, seasons, etc.). It shows the prevailing wind direction. Although many variations exist, the most common graphic form consists of a circle from which sixteen lines emanate, each line corresponding with a compass point. The length of each line is proportioned to the frequency (indicated in percentages of total time) of wind from that direction. The frequency of calm conditions is usually shown in the center. In constructing or interpreting wind roses it is important to remember the meteorological convention that *wind direction refers to the direction from which the wind is blowing.* Therefore, a bar, line or projection which extends to the west on a wind rose diagram indicates the frequency of wind blowing FROM the west and NOT the frequency of wind TOWARDS the west.

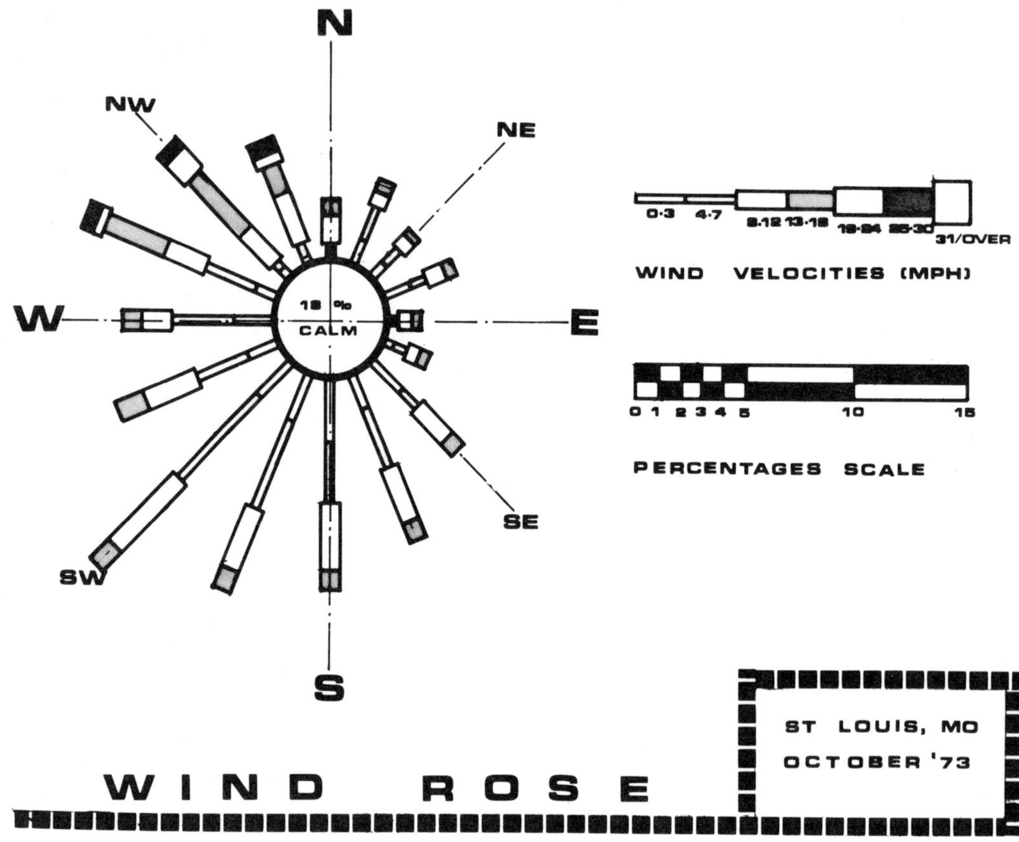

Figure C-1

Appendix D

ECONOMIC FEASIBILITY STUDY

Although Economic Feasibility Studies and Reports constitute a specialized portion of economic analyses, the following example provides a generalized outline of sequential steps which could be followed in developing a simple feasibility evaluation.

1. Building Areas

    1.1   Gross Area

    1.2   Standard Net Assignable or Rental (area/units/etc.) ____________________

    1.3   Plan Efficiency Ratio ____________________

2. Estimated Project Cost
(see Sub-Chapter 8.4/Figure 8.33) subtotal ____

3. Financing

    3.1   Estimated Project Cost (from 2) ________

    3.2   Mortgage at (x%) (add) ____________

        3.20 Equity Required ____________

        3.21 Land Cost (subtract) ____________

    3.3   Cash Required subtotal ____________

4. Annual Gross Income

    4.1   Overall Income
    (Standard Net Assignable or Units × rental rate) = ____________________

    4.2   Vacancy Percentage (Y%) (subtract) ____

    4.3   Estimated Annual Gross
    Income subtotal ____________________

5. Annual Operating Cost

    5.1   Operating Expenses ____________________

    5.2   Taxes (add) ____________________

    5.3   Insurance (add) ____________________

    5.4   Maintenance and Miscellaneous (add) ____

    5.5   Estimated Annual Operating
    Cost subtotal ____________________

6. Net Income

    6.1   Estimated Annual Gross
    Income (subtotal 4.3) ____________

    6.2   Estimated Annual Operating
    Cost (subtotal 5.5) (subtract) ____________

    6.3   Estimated Annual Net Income subtotal ____

7. Return

    7.1   Estimated Annual Net
    Income (subtotal 6.3) ____________

    7.2   Annual Amortization,
    Interest (subtract) ____________

    7.3   Estimated Annual Return[D1] for the
    Duration of Mortgage Time subtotal ______

[D1]Establish the percentage of equity this return subtotal represents during mortgage amortization and afterwards to effectively evaluate return percentages. It is also convenient to establish the rate of return if projects are owner-financed by establishing the return percentage an estimated annual net income (6.3) represents against an investment of total project cost, as well as how long it would take to amortize such investment at a desired percentage of return.

Appendix E

POPULATION PROJECTIONS

Population projections can be done in several ways, each one of which offers specific advantages and disadvantages under different conditions and as applied to different situations.

A brief summary of several of these methods is presented below:

1. Mathematical projections (arithmetical or geometric)

    1.1 Compute the average NUMERICAL population change and project such increase into the future.

    1.2 Compute the average RATE of population change (generally a percentage of the total) and project this rate into the future.

    1.3 Plot a curve of past population growth on a semi-logarithmic scale and develop a corresponding exponential equation which fits the curve. Projections can be obtained then from such equation.

    1.4 Fit a specific mathematical curve to the curve of past population's growth and determine the population projection from this mathematical curve.[E1]

Generally arithmetical projections done by these means will provide MINIMUM forecasts. while geometric projections will provide MAXIMUM forecasts. It is recommendable, therefore, to use both as an aid in establishing these parameters.

Mathematical projections offer the advantages of relative dependability for SHORT TERM projections and ease of execution, as well as the fact that they are generally adequate for areas which have experienced a more or less constant rate of change or growth of population. The disadvantages of these projections lie in one essential issue: the assumption of constant factors and its consequences in the development of growth and change. Undoubtedly, in view of the recent developments at almost every level of the sociological scale, this assumption is extremely unreliable.

2. Natural Projections

    2.1 Net migration. Study past migrations by determining their volume, direction and composition as well as their relationships to conditions existing at such times in a specific area. past or foreseeable changes and their interrelationships are then established and evaluated. The developments of these appraisals can then be determined as related to the net migration for a specific period. With these studies it is possible to develop minimum/maximum levels of growth and change for future periods of time as related to past equivalences.

    2.2 Natural increases. The principal issues to consider in this procedure are the population composition factors: race, sex and age. The process is as follows:

    2.21 By subdividing the population in 5-year or 10-year age groups, the survivors of the resident population for each sex are established. Such survival rate can be easily projected through the use of mortality tables and trends such as the *Population Pyramid.*[E2]

2.22 The total net migration for the projected date is either added or subtracted (depending on the migration direction) from the surviving residents in each age group.[E3]

2.23 Project birth rates by age of mothers (residents and migrants) by establishing age, birth rates and multiplying by the average number of women[E4] in age groups within childbearing ages.

2.24 Establish survivors of projected birth rates by using death rate factors for young children.

2.25 To the subtotal figure of former resident population and their projected babies, the subtotals of net migration and their projected babies must be added or subtracted, this, once again, depends on the established migratory direction from where a final population projection is obtained.

3. Equivalency Projections

Relationships to other areas. Since the population growth or change in a specific location can be influenced by economic or social factors affecting the region or state, the population projection for the areas may be closely related to or affect the growth or change in a smaller location. For this reason, economic regions or state projections can become extremely useful guides in the development of population projections for a specific location.

In developing equivalency projections, one simple method is to determine the percentage of population which the specific location represents in its economic region, state, etc., and by using the systems outlined in the mathematical procedures establish an equivalency from which the smaller projections can be easily obtained.

It is noteworthy that the basis for these methods assumes a constant relationship between local communities and larger areas and this may not always be the case. Furthermore, unforeseeable changes in local, as well as national levels may greatly affect a particular community and automatically invalidate the results of a population projection based on an equivalency approach.

On the other hand, the system has the advantage of offering a broader knowledge of the national, state or regional factors and trends affecting the population changes in specific locations making it easier and sometimes even more accurate to develop population projections at the smaller scale level.

4. Specialized Projections.

Other types of population projections include the more definite grouping of social elements by age, sex, race, educational groups, births and deaths by age groups, etc. Most of the systems previously outlined can be easily adapted to fit each one of these specialized studies. In each instance, the statistics department at state and local levels, as well as the United States Department of Commerce, Bureau of the Census, can offer excellent guides in the analysis and determination of most of these procedures, developments and data.

[E1]Such as the curve developed by P. F. Verhulst in 1838 based on a "Law of Growth for a Limited Area."

[E2]Refer to United States *Population Pyramid,* United States Department of Commerce, Bureau of the Census, 1950-1960.

[E3]Note that sex and age of migrant populations, as well as the race in specific cases, will be generally quite different from resident populations. These state or local population composition factors should be carefully analyzed and used as guidelines in projecting the race, sex and age distribution of any net migration.

[E4]Average figures are usually obtained by a simple interpolation or by averaging the number of women at the beginning and end of the forecast period for each childbearing age group.

Bibliography

ANTILL, James M. and WOODHEAD, Ronald W., *Critical Path Methods in Construction Practice.* New York: John Wiley and Sons, Inc., 1965.

AYERS, Chesley, *Specifications: For Architecture, Engineering and Construction.* New York: McGraw-Hill Book Company, 1975.

CALLENDER, John Hancock, (Ed.), *Time-Saver Standards, A Handbook of Architectural Design.* New York: McGraw-Hill Book Company, 1966. Fourth edition.

CAUDILL, William W., *Toward Better School Design.* New York: F. W. Dodge Corporation, 1954.

COWGILL, Clinton H. and SMALL, Ben John, *Architectural Practice.* New York: Reinhold Publishing Corp., 1959. Third edition.

DAVIS, Gordon B., *Computer Data Processing.* New York: McGraw-Hill Book Company, 1973. Second edition.

DEASY, C. M., *Design for Human Affairs.* New York: John Wiley and Sons, 1974.

DE CHIARA, Joseph and KOPPELMAN, Lee, *Planning Design Criteria.* New York: Van Nostrand Reinhold Company, 1969.

DIFFRIENT, Niels, et al, *Human Scale.* Cambridge, Mass: The M.I.T. Press, 1974.

DRUCKER, Peter F., *Management: Tasks, Responsibilities, Practices.* New York: Harper and Row, 1973.

DUFFY, Francis, CAVE, Colin and WORTHINGTON, John, *Planning Office Space.* London, England: The Architectural Press Ltd. New York: Nichols Publishing, 1976.

EVANS, Benjamin H., and WHEELER, Jr., C. Herbert, *Emerging Techniques. 2 Architectural Programming.* Washington, D.C.: The American Institute of Architects, 1969.

FEREBEE, Ann, (Ed.), *Proceedings of the First National Conference on Urban Design.* New York: R C Publications, Inc., 1978.

FOXHALL, William B., *Professional Construction Management and Project Administration.* New York: Architectural Record and The American Institute of Architects, 1972.

GUTMAN, Robert, (Ed.), *People and Buildings.* New York: Basic Books, 1972.

GUTTRIDGE, Bryan and WAINWRIGHT, Jonathan R., *Computers in Architectural Practice.* New York: Halsted Press/John Wiley and Sons, Inc., 1973.

HALL, Edward T., *The Hidden Dimension.* New York: Doubleday and Co., 1969.

HANDLER, A. B., *Systems Approach to Architecture.* New York: American Elsevier Publishing Co., 1970.

HARRIS, Cyril M., *Dictionary of Architecture and Construction.* New York: McGraw-Hill, Inc., 1975.

HEIMSATH, Clovis, *Behavioral Architecture: Towards an Accountable Design Process.* New York: McGraw-Hill, Inc., 1977.

HUBBARD, Henry V. and KIMBALL, Theodora, *An Introduction to the Study of Landscape Design.* Boston: Hubbard Educational Trust, 1967.

HUNT, Jr., William Dudley, (Ed.), *The American Institute of Architects Comprehensive Architectural Services — General Principles and Practice.* New York: McGraw- Hill Book Company, 1965.

HUNT, Jr., William Dudley, (Ed.), *The American Institute of Architects Creative Control of Building Costs.* New York: McGraw-Hill Book Company, 1967.

LAPIDUS, Morris, *Architecture as a Profession and a Business.* New York: Van Nostrand Reinhold, 1967.

LASEAU, Paul, *Graphic Problem Solving for Architects and Builders.* Boston: Cahners Publishing Company, Inc., 1975.

MITCHELL, William J., *Computer-Aided Architectural Design.* New York: Petrocelli/Charter, 1977.

MOORE, Gary T., (Ed.), *Emerging Methods in Environmental Design and Planning.* Cambridge, Mass: The M.I.T. Press, 1973.

O'BRIEN, James J., *Value Analysis in Design and Construction.* New York: McGraw-Hill Book Company, 1976.

OLGYAY, Victor, *Design With Climate.* Princeton, New Jersey: Princeton University Press, 1973. Fourth printing.

PAINTER, Dean E., *Air Pollution Technology,* Reston, VA: Reston Publishing Co./A Prentice-Hall Company, 1974.

PEÑA, William, *Problem Seeking, An Architectural Programming Primer.* Boston: Cahners Books International, 1977.

PEVSNER, Nikolaus, FLEMING, John, and HONOUR, Hugh, *A Dictionary of Architecture.* Woodstock, New York: The Overlook Press, 1976.

PILE, John, *Open Office Planning.* New York: Whitney Library of Design, 1978.

PORTEOUS, J. D., *Environment and Behavior, Planning and Everyday Urban Life.* Reading, Mass: Addison-Wesley Publishing Co., 1977.

PRESIER, Wolfgang F. E., (Ed.), *Facility Programming: Methods and Applications.* Stroudsburg, Pa: Dowden, Hutchinson and Ross, 1978.

PULVER, H. E., *Construction Estimates and Costs.* New York: McGraw-Hill Book Company, 1969. Fourth edition.

RAPOPORT, Amos, (Ed.), *The Mutual Interaction of People and Their Built Environment.* Chicago: Aldine Press, 1976.

RAU, J. G. and WOOTEN, D. C., (Eds.), *Environmental Impact Analysis Handbook.* New York: McGraw-Hill, 1980.

RUBENSTEIN, Harvey M., *A Guide to Site and Environmental Planning.* New York: John Wiley and Sons, Inc., 1969.

RYLE, Gilbert, *The Concept of Mind.* New York: Barnes and Noble Books/Harper and Row, Publishers, 1949.

SANOFF, Henry, *Methods in Architectural Programming.* Stroudsburg, Pa: Dowden, Hutchinson and Ross, 1977.

SHIRK, Evelyn, *The Ethical Dimension.* New York: Appleton-Century-Crofts/Meredith Publishing Company, 1965.

UNIFORM BUILDING CODE 1979 Edition. International Conference of Building Officials, Whittier, Cal., 1979. Second printing.

WARD, Barbara, *The Home of Man.* New York: W. W. Norton and Company, Inc., 1976.

WEIZENBAUM, Joseph, *Computer Power and Human Reason: From Judgment to Calculation.* San Francisco: W. H. Freeman and Co., 1976.

WELS, Byron G., *Personal Computers, What They Are and How to Use Them.* Englewood Cliffs, New Jersey: Trafalgar House Publishing, Inc., 1978.

ZEISEL, John, *Sociology and Architectural Design.* New York: Russell Sage Foundation, 1975.

Index